CURRENT PRACTICES FOR INTERPRETING ENGINEERING DRAWINGS

CURRENT PRACTICES FOR INTERPRETING ENGINEERING DRAWINGS

Edward A. Maruggi, Ph.D.
PROFESSOR EMERITUS
ROCHESTER INSTITUTE OF TECHNOLOGY

West Publishing Company
MINNEAPOLIS/ST. PAUL   NEW YORK   LOS ANGELES   SAN FRANCISCO

WEST'S COMMITMENT TO THE ENVIRONMENT

In 1906, West Publishing Company began recycling materials left over from the production of books. This began a tradition of efficient and responsible use of resources. Today, up to 95 percent of our legal books and 70 percent of our college and school texts are printed on recycled, acid-free stock. West also recycles nearly 22 million pounds of scrap paper annually—the equivalent of 181,717 trees. Since the 1960s, West has devised ways to capture and recycle waste inks, solvents, oils, and vapors created in the printing process. We also recycle plastics of all kinds, wood, glass, corrugated cardboard, and batteries, and have eliminated the use of Styrofoam book packaging. We at West are proud of the longevity and the scope of our commitment to the environment.

Production, Prepress, Printing and Binding by West Publishing Company.

British Library Cataloguing-in-Publication Data. A catalogue record for this book is available from the British Library.

Cover image courtesy of ICEM Technologies, a division of Control Data Systems, Inc.

Interior text design by Dapper Design.

Copyediting and composition by Lachina Publishing Services, Inc.

Accuracy checking by Gregory Rodgers.

All photos not specifically credited are courtesy of Mike Spencer, Rochester Institute of Technology.

COPYRIGHT © 1995   By WEST PUBLISHING COMPANY
                                  610 Opperman Drive
                                  P.O. Box 64526
                                  St. Paul, MN 55164-0526

All rights reserved

Printed in the United States of America

02 01 00 99 98 97 96 95   8 7 6 5 4 3 2 1 0

Library of Congress Cataloging-in-Publication Data

Maruggi, Edward A.
    Current practices for interpreting engineering drawings / Edward A. Maruggi.
        p.   cm.
    Includes index.
    ISBN 0-314-04576-7
      1. Engineering drawings. I. title.
T379.M29 1995
604.2—dc20   94-24210
                                                  CIP

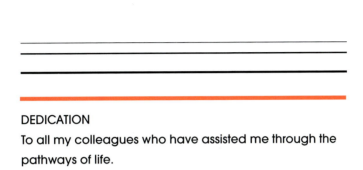

DEDICATION

To all my colleagues who have assisted me through the pathways of life.

CONTENTS

Preface   xiii

SECTION ONE

Introduction to Current Practices for Interpreting Engineering Drawings   1

1.1   Rationale For Interpreting Drawings   1
1.2   Some Ancient Practices   3
1.3   Current Practices   3
1.4   Summary   5
Technical Terms for Study   5
Competency Quiz   7
Part A—Comprehension   7
Part B— Technical Terms   9

SECTION TWO

Measuring and Measurement Systems   11

2.1   Introduction   11
 2.1.1   THE DECIMAL-INCH SYSTEM   12
 2.1.2   THE INTERNATIONAL SYSTEM OF UNITS   13
 2.1.3   THE DUAL-DIMENSIONING SYSTEM   14
2.2   Using Basic Measuring Tools   14
 2.2.1   THE STEEL RULE   15
 2.2.2   THE CALIPER   15
 2.2.3   THE PROTRACTOR   21
 2.2.4   THE MICROMETER   21
2.3   Summary   24
Technical Terms for Study   24
Competency Quiz   27
Part A—Comprehension   27
Part B—Technical Terms   31
Part C—Technical Mathematics   32

SECTION THREE

Drawing Media and Formats   35

3.1   Introduction   35
3.2   Drawing Media   35
 3.2.1   STANDARD DRAWING SIZES   36

3.3   Parts of a Drawing   36
 3.3.1   DRAWING FIELD   37
 3.3.2   TITLE BLOCK   37
 3.3.3   REVISION AREA   39
 3.3.4   PARTS LIST   39
 3.3.5   PART IDENTIFICATION   39
 3.3.6   PART-NUMBERING SYSTEM   40
3.4   Summary   40
Technical Terms for Study   41
Competency Quiz   43
Part A—Comprehension   43
Part B—Technical Terms   45
Part C—Drawing Interpretation   47

SECTION FOUR

Line Conventions and Text Presentation   49

4.1   Introduction   49
4.2   Line Convention   49
4.3   Text Presentation   51
 4.3.1   FREEHAND LETTERING TECHNIQUES   51
 4.3.2   STYLES OF LETTERING   51
 4.3.3   MECHANICAL LETTERING AIDS   53
 4.3.4   COMPUTER-GENERATED TEXT   55
4.4   Summary   56
Technical Terms for Study   56
Competency Quiz   57
Part A—Comprehension   57
Part B—Technical Terms   59
Part C—Lettering Activity   60

SECTION FIVE

Technical Sketching   61

5.1   Introduction   61
5.2   Types of Technical Sketches   62
 5.2.1   MULTIVIEW SKETCH   62
 5.2.2   ISOMETRIC SKETCH   63
 5.2.3   OBLIQUE SKETCH   63
 5.2.4   PERSPECTIVE SKETCH   64
5.3   Sketching Squares, Rectangles, and Triangles   66
5.4   Sketching Circles and Arcs   69
5.5   Suggestions for Producing Technical

Sketches 69
5.6 Summary 70
Technical Terms for Study 70
Competency Quiz 73
Part A—Comprehension 73
Part B—Technical Terms 75
Part C—Sketching 76

SECTION SIX

Principles of Orthographic Projection 81

6.1 Introduction 81
   6.1.1 THIRD-ANGLE PROJECTION 81
   6.1.2 FIRST-ANGLE PROJECTION 82
6.2 The Multiview Drawing 82
6.3 The Selection and Placement of Views 84
6.4 The One-View Drawing 84
6.5 The Two-View Drawing 85
6.6 The Three-View Drawing 85
6.7 The Projection of Points, Lines, and Surfaces 85
   6.7.1 POINTS 85
   6.7.2 LINES 87
   6.7.3 SURFACES 89
6.8 The Precedence of Lines 89
6.9 Summary 89
Technical Terms for Study 90
Competency Quiz 93
Part A—Comprehension 93
Part B—Technical Terms 96
Part C—View and Surface Identification 97
Part D—Point, Line, and Surface Identification 99
Part E—Technical Mathematics 101

SECTION SEVEN

Sectional and Auxiliary Views 103

7.1 Introduction 103
7.2 Sectional Views 104
   7.2.1 THE CUTTING PLANE LINE 104
   7.2.2 CONVENTIONS FOR SECTIONING 104
7.3 Types of Sectional Views 106
   7.3.1 THE FULL SECTION 106
   7.3.2 THE HALF SECTION 106
   7.3.3 THE BROKEN-OUT SECTION 106
   7.3.4 THE REVOLVED SECTION 109
   7.3.5 THE OFFSET SECTION 109
   7.3.6 THE REMOVED SECTION 110

   7.3.7 SECTIONS THROUGH WEBS, SPOKES, AND RIBS 110
   7.3.8 THE UNLINED SECTION 110
7.4 Breaks in Elongated Objects 112
7.5 Auxiliary Views 112
   7.5.1 TYPES OF AUXILIARY VIEWS 113
7.6 Summary 117
Technical Terms for Study 119
Competency Quiz 121
Part A—Comprehension 121
Part B—Technical Terms 124
Part C—Sectional View Identification 125
Part D—Auxiliary View Identification 129

SECTION EIGHT

Dimensioning 131

8.1 The Purpose of Dimensions 131
8.2 Fundamental Rules of Dimensioning 132
8.3 Types of Dimensioning Systems 132
   8.3.1 THE UNIDIRECTIONAL SYSTEM 132
   8.3.2 THE ALIGNED SYSTEM 132
8.4 Basic Dimensioning Terms 133
8.5 Application of Dimensions 135
   8.5.1 THE DIMENSION LINE 135
   8.5.2 THE EXTENSION LINE 136
   8.5.3 THE LEADER 136
   8.5.4 DIMENSIONS NOT TO SCALE 137
8.6 Dimensioning of Features 137
   8.6.1 DIAMETERS 137
   8.6.2 RADII 138
   8.6.3 CHORDS, ARCS, AND ANGLES 139
   8.6.4 ROUNDED ENDS 140
   8.6.5 HOLES 140
   8.6.6 THE CHAMFER 143
   8.6.7 THE KEYWAY 144
8.7 Summary 145
Technical Terms for Study 146
Competency Quiz 149
Part A—Comprehension 149
Part B—Technical Terms 151
Part C—Dimension Identification 152
Part D—Technical Mathematics 158

SECTION NINE

Tolerancing 161

9.1 The Need for Tolerances 161
9.2 The Purpose of Tolerancing 161
9.3 Tolerancing Terms 162

9.4 Tolerancing Methods 162
9.5 Interpretation of Limits 163
    9.5.1 SINGLE LIMITS 163
9.6 The Accumulation of Tolerances 164
9.7 Symbology 165
    9.7.1 THE GEOMETRIC CHARACTERISTIC SYMBOL 165
    9.7.2 THE FEATURE CONTROL FRAME 165
    9.7.3 THE DATUM IDENTIFICATION SYMBOL 166
    9.7.4 THE BASIC DIMENSION AND THE DATUM TARGET SYMBOL 167
    9.7.5 THE TOLERANCE ZONE 167
    9.7.6 TOLERANCING MODIFIERS 167
9.8 The Application and Interpretation of Geometric Characteristic Symbols 168
    9.8.1 STRAIGHTNESS 168
    9.8.2 FLATNESS 169
    9.8.3 ROUNDNESS 170
    9.8.4 CYLINDRICITY 170
    9.8.5 LINE PROFILES AND SURFACE PROFILES 171
    9.8.6 ANGULARITY 171
    9.8.7 PERPENDICULARITY 174
    9.8.8 PARALLELISM 175
    9.8.9 POSITIONAL TOLERANCE 176
    9.8.10 CONCENTRICITY 176
    9.8.11 RUNOUT 176
9.9 Summary 178
Technical Terms for Study 179
Competency Quiz 183
Part A—Comprehension 183
Part B—Technical Terms 185
Part C—Tolerancing Application 187
Part D—Tolerancing Interpretation 191

SECTION TEN

Fasteners and Joining Methods 193

10.1 Introduction 193
10.2 The Screw Thread 193
10.3 Screw Thread Terminology 195
10.4 Drawing Application of Screw Threads 196
    10.4.1 THREAD REPRESENTATION 196
    10.4.2 DECIMAL-INCH THREAD DESIGNATION 196
    10.4.3 METRIC THREAD DESIGNATION 198
10.5 Fasteners 198
    10.5.1 THE THREADED FASTENER 199
    10.5.2 THE UNTHREADED FASTENER 202
10.6 Joinings 205
    10.6.1 WELDING 206
    10.6.2 RESISTANCE WELDING 208
    10.6.3 BRAZING 211
    10.6.4 SOLDERING 213
    10.6.5 ADHESIVE BONDING 213

10.7 Summary 213
Technical Terms for Study 214
Competency Quiz 217
Part A—Comprehension 217
Part B—Technical Terms 221
Part C—Interpreting Fasteners 222
Part D—Interpreting Joining Methods 227

SECTION ELEVEN

Power Transmission Elements 233

11.1 Introduction 233
11.2 Gears 233
    11.2.1 THE SPUR GEAR 234
    11.2.2 THE BEVEL GEAR 236
    11.2.3 THE WORM AND WORM GEAR 239
11.3 Cams 240
    11.3.1 TYPES OF CAMS 242
    11.3.2 CAM LAYOUT AND NOMENCLATURE 242
11.4 Belts 244
    11.4.1 TYPES OF BELT DRIVES 246
    11.4.2 TYPES OF BELTS 246
    11.4.3 FLAT BELT PULLEYS 246
    11.4.4 V-BELTS 248
11.5 Chain 249
    11.5.1 TYPES OF CHAIN 250
    11.5.2 THE SPROCKET 251
11.6 Summary 253
Technical Terms for Study 253
Competency Quiz 257
Part A—Comprehension 257
Part B—Technical Terms 260
Part C—Interpreting Gears 263
Part D—Interpreting Cams, Pulleys, and Sprockets 269
Part E—Technical Mathematics 274

SECTION TWELVE

Surface Texture and Protective Coatings 277

12.1 Introduction 277
12.2 Surface Texture Definition and Nomenclature 278
    12.2.1 SURFACE CONTROL SYSTEM 279
    12.2.2 ROUGHNESS 280
    12.2.3 WAVINESS 281
    12.2.4 LAY 281
    12.2.5 FLAWS 282
12.3 Application of Surface Texture Symbols 282

12.4 Protective Coatings 282
    12.4.1 METALLIC COATINGS 283
    12.4.2 NONMETALLIC COATINGS 285
    12.4.3 POLYMER COATINGS 286
    12.4.4 FINISH BLOCK ENTRIES ON AN
    ENGINEERING DRAWING 286
12.5 Summary 287
    Technical Terms for Study 288
    Competency Quiz 291
    Part A—Comprehension 291
    Part B—Technical Terms 294
    Part C—Application of Surface Texture
    Symbols 295

SECTION THIRTEEN

Detail and Assembly Drawings 299

13.1 Introduction 299
13.2 Types of Detail and Assembly Drawings 299
    13.2.1 THE DESIGN LAYOUT DRAWING 300
    13.2.2 THE DETAIL DRAWING 300
    13.2.3 THE ASSEMBLY DRAWING 301
13.3 Summary 305
    Technical Terms For Study 305
    Competency Quiz 313
    Part A—Comprehension 313
    Part B—Technical Terms 315
    Part C—Interpreting Detail Drawings 317
    Part D—Interpreting Assembly Drawings
    321

SECTION FOURTEEN

Development and Sheet Metal Drawings 327

14.1 Introduction 327
14.2 Types of Development Drawings 329
    14.2.1 PARALLEL LINE DEVELOPMENT 330
    14.2.2 RADIAL LINE DEVELOPMENT 333
    14.2.3 TRIANGULATION DEVELOPMENT 335
14.3 Sheet Metal Drawings 335
    14.3.1 THE DEVELOPED LENGTH 337
    14.3.2 DETAILS OF SHEET METAL PARTS 338
14.4 Summary 340
    Technical Terms for Study 341
    Competency Quiz 343
    Part A—Comprehension 343
    Part B—Technical Terms 345
    Part C—Development 346
    Part D—Sheet Metal Drawing 350

SECTION FIFTEEN

Computer-Generated Drawings 353

15.1 Introduction 353
15.2 The Basic CAD Concept 354
15.3 System Components 354
    15.3.1 THE INPUT DEVICE 355
    15.3.2 THE OUTPUT DEVICE 358
    15.3.3 THE PROCESSING DEVICE 360
15.4 Types of Computer Systems 361
    15.4.1 THE MAINFRAME-BASED SYSTEM 361
    15.4.2 THE MINICOMPUTER-BASED SYSTEM 362
    15.4.3 THE MICROCOMPUTER-BASED
    SYSTEM 362
    15.4.4 THE LAPTOP COMPUTER 362
15.5 CAD-Based Software for Producing
    Engineering Drawings 363
15.6 CAD Applications 363
    15.6.1 PREPARING FOR GRAPHICS
    PRODUCTION 363
    15.6.2 THE MENU AND COMMAND SYSTEM 364
    15.6.3 CREATING A DESIGN FILE 365
    15.6.4 DRAWING LINE FORMS 365
    15.6.5 CREATING TEXT 366
    15.6.5 ESTABLISHING LAYERS 366
    15.6.7 PLOTTING 366
15.7 Summary 370
    Technical Terms for Study 370
    Competency Quiz 373
    Part A—Comprehension 373
    Part B—Technical Terms 376

SECTION SIXTEEN

Supplementary Competency Exercises 379

16.1 Introduction 379
    Part A—Detail Drawing Interpretation 381
    Part B—Assembly Drawing
    Interpretation 384

APPENDIX A
    Hole/Drill Size Chart 385
    Inches/Metric Decimal Equivalents 386
APPENDIX B
    Tap Drill Sizes for Machine Screws 387
APPENDIX C
    Clearance Hole Sizes for Machine
    Screws 388
APPENDIX D
    Word Abbreviations on Drawings 389

APPENDIX E

Standard Screw Thread Chart—External
Threads   393
Standard Screw Thread Chart—Internal
Threads   395

APPENDIX F

Square End Straight Dowel Pin   397
Clevis Pin   397
Cotter Pin   398
Taper Pin   398
Grooved Pin—ASA   399
Grooved Pin—ABA   400

APPENDIX G

Bends in Sheet Metal—90-Degree
Developed Length   401
Bends in Sheet Metal—Greater Than 90-
Degree Developed Length   402

APPENDIX H

Sheet Metal Gages   403

APPENDIX I

Square, Flat, Plain Taper, and Gib Head
Keys   404
Woodruff Keys   405
Pratt and Whitney Round-End Keys   406

INDEX   407

PREFACE

Current Practices for Interpreting Engineering Drawings is a worktext of educational experiences whose purpose is to provide the learner with the knowledge and ability to read and understand engineering, working, and production drawings used in industry today. It also provides the basic skills needed to produce a freehand sketch.

The rationale for the worktext is that there will be an increasing demand for workers with technical skills in the manufacturing sector throughout the world in the 1990s and well into the 2000s. The primary vehicle for communication in business and industry is the drawing (working drawings, technical drawings and sketches, production drawings, graphics, and CAD/CAM drawings) and various types of diagrams. The form that drawings eventually take is the *blueprint, whiteprint,* or some other type of visual *copy,* in the form of an image or hard copy. Technical personnel must be competent enough in the workplace to be able to read and accurately interpret the many different kinds of copy.

The worktext includes an introduction to the topic of and rationale for reading and interpreting various kinds of drawings used in the manufacturing process, and then moves to sections on how to use measuring tools available in the shop; different types of media and drawing formats; lettering techniques, text presentation, and line conventions; the types of technical sketches; and the sketching and interpretation of single-view, two-view, and three-view drawings, including orthographic projection. In addition, sectional and auxiliary views, surface texture, dimensioning systems, tolerancing methods, fasteners and joining methods, power transmission elements, detail and assembly drawings, development and sheet metal drawings, and computer-generated drawings are covered. Exercises in technical mathematics appropriate to specific topical areas are included. Because of the rapidly increasing rate at which drawings are being produced using CAD systems, all mechanical drawings in the worktext were produced using Claris CAD software, Macintosh IIsi hardware, and a Hewlett Packard HP IIIP laser jet printer. All drawings conform to the standards of the American National Standards Institute (ANSI), the United States Department of Defense (DOD), the International Organization of Standards (ISO), and other relevant, accepted military and industrial organizations.

The worktext illustrates up-to-date examples as well as current practices in each of its sections. Many illustrations are used to facilitate the application of each described example.

This work is intended to be a core text along with its instructional guide and solutions manual. As is usually true with materials of this type, the text organization allows the individual learner to progress from basic, simple concepts and exercises to more challenging work. Each section is prefaced by clearly stated learner outcomes and performance is measured against these experiences by competency exercises at the end of each section. A technical glossary is also included at the end of each section, with concise definitions of important terms. The text is sufficiently illustrated to permit its use in a self-paced mode, or the instructor may choose to lecture on each topic. The worktext is self-contained; that is, no additional materials or references are needed to successfully complete the coursework. The many tables and charts in the appendixes ensure that no additional references are required. The tear-out page format of the book allows the student to work directly on a table or a drafting board to complete all competency exercises.

The educational experiences provided are intended to serve learners with varying backgrounds. This worktext is appropriate for students in high school technical programs, community colleges, technical institutes, vocational programs, industrial training programs, machine and mechanical technology programs, industrial technology programs, manufacturing processes programs, engineering technology courses, quality control programs, and inspection programs. Specific personnel in the workplace who would benefit from its contents include apprentices, journeyperson machinists, tool and diemakers, mold makers, pattern makers, instrument makers, machine builders, and mechanical technicians, as well as those thinking about a career in the technical industries. In short, anyone who wants or needs to read and accurately interpret engineering drawings (and draw basic freehand sketches) will benefit from the contents of this worktext.

ACKNOWLEDGMENTS

No textbook can be written without the help of the many people who have affected the author in some positive way. Colleagues from industry (particularly General Dynamics Corporation) and those from higher education (the University of Minnesota and the College of the National Technical Institute for the Deaf at Rochester Institute of Technology) provided me with the knowledge, ability, confidence, and technical expertise to produce this work. Specific technical advice and support was provided by Eder Benati, Will Yates, and Jorge Samper of NTID at RIT. I am very appreciative also for the technical expertise provided by my son, Edward P. Maruggi, and the gentle nudging and encouragement from my wife, Carolyn.

Sincere thanks and appreciation also go out to the L. S. Starrett Company; Boston Gear, a Division of Incom International; Paper Machinery Corporation; Boker's Incorporated; International Business Machines; Hewlett-Packard; Apple Computer; Xerox Corporation; and Rotor Clip Company, Emerson Power Transmission Corporation, Epson America, and Canon Computer Systems for allowing the use of photos and drawings of products they manufacture and provide to customers around the world.

My editor, Christopher Conty, Stefanie Reardon of West Publishing Company, and Lachina Publishing Services deserve a special thanks for their dedication to producing a quality product. The textbook reviewers provided by West Educational Publishing Company deserve a special acknowledgement for taking the time to critique this publication and to make it acceptable for its intended use.

REVIEWERS

Thomas Acuff	*Black Hawk College (IL)*
Robert Benware	*Daytona Beach Community College (FL)*
William Denardo	*Lansing Community College (MI)*
Peter Fricano	*Triton College (IL)*
Jerry Hayes	*Texas State Technical College, Abilene*
Roely Hellinga	*George Brown College AA&T (Ontario, Canada)*
John Hoffman	*Delta College (MI)*
Albert Mueller	*Waukesha County Technical College (WI)*
Greg Rodgers	*George Brown College AA&T (Ontario, Canada)*
Paul L. Rosengren	*Gulf Coast Community College (FL)*
Jeff Woodson	*Columbus State Community College (OH)*
Bruce R. Wright	*State University of New York, College of A&T, Cobleskill*

ABOUT THE AUTHOR

Edward Albert Maruggi has extensive experience in both industry and education. He began his career as a drafter of mechanical parts for industrial laundry equipment at the American Laundry Machinery Corporation in Rochester, New York. In addition, he was employed by the Mixing Equipment Company, a division of General Signal Corporation, as a drafter of components for various size mixers and agitators used primarily in the chemical and petroleum industries. He also worked for the Tenna Corporation as a mechanical designer of automobile radios and antennas.

Most of his industrial experience, however, was as a mechanical engineer, section head, and engineering manager at the General Dynamics Corporation, Electronics Division, Rochester, New York, where he worked on various engineering projects under contract for the United States Government. Design projects included sonar, radar, and transistorized battle announce systems for the Navy and an Infrared test station for the F-111 Aerospace Ground Equipment program for the Air Force.

For many years, Dr. Maruggi served higher education at the Rochester Institute of Technology, College of the National Technical Institute for the Deaf, in various faculty and administrative roles at the undergraduate and graduate level including professor, Chairperson, Director, and Assistant Dean for Engineering Technology programs for hearing impaired students. He was accorded the rank of Professor Emeritus in 1992.

Among the awards Dr. Maruggi received through the years include an Education Professions Development Act Fellowship at the University of Minnesota for a doctoral program in vocational education (1972/1973). He was selected by the Rotary Foundation for the Teacher of the Handicapped Award to study and conduct research in Italy for one academic year (1980/1981), and also recognized as a Pioneer in Engineering Technology in New York State by the Council of Engineering Technology (1993).

In addition to earning an engineering technology degree from Rochester Institute of Technology, Dr. Maruggi has both undergraduate and graduate degrees in Vocational-Technical Education from the State University of New York at Oswego and a Ph.D. in Vocational Education Administration from the University of Minnesota, Minneapolis, Minnesota.

Dr. Maruggi has presented and published extensively having presented more than twenty-five papers and several articles in the field of technical education, industrial management, and career education. In addition, he is a textbook author. His publications include Machine Shop Manual (1972), privately produced; Technical Graphics; Electronics Worktext (1986) for Charles E. Merrill (Prentice-Hall); second edition (1991); The Technology of Drafting (1991) for Charles E. Merrill (Prentice-Hall); Disegno Electronico (1989); second edition (1993) by Jackson Libri, Milan, Italy.

SECTION 1

Introduction to Current Practices for Interpreting Engineering Drawings

LEARNER OUTCOMES

You will be able to:

- State the purpose for a drawing.
- Describe the value of being able to interpret a drawing.
- Explain the diazo copy process.
- Identify nine common technical drawings required by industry.
- Explain the form that a drawing takes in an industrial community.
- Describe two methods for producing a hard copy.
- Define ten technical terms.
- Demonstrate your learning through the successful completion of competency exercises.

1.1 RATIONALE FOR INTERPRETING DRAWINGS

In today's marketplace, many American consumer, commercial, and military products are manufactured in such countries as Japan, Taiwan, Korea, China, and the like. Automakers from Japan and Germany are producing cars in the United States. Large U.S. multinational companies have joint partnerships with foreign industries. An American military contractor of fighter aircraft produces its airframes in the United States, but hundreds of its assemblies, subassemblies, and parts are made in other countries. The universal language that ties industrialized nations together for the production of goods throughout the world is the engineering drawing. The form that the drawing takes in the industrial community is the print. The word *Print* refers to the *blueprint, whiteprint, sepia, brown copy,* or *Xerox,* which are some of the many names given to *hard copy* of a drawing. Because of the wonder of today's computer-aided drafting (CAD) technology, the first image of a drawing, very often, is a visual image on a computer screen, or monitor. In this instance, a hard copy is produced on a printer or a plotter from the visual form. Examples of a monitor and a plotter are shown in Figures 1.1 and 1.2.

If a drawing is produced through traditional means by a drafter or designer on a drafting board, the hard copy is usually produced on a blueline diazo process machine. The diazo process of reproducing engineering drawings has been the primary method of making

FIGURE 1.1
A monitor produces a visual image

FIGURE 1.2
Printers produce hard copy

prints since the 1950s. The dry diazo process is based on the reaction of light-sensitive paper to ultraviolet light. When exposed to this light, this paper has the characteristic of decomposing into a colorless substance. Ammonia is usually used as the developing liquid in this process, and its odor can be a problem if the equipment is not vented properly. This type of printing machine produces blueline prints, referred to as *blueprints*. Figure 1.3 illustrates the flow of a drawing through the blueline diazo process machine.

The purpose of the engineering drawing is to provide visually oriented data that is usable by technical, engineering, and manufacturing personnel to assist in the production of goods and services. The engineering drawing, which, as the name implies, was produced in the engineering department, is the primary vehicle for this language. Upon release from the engineering department, it is frequently referred to as a **production drawing**, or **working drawing**.

The drawing, therefore, is the language of industry and is a method of communication.

This graphic technical language is used in engineering departments and manufacturing facilities, between contractors and customers, and between buyers and vendors. To interpret, describe, and analyze a drawing, a person must be able to understand its symbols, lines, shapes, views, surfaces, dimensions, and detailed information, which take the place of many words. One in possession of these skills will become an intelligent and productive member of a manufacturing or production group or team. (Please see Figure 1.4.)

The industrial sector requires many types of drawings and diagrams for the various applications and uses of consumer, commercial, and military products

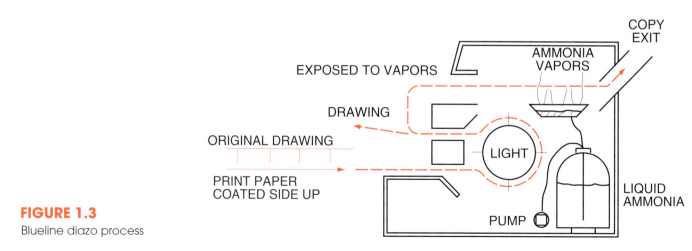

FIGURE 1.3
Blueline diazo process

Section 1: *Introduction to Current Practices for Interpreting Engineering Drawings* 3

FIGURE 1.4
Two people interpreting a drawing

produced throughout the world. Some types of drawings are required more often than others, depending on the nature of the particular industry's product or product line, and on the structure and size of the company. The following represents a list of the more common types of mechanical and related drawings and diagrams required by industry today:

- The assembly drawing
- The control drawing
- The design layout drawing
- The detail drawing
- The erection drawing
- The expanded assembly drawing, also referred to as the *exploded view drawing*
- The installation drawing
- The kit drawing
- The mechanical schematic drawing
- The numerical control (N/C) drawing
- The packaging drawing
- The piping drawing
- The proposal drawing
- The tabulated assembly drawing
- The tube bend drawing

Several of these drawings will be covered in subsequent sections in this worktext.

1.2 SOME ANCIENT PRACTICES

For thousands of years, various forms of drawings have been produced to illustrate designs, ideas, and concepts, They have been helpful in the construction of various structures since very early times. Primitive humans made sketches and drawings with crude tools, such as rocks, sticks, and other available sharp instruments. A graphic language called *hieroglyphics* was used by the Egyptians between 4000 and 3000 B.C. to portray and represent objects. The papyrus plant provided a parchment type of writing and drawing surface, like paper, that allowed more advanced and elaborate sketches to be produced. An example of this language is shown in Figure 1.5. The characters shown represent several different areas, including floor plans; health, medical, and religious symbols; and the environment (the sun rising).

1.3 CURRENT PRACTICES

Since these more primitive methods of producing drawings were used, tremendous changes have taken place. Manual drafting practices had predominated until the advent of computer-aided drafting. Manual drafting can be performed with a minimal investment in equipment and tools, or one can spend large sums of money in outfitting a drafting laboratory. Modern

FIGURE 1.5
Egyptian hieroglyphics

drafting rooms are equipped with large, full-size manual drafting stations that include drafting boards and drafting machines designed to move in any direction quickly and accurately to accommodate the technical needs of the drafter. In addition, drafting chairs are comfortable, supportive, and adjustable to various heights. The manual drafting station usually includes a reference table located behind or to the side of the drafter. This table is used for reference materials the drafter needs, which may include drawings, layouts, drafting manuals, reference books, and catalogs. Figure 1.6 illustrates a typical manual drafting station used in a modern drafting room.

The automated drafting station is called the *computer-aided drafting (CAD) workstation*. It is used for producing drawings and consists of two basic elements: **hardware** and **software**. *Hardware* may be defined as the physical components of the system, including electrical, electronic, electromechanical, magnetic, and mechanical devices. Software consists of sets of procedures, programs, and related documentation that direct operation of the system to produce graphics and related text material. Hardware for a basic CAD system always consists of a **central processing unit (CPU)**, a **memory device**, an **input device**, an **output device**, and a storage facility for programs and files. The system illustrated in Figure 1.7 is a complete **interactive** computer system for two- or three-dimensional drafting. In-depth coverage of computer-generated drawings is presented in Section 15.

Current practices in drafting may be found in "Dimensioning and Tolerancing for Engineering Drawings," a specification of the American National Standards Institute (ANSI), revised in 1982 and often called ANSI Y 14.5 M. This document establishes and illustrates current methods to be used in specifying dimensions and tolerances on drawings. Also,

FIGURE 1.6
Modern manual drafting station

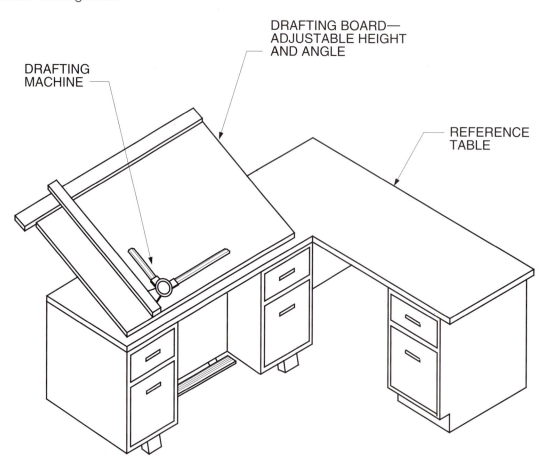

FIGURE 1.7
CAD workstation

there is apparently a new awareness of, and a need for, moving more rapidly and directly to the metric system of measurement. Major multinational industries have converted to SI units *(Système international d'unités)* in an effort to support their worldwide operations in dealing with countries committed to the metric system.

1.4 SUMMARY

Understanding and interpreting engineering drawings in today's marketplace is an important skill to possess. The print is the language of industrialized nations throughout the world. There are many kinds of prints used in industry and business. Both visual images and hard copy print were explained in this section, and several types of technical drawings required by industry were identified. Some ancient drawing techniques were discussed. Current practices were outlined and equipment for both manual drafting and computer-aided drafting was listed and defined.

TECHNICAL TERMS FOR STUDY

blueline diazo process A method of producing a print (hard copy) of a drawing through the process of exposing light-sensitive paper to ultraviolet light and then feeding it through a developing solution, primarily ammonia.

central processing unit (CPU) Considered the brain of a computing system. Integrated-circuit chips form an important part of the microprocessor portion of the computer, which accomplishes the logical processing of data. The CPU contains the arithmetic, logic, and control circuits of the system.

computer-aided drafting (CAD) An automated electronic graphics tool that replaces traditional equipment and drafting tools. It produces computer-generated drawings.

engineering drawing The art of mechanical drawing as it pertains to the field of engineering; communicating through the use of lines, symbols, shapes, views, dimensions, and detailed information, which take the place of many words.

hardware The physical components of a computer system, including the electrical, electronic, electro-mechanical, magnetic, and mechanical devices.

input device A unit that provides data and instructions to the computer. An input device takes information from the operator and prepares it for processing by the CPU.

interactive Describes a CAD system in which there is communication or interaction among the various components of the system as well as between the system and the operator.

memory The place where data is stored in the computer. The size of the memory is determined by the amount of information it can store.

monitor A unit that displays a visual image on a video screen. Also known as a CRT, a graphics display, or a screen.

output device A device that produces output. The reason for inputting data into the computer is to receive output. Output has two forms, visual copy or hard copy.

plotter An output device that draws mechanically, with a pen, or electrostatically on paper, vellum, or polyester film. It produces a hard copy.

print The form that a drawing takes when a hard copy is made.

production drawing A form of technical drawing that presents information, ideas, and instructions in pictorial and text form. Production drawings are usually classified into two groups: detail drawings and assembly drawings. Also called a **working drawing**.

software A collection of programs used to control the internal functioning of the hardware of a computer system.

working drawing *See* **production drawing**.

Section 1: *Introduction to Current Practices for Interpreting Engineering Drawings*

Student _____   Date _____

Section 1: Competency Quiz

PART A COMPREHENSION

1. What is the purpose of an engineering drawing? (6 pts)

2. Why is it valuable to be able to interpret an engineering drawing? (6 pts)

3. What kinds of equipment produce a hard copy? (12 pts)

4. What is a monitor used for? (6 pts)

5. Identify six different drawings and diagrams used by industry. (12 pts)

6. What was the graphics language used by the early Egyptians? (6 pts)

7. Name four pieces of equipment used in a traditional drafting room. (8 pts)

8. Identify four pieces of basic equipment required on a CAD system. (8 pts)

©1995 West Publishing Company

8   Section 1: *Competency Quiz*

9. Define the term *hardware*. (6 pts)

10. Define *software*. (6 pts)

11. What is the ANSI specification for dimensioning and tolerancing engineering drawings? (6 pts)

12. What do the letters *CPU* stand for? (6 pts)

13. What is the function of a CPU? (6 pts)

14. Define *production drawing*. (6 pts)

Section 1: *Introduction to Current Practices for Interpreting Engineering Drawings*   **9**

Student _____   Date _____

PART B TECHNICAL TERMS

For each definition, select the correct technical term from the list on the bottom of the page. (10 pts each)

1. _____ The form that a drawing takes when a hard copy is made.

2. _____ A collection of programs used to control the internal functioning of the hardware of a computer system.

3. _____ An output device that draws mechanically, with a pen, or electrostatically on paper, vellum, or polyester film.

4. _____ The physical components of a computer, including the electrical, electronic, electromechanical, magnetic, and mechanical devices.

5. _____ A term describing a CAD system in which there is communication or interaction among the various components of the system as well as with the operator.

6. _____ The place where data is stored in the computer.

7. _____ An output device that displays a visual image on a video screen.

8. _____ An automated electronic graphics tool that replaces traditional drafting equipment and tools.

9. _____ A method of producing a print through the process of exposing light-sensitive paper to ultraviolet light and then feeding it through a developing solution.

10. _____ Considered the brain of a computer system. It contains the arithmetic, logic, and control systems.

A. blueline diazo process
B. CAD
C. central processing unit
D. engineering drawing
E. hardware
F. input device
G. interactive
H. memory
I. monitor
J. plotter
K. print
L. production drawing
M. software
N. working drawing

©1995 West Publishing Company

SECTION 2

Measuring and Measurement Systems

LEARNER OUTCOMES

You will be able to:

- Describe the difference between the terms *measuring* and *measurement systems*.
- Identify two basic measurement systems.
- Define the dual-dimensioning system.
- Identify five precision and nonprecision measuring devices.
- Identify correct measurements using both the decimal-inch and metric scales.
- State the purpose of the spring caliper.
- Identify the basic component parts of the vernier caliper.
- Identify the basic component parts of the micrometer.
- Define ten technical terms.
- Demonstrate your learning through the successful completion of competency exercises.

2.1 INTRODUCTION

Working with measuring tools and measurement systems is part of every technician's job in the manufacturing sector. In the performance of machining operations, the tradesperson is obligated to use the tools of the trade, which include measuring tools and equipment. For the needs of persons working in machining occupations, **measuring** may be defined as determining or estimating the size or shape of an object by comparing it with some previously established standard. Measuring or measurement can take several forms. For example, a pound or a kilogram is a measure of weight. A quart or liter is a measure of liquid. The number of square feet in a building is a measure of area. The contents of a closed cylinder in cubic centimeters is a measure of volume. Using a measuring device to determine the length, width, or height of a machined part are examples of the act of measuring. This worktext deals primarily with dimensional measurements. A **dimensional measurement**, which is a means of expressing the distance between two points, can be either linear or angular. It is a numerical value expressed in appropriate units of measure, such as inches, millimeters, yards, meters, degrees, minutes, and so forth. These are increments used by industrial nations and conform to some national or international standard.

There are a significant number of multinational companies in existence today, which means that the same part or group of assembled parts can conceivably be produced in several different locations at the same time; therefore, the precision measurement and consistency in producing mechanical parts is of utmost importance. In today's industrial environment, the term used to describe the science of weights and measures is **metrology**. Metrology is an important aspect of the ability to produce quality products in the United States or elsewhere.

On a drawing, a **dimension** is a numerical value used to describe the size, shape, and features of a part or an object. Measuring accurately is a requirement for the successful production of quality goods in business and industry. The industrial worker must know about measurement and how it is applied in the workplace. In this section, various types of shop measuring tools and devices will be described and their uses explained.

Measuring tools must be used with a measurement system. **Measurement systems** are standardized methods used throughout the world for measurement. There are two basic types of measuring systems. The one used predominantly in the United States is the **decimal-inch system**. The other, which most of the world follows, is the **International System of Units**, or **SI** (from the French term *Système international d'unités*). This system is an expanded version of the metric system. In the late nineteenth century, the SI (also called the international metric system) was legalized in the United States, but even today its use is not mandatory. It is used primarily by large corporations and those industries that have overseas subsidiaries.

■ 2.1.1 The Decimal-Inch System

The basis for the decimal-inch system is a two-place decimal unit using an increment of .02 inch (¹⁄₅₀ inch) or multiples of .02 inch, such as .800, 1.600, and 1.260 inch, as illustrated in Figure 2.1.

Decimal-inch dimensions which are not multiples of .02 inch, such as .15, .03, and .05 inch, for example, are only used when it is necessary to meet some design requirement. When greater accuracy is required for close-tolerance work and fits, three- or four-place decimals are used, as shown in Figure 2.2. (Note that for values less than one, the zero does not precede the decimal point.) Appendix A, "Inches/Metric Decimal Equivalents," identifies common decimal equivalents for the decimal-inch system of measurement.

FIGURE 2.1
Decimal-inch measuring unit

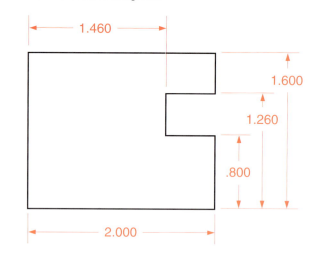

FIGURE 2.2
Decimal-inch dimensions

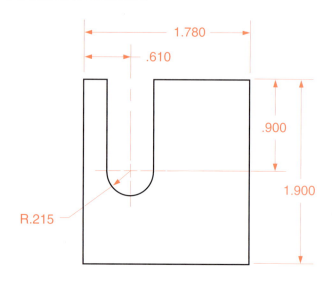

Because of our general lack of knowledge of other measurement systems, we in the United States are primarily familiar with the standard system of linear units shown here:

12 inches (")	= 1 foot (ft)
36 inches (")	= 1 yard (yd)
3 feet (ft)	= 1 yard (yd)
5,280 feet (ft)	= 1 mile (mi)
1,760 yards (yds)	= 1 mile (mi)

FIGURE 2.3
Angular units

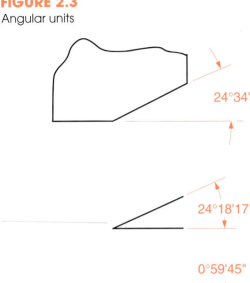

In this system, all dimensions are expressed as decimals except angular dimensions, which continue to be expressed in degrees, minutes, or seconds or as decimals. These dimensions are demonstrated in Figure 2.3.

■ 2.1.2 The International System of Units

In the late eighteenth century, the French established the **meter** for measurement purposes, thus launching the metric system. It was determined that the meter represented one ten-millionth (1/10,000,000) of the distance from the earth's equator to the pole. The meter, therefore, became 39.37 inches in length, or approximately 1.1 yards.

Probably the greatest advantage of the metric system is its ease in computation. The system is based on the use of the numeral 10 or multiples of 10, and it is much easier to use than the decimal-inch system. For example:

1 mm = 1 millimeter (¹/₁₀₀₀ of a meter)
1 cm = 1 centimeter (¹/₁₀₀ of a meter)
     = 10 mm
1 dm = 1 decimeter (¹/₁₀ of a meter)
     = 10 cm
     = 100 mm
1 m = 1 meter
     = 100 cm
     = 1000 mm
1 km = 1 kilometer
     = 1000 m
     = 100000 cm
     = 1000000 mm

The **millimeter (mm)** is the very basic increment used in industry for both linear and angular measurements. For students in the machine trades who are unfamiliar with metric units, it is extremely important to memorize the following conversions:

1 inch = **25.4 millimeters**
(decimal-inch system)   **(metric system)**

.03937 inch = **1 millimeter**
(decimal-inch system)   **(metric system)**

The following guidelines are normally observed for millimeter dimensions on drawings and are depicted in Figure 2.4.

1. When the dimension is less than one millimeter, a zero precedes the decimal point (Figure 2.4a).
2. When the dimension is a whole number, neither the decimal point nor a zero is shown (Figure 2.4b).
3. When the dimension exceeds a whole number by a decimal fraction of one millimeter, the last digit to the right of the decimal point is not followed by a zero (Figure 2.4c).
4. Neither commas nor spaces are used to separate digits into groups in specifying millimeter dimensions on drawings (Figure 2.4d).

FIGURE 2.4
Guidelines for millimeter dimensions

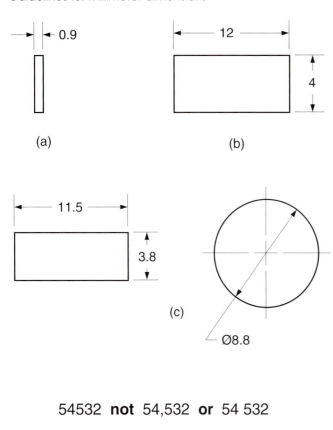

FIGURE 2.5
Acceptable dual-dimensioning practices

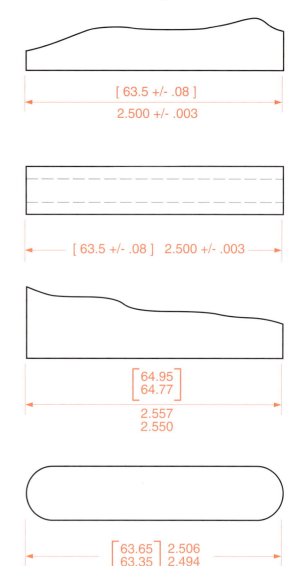

2.1.3 The Dual-Dimensioning System

Because the metric system is not universally accepted in the United States but used throughout the rest of the world, many industries use both the decimal-inch and metric systems simultaneously. This combined system is called the **dual-dimensioning system**. When this system is used, both metric and decimal-inch values are shown on the same drawing. For example, in a method approved by the **American National Standards Institute (ANSI)**, metric dimensions may be shown in brackets, [ ], and placed above the decimal-inch dimension, or they may be located side by side, as Figure 2.5 illustrates. The dimension line may be used to separate the decimal-inch and metric dimensions when conditions warrant.

Note: Only one of the methods shown in Figure 2.5 is permitted on a single drawing.

When a drawing includes dual-dimensioning practices, it is clearly identified in a note close to the title block of the drawing and is normally displayed in one of the ways shown in Figure 2.6.

An alternate method of dual-dimensioning takes the form of a conversion table of all dimensional values. The usual location on a drawing for this table is in the upper-left-hand corner of a drawing, as portrayed in Figure 2.7.

2.2 USING BASIC MEASURING TOOLS

Over the years, producing high-quality, tight-tolerance, precise assemblies for industry and for NASA has become more and more complicated. Machinists

Section 2: *Measuring and Measurement Systems* **15**

FIGURE 2.6
Dual-dimensioning identification

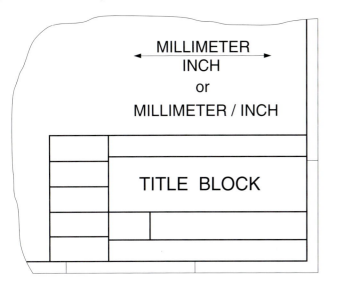

FIGURE 2.7
Alternate identification of dual dimensions

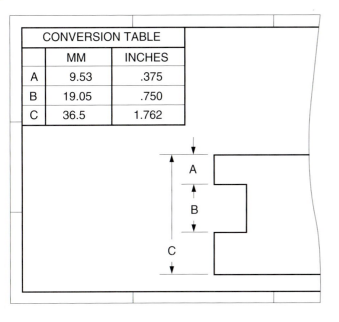

operations and also for inspecting completed workpieces. In addition to understanding both the decimal-inch and metric measurement systems, the ability to perform technical mathematic calculations is of extreme importance.

Measuring devices are generally categorized into two groups: **nonprecision** and **precision measuring tools**. The nonprecision types include the tape measure; the steel rule (scale); certain kinds of calipers (inside and outside); architects, metric, and engineers scales; and the protractor. The precision types are those tools that, through the use of an adjusting screw or other device, are able to perform very exact measurements, often to .0001 inch. Examples of these include the micrometer (inside and outside), vernier caliper, height and depth gages, dial indicator, gage blocks, and sine tools. The remainder of this section will cover a sampling of both nonprecision and precision measurement tools and devices.

■ 2.2.1 The Steel Rule

The steel rule is a nonprecision measuring tool available in several different shapes and sizes. High-quality rules are made of hardened and tempered alloy steel or stainless steel. Rule manufacturers produce them in rigid or flexible form and in lengths from 1 to 48 inches. The most common steel rule is a rigid one that is 6 inches in length and .75 inch wide. Rules are available in decimal-inch, metric, or dual measurement systems. Figures 2.8a and 2.8b show two rigid steel rules and illustrate the graduations on both for the decimal-inch and metric systems.

When reading a steel rule, it is necessary to be able to identify the major rule divisions, such as 1 inch, 2 inches, 10 millimeters, 20 millimeters, and so on, and the subdivisions as well. Sample readings are shown in Figures 2.9 and 2.10.

■ 2.2.2 The Caliper

The **caliper** is a device used to measure the inside or the outside diameter of a workpiece. It is also used at times to measure other dimensions. There are several different types of calipers used in industry. They include the *spring*, *vernier*, *dial*, *electronic*, *digital*, and *hermaphrodite* types. Examples of the various types are shown in Figure 2.11.

and shop personnel have had to work to closer tolerances and as a result, better quality-control measures have been developed. To be an effective technician in the modern machine trades industry, a worker must be qualified to understand and use all the common measuring and layout tools available. These types of tools are used when performing bench and machine

16 Section 2: *Measuring and Measurement Systems*

FIGURE 2.8a
Steel rule showing decimal-inch graduations

FIGURE 2.8b
Steel rule showing metric graduations

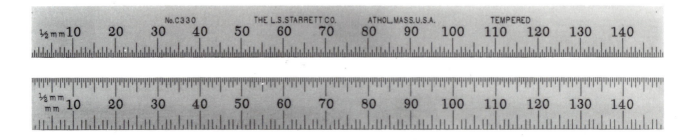

FIGURE 2.9
Reading a decimal-inch rigid steel scale

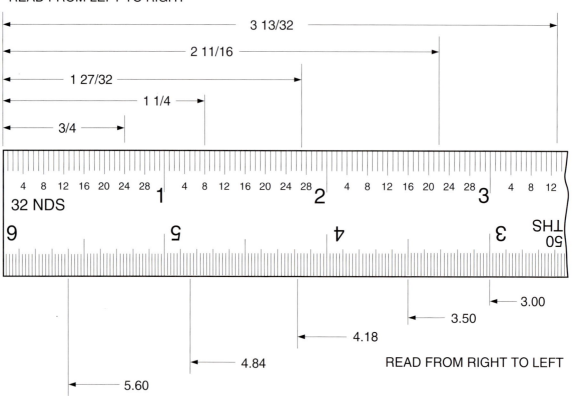

FIGURE 2.10
Examples of metric rule readings

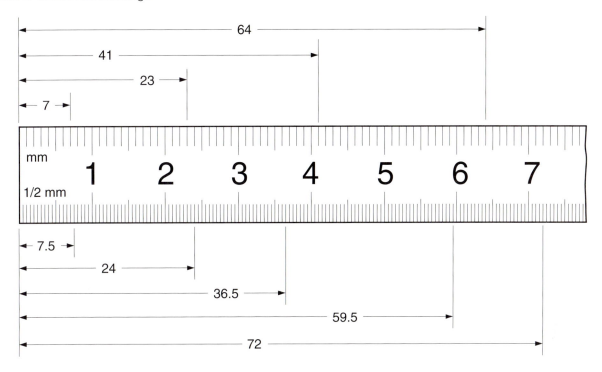

FIGURE 2.11
Different types of calipers used in industry

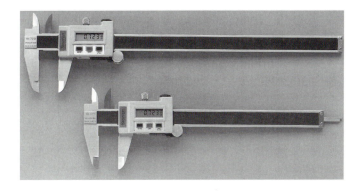

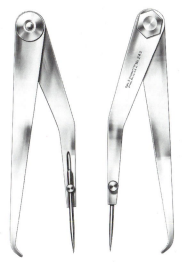

The most commonly used caliper is the **spring caliper**. It is considered a nonprecision transfer measurement tool for performing inside measurements on holes, slots, and grooves, and outside measurements on shafts, cylinders, and pipes. The spring caliper is so named because it has a circular spring ring located above the pivot point, which allows the legs of the caliper to move inward or outward. A screw-and-nut mechanism is used to adjust the caliper to the required setting. Figure 2.12 illustrates the various components of the spring caliper.

FIGURE 2.12
Component parts of a spring caliper

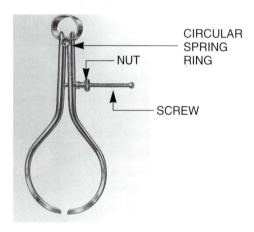

Calipers range in size from 2 inches to 8 inches. The 2-inch size indicates the maximum space the 2-inch caliper is capable of measuring. The spring caliper is also called a comparison measurement tool because after taking the measurement of a workpiece, the accuracy of the measurement must be determined by comparing the caliper setting to a scale, rule, or some other device to assess its accuracy. When using a caliper for the first time, follow these steps for determining the outside diameter of a workpiece:

1. Set one jaw of the caliper on the workpiece, as shown in Figure 2.13a.
2. Using this point as a pivot point, open the caliper until the free jaw barely clears the diameter to be measured.
3. Swing the jaw over the maximum diameter a few times while turning the adjust screw until the feel of a slight pressure exists between the jaws and the workpiece. Do not overtighten the adjusting screw.
4. When satisfied that the measurement is as accurate as possible, remove the caliper and compare it to a steel rule to determine the correct reading for the measurement (see Figure 2.13b). The procedure for the inside caliper is similar.

The **vernier caliper**, which is a precision measuring device, can be used to measure accurately in increments of .001 inch for internal, external, and depth dimensions. Because of this flexibility, this tool has a wide range of applications in the shop. The term **vernier** refers to a scale used for measuring a fractional part of one of the divisions of a longer scale. It is named after Pierre Vernier, a French mathemati-

FIGURE 2.13a
Setting the spring caliper for measurement

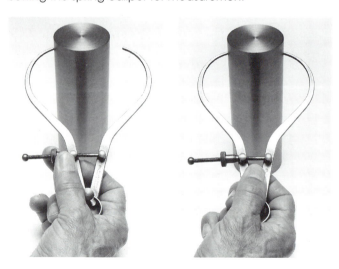

FIGURE 2.13b
Determining the correct measurement

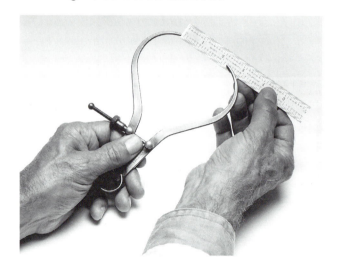

FIGURE 2.14
Uses of the vernier caliper

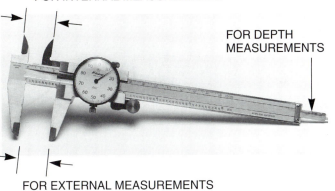

FIGURE 2.15
Component parts of the vernier caliper

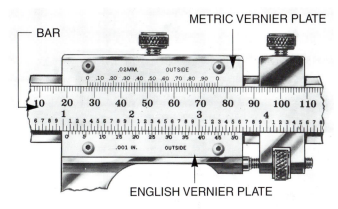

cian (1580-1637). Figure 2.14 illustrates the three main purposes of the vernier caliper. Note that the left-hand end is for measuring inside and outside dimensions, while the right-hand side is for measuring depth.

The most-used vernier calipers are perhaps the 25- and 50-division models. On the 25-division type, the *vernier beam* or *bar* scale is divided into sections of 1 inch, .1 inch, and minor divisions of .025 inch. The *vernier plate*, however, is divided into 25 equal parts that, when combined, are the same length as the 24 divisions on the beam. The vernier plate divisions are, therefore, .001 inch shorter than the divisions on the beam. This allows measurements to be accurate within $1/1{,}000$ of an inch. The only exception is when the zero on the vernier scale is precisely aligned with a line on the beam, in which case the line numbered 25 also lines up with a line on the beam. Figure 2.15 identifies the various important components of the vernier caliper.

When reading the vernier caliper, the *movable jaw* and *fixed jaw* should contact the workpiece firmly, but not with excessive pressure. Use the fixed jaw as the reference point. If possible, try to make readings while the tool is in contact with the workpiece. Also, take measurements more than once to ensure the accuracy of the measurement. To read the vernier caliper, count all the readings to the left of the zero on the vernier scale according to the following steps (as shown in Figure 2.16):

1. On the true scale (beam), the largest whole number to the left of the vernier scale equals

    1.000"

2. There are 4 numbered increments of .100 inch each to the left of the zero on the vernier scale.

    4 x .100" = .400"

3. There are two whole divisions to the left of the zero on the vernier scale.

    2 x .025" = .050"

4. The number 9 on the vernier scale is the very first line aligned directly with a line on the true scale. Since each increment on the vernier scale is equal to .001 inch, multiply .001 inch by 9.

    9 x .001" = .009"

5. Therefore, the total accurate reading in this case equals

    1.000"
    .400"
    .050"
    .009"
    ─────
    1.459"

An additional vernier caliper reading example appears in Figure 2.17.

20 Section 2: *Measuring and Measurement Systems*

FIGURE 2.16
Vernier caliper reading of 1.459 inch

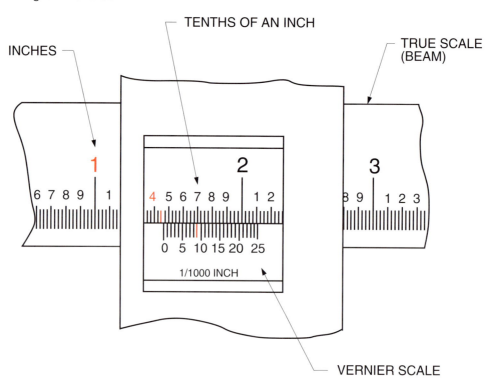

FIGURE 2.17
Vernier caliper reading of 2.875 inches

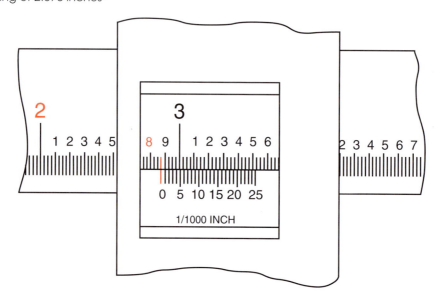

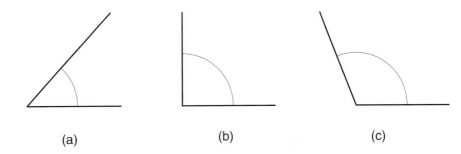

FIGURE 2.18
Angles: (a) acute, (b) right, (c) obtuse

■ 2.2.3 The Protractor

For success in the workplace, machine trades personnel should be able to measure angles accurately. The **protractor** is an angle-measuring instrument that allows a technician to determine angles on mechanical parts. An **angle** is formed when two lines originate from a common point. There are three types of angles: the *acute angle*, whose space occupies less than 90 degrees; the *right angle*, which is exactly 90 degrees; and the *obtuse angle*, which occupies more than 90 degrees. An example of each is shown in Figure 2.18.

In the decimal-inch system, the unit of angular measure is the **degree**. A few basic angular equivalents are listed here:

1 minute   = 60 seconds (1' = 60")
1 degree   = 60 minutes (1° = 60')
360 degrees = a circle

The metric system uses the **radian** as its basic unit of angular measurement. Because the radian is not frequently used in regular machining occupations, it will not be covered here, except to indicate that a radian is the length of an *arc* (on the *circumference* of a circle) that is equal to the *radius* of the circle. Because the circumference of a circle is equal to $2\pi r$, there are two radians in a circle. When converting radians to degrees, the following is true:

one radian = $360°/2\pi r$ = $57°17'44''$

There are several different types of protractors in use today. They include the *plate, bevel, universal bevel, vernier*, and *combination set bevel protractor*. The bevel protractor is usually part of a machinist's combination set. It is so named because there are two main components to the assembly, a protractor and a steel rule that work in combination. The steel rule is available in 6-, 12-, or 24-inch lengths. The protractor is used jointly with the rule to lay out angles or to make measurements on a workpiece during machining operations. As shown in Figure 2.19, the protractor is made to slide along the edge of the rule, which has a groove in it and locks into any position through the use of a lock nut. Note that the protractor has a flat base that allows it to rest squarely on the workpiece. Most protractors are graduated in 1-degree increments, from 0 degrees to 180 degrees, and are considered a nonprecision measuring instrument. Both the scale and the protractor are direct reading devices.

FIGURE 2.19
The combination set bevel protractor

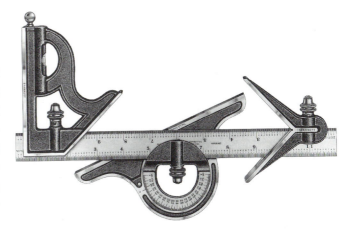

■ 2.2.4 The Micrometer

The **micrometer** is considered a precision measuring device. It was developed more than 350 years ago. It came into prominence about the time of the Civil War and has grown in popularity over the years as machined products were required to meet more precise manufacturing standards. The micrometer takes many forms and is the most commonly used precision measuring instrument in the workplace. The various types of micrometers include the *blade, disc, hub, outside, screw thread,* and *digital electronic micrometer.* They are used to indicate outside and inside diameters as well as depth. Representative examples of the various micrometers used in industry are illustrated in Figure 2.20.

The type that will be covered in this section is the outside-measuring micrometer. To understand the workings of a micrometer, one must know the individual parts and their function within the micrometer assembly. Figure 2.21 describes each part of the micrometer assembly.

FIGURE 2.20
Examples of several types of micrometers

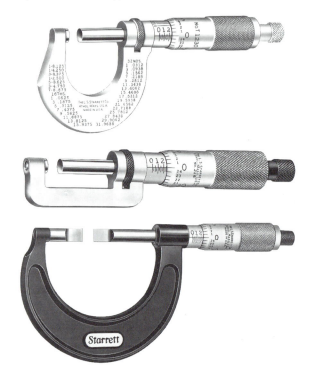

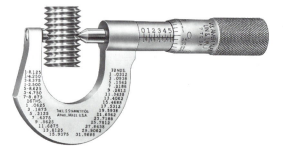

FIGURE 2.21
Component parts of the outside-measuring micrometer

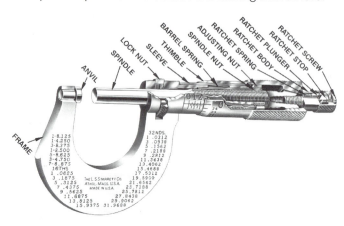

As the *ratchet screw* of a micrometer is rotated clockwise, the *spindle* will advance a precise distance toward the *anvil*, therefore closing the distance between the spindle and the anvil. Note in Figure 2.21 the location of the *thimble*. It has a beveled section that is divided into 25 equally spaced divisions around its circumference. The thimble rides over the *sleeve*, which is graduated into 10 equal major divisions, each with 4 smaller increments. When the thimble is rotated one revolution, the spindle, which has 40 threads per inch, advances p which is equal to $\frac{1}{40}$ of an inch, or 0.025 inch. The pitch (p) is represented in thousandths of an inch, as shown here:

$$p = \frac{1}{\text{number of threads per inch}}$$

$$p = \frac{1}{40}$$

$$p = 0.025"$$

Graduations for the inch micrometer are shown in Figure 2.22. Note that the reading of the micrometer in Figure 2.22 is precisely 1.000 inch.

FIGURE 2.22
Graduations for the inch micrometer

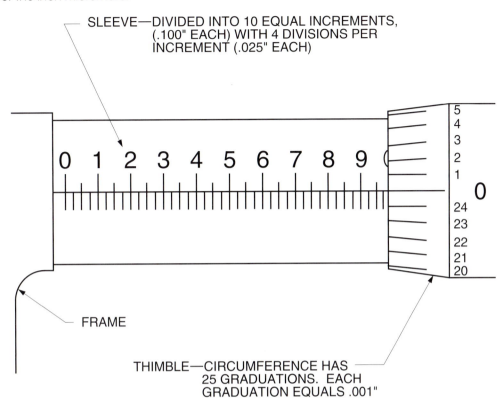

Because the micrometer is considered a precision tool, care must be taken when using it. Excessive pressure when measuring a workpiece will not only provide an inaccurate reading but may also place a strain on the frame. Tradespersons have to learn the certain "feel" involved in determining the correct amount of pressure when using this tool. When measuring round or cylindrical pieces, at least two readings should be made because parts may not be perfectly round (concentric). Reading the micrometer requires observing the graduations on the sleeve as well as the ones on the thimble. The steps used to determine the sample reading identified in Figure 2.23 is as follows:

1. Note that increment 6 is the last whole number shown on the sleeve. The distance between each of these numbers is equal to .100 inch.

$$6 \times .100" = .600"$$

FIGURE 2.23
Micrometer reading of .673 inch

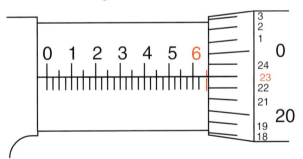

2. In addition, two smaller divisions are visible on the sleeve. Each division equals .025 inch.

$$2 \times .025" = .050"$$

3. The number 23 is directly in line with the long horizontal line (called the index line) on the sleeve. The distance between each increment on the thimble is .001 inch.

$$1 \times .023" = .023"$$

4. Therefore, the reading in Figure 2.23 equals

.600"
.050"
.023"
―――
.673"

An additional example of a micrometer reading is shown in Figure 2.24.

FIGURE 2.24
Micrometer reading of .442 inch

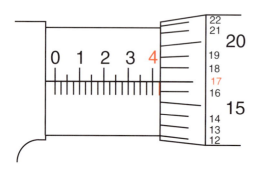

2.3 SUMMARY

This section emphasized the importance of measuring tools and measurement systems and their relationship to the machine trades industry. Two basic measurement systems, the decimal-inch system and the International System of Units, were covered in depth. The one predominately used in the United States is the decimal-inch system, while most of the world uses the metric system. Illustrations and examples of measurements made with each system were provided. The dual-dimensioning system, which is used mostly by multinational corporations was discussed, and its use and value were explained. The use of basic measuring tools and devices was outlined, and their value to the worker was stressed. The tools detailed are those used primarily in performing bench and machine operations and also for inspecting completed workpieces. The two categories of measuring tools are nonprecision and precision types. Nonprecision types are used to perform measurements of a fractional nature. Discussed were the steel rule, the spring caliper, and the protractor. The precision types covered were the vernier caliper and the micrometer. The precision types are used, very often, to measure in increments of .0001 inch. Major components of these devices were illustrated and explained, and in-depth instructions on how to read these instruments were given.

TECHNICAL TERMS FOR STUDY

American National Standards Institute (ANSI) A professional body that has developed standards for common procedures in many areas of industry.

angle Formed when two lines originate from a common point. Three types of angles exist: the acute, the obtuse, and the right angle. Sometimes, a straight angle is referred to as being one with 180 degrees.

caliper A device for measuring the inside or outside diameter of a workpiece. There are several types, including the spring, vernier, dial, electronic, and hermaphrodite (so called because of its unusual leg design).

decimal-inch system A measurement system based on the use of a two-place decimal increment of .02 inch ($1/50$ inch) or multiples of .02 inch.

degree A unit of angular measurement in the decimal-inch system.

dimension A numerical value on a drawing used to describe the size, shape, and features of an object or a part.

dimensional measurement A means of expressing the distance between two points. A numerical value that can be linear or angular; expressed in appropriate units of measure, such as inches, yards, millimeters, meters, degrees, minutes, and so forth.

dual-dimensioning system A measurement system combining the decimal-inch and the metric system of dimensioning on the same drawing.

International System of Units A measurement system used throughout most of the world; an updated version of the metric system established in

Section 2: *Measuring and Measurement Systems* **25**

France in the nineteenth century. It is based on the use of the numeral 10 or multiples of 10, and it is much easier to work with than the decimal-inch system. It uses the meter for measurement purposes.

measurement systems The methods used throughout the world for measurement.

measuring The act of performing a measurement.

meter A unit of measurement determined by the French to be equal to one ten-millionth ($^1/_{10,000,000}$) of the distance from the earth's equator to the pole. A meter is equal to 39.37 inches, or approximately 1.1 yards.

metrology The science of weights and measures.

micrometer The most commonly used precision measuring device in the machine trades. It was developed more than 350 years ago. Various types include the blade, disc, hub, indicating, direct reading, screw thread, and digital electronic.

millimeter The basic increment used in industry for both linear and angular metric measurements.

nonprecision measuring tool A device used to measure in fractional increments on workpieces. Examples include the tape measure, steel rule, drafting scales, certain kinds of calipers, and the protractor.

pitch *(p)* A measurement that equals 1 divided by the number of threads per inch.

precision measuring tool A measurement tool capable of performing very exact measurements, often within .0001 inch, whose fine adjustment is accomplished through the use of an adjusting screw.

protractor An angle-measuring instrument for use on mechanical parts. Often part of a machinist's combination set, which incorporates a protractor and a steel rule that work in combination with each other.

radian The basic unit of angular measurement in the metric system. A radian is the length of an arc (on the circumference of a circle) equal to the radius of the circle.

SI An abbreviation for the French term *Système international d'unités*. Same as the International System of Units.

spring caliper One of the types of nonprecision measuring devices for performing inside measurements on holes, slots, and grooves, and outside measurements on shafts, cylinders, and pipes. It is so named because it has a spring ring located above the pivot point, which allows the legs of the caliper to move inward or outward. A screw-and-nut mechanism are used to adjust the caliper to the required setting.

steel rule A nonprecision measuring tool made of hardened and tempered steel or stainless steel and available in various sizes and shapes. Steel rules are made in flexible or rigid form and in lengths up to 48 inches.

vernier A scale used to measure a fractional part of one of the divisions of a longer scale. Named after a French mathematician, Pierre Vernier.

vernier caliper A precision measuring device that can be used to measure accurately to within .001 inch for internal, external, or depth dimensions.

Section 2: *Measuring and Measurement Systems*   **27**

Student _____   Date _____

Section 2: Competency Quiz

PART A COMPREHENSION

1. Name two types of basic measurement systems. (4 pts)

2. Identify four guidelines normally observed when millimeter dimensions are specified on a drawing. (6 pts)

3. What does the term *dual-dimensioning* mean? (3 pts)

4. Name two groups of measuring tools. (4 pts)

5. Identify four different nonprecision measuring tools. (8 pts)

6. Identify four different kinds of precision measuring tools. (8 pts)

©1995 West Publishing Company

Section 2: *Competency Quiz*

7. In the shop, what is the purpose of a caliper? Name four different types. (11 pts)

8. Why is the vernier caliper considered a precision measuring device? (3 pts)

9. Why is the bevel protractor considered a combination set? (3 pts)

10. What is the most commonly used precision measuring device in a typical shop? (3 pts)

11. What is the name of the following measuring tool? Fill in the blank spaces with the correct measurements. (15 pts)

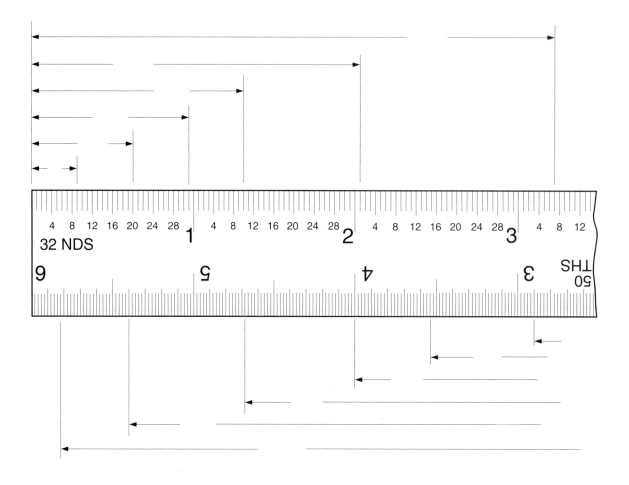

Section 2: *Measuring and Measurement Systems* 29

Student _____   Date _____

12. What is the name of the following measuring tool? Fill in the blank spaces with the correct measurements. (17 pts)

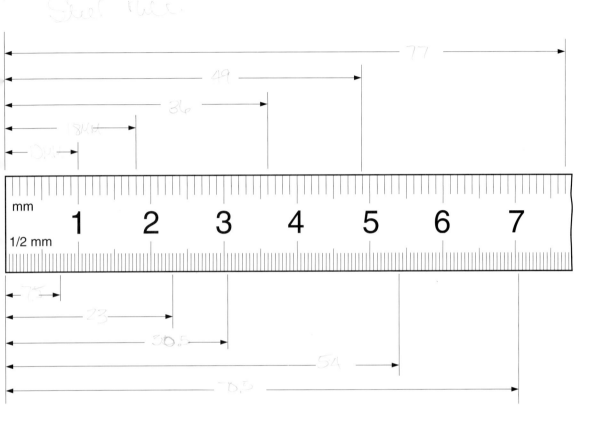

13. What are the correct readings on the two vernier caliper drawings that follow? (6 pts)

a. _____

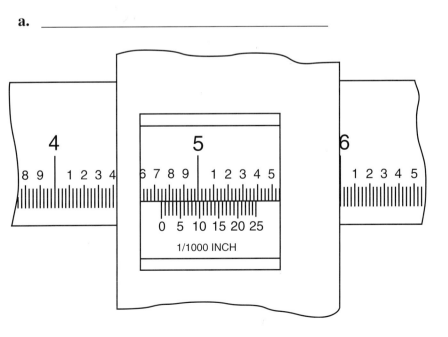

©1995 West Publishing Company

b. _____

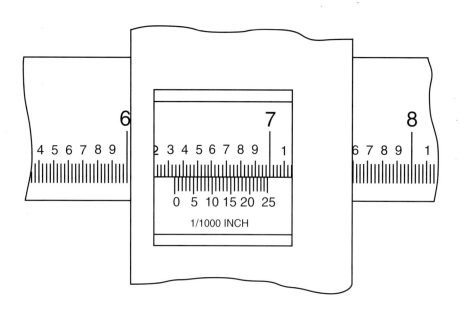

14. What are the correct readings on the three micrometer drawings that follow? (9 pts)

a. _____   **b.** _____

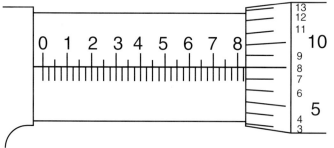   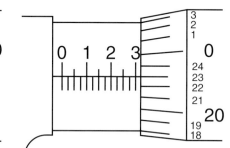

c. _____

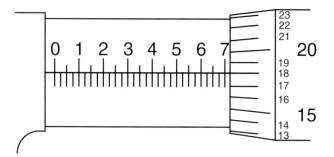

Section 2: *Measuring and Measurement Systems* **31**

Student _____ Date _____

PART B TECHNICAL TERMS

For each definition, select the correct technical term from the list on the bottom of this page. (10 pts each)

1. _Degree_____ The unit of angular measurement in the decimal-inch system.

2. _Steel rule_____ A nonprecision measuring tool made of hardened and tempered steel or stainless steel and available in flexible or rigid form up to 48 inches in length.

3. _radian_____ The basic unit of angular measurement in the metric system.

4. _Metrology_____ The science of weights and measures.

5. _dual-dimensioning_ A system combining the decimal-inch and the metric system of dimensioning on the same drawing.

6. _p - pitch_____ Equals 1 divided by the number of threads per inch.

7. _micrometer_____ The most commonly used precision measuring tool in the machine trades, developed more than 350 years ago.

8. _protractor_____ An angle-measuring instrument for use on mechanical parts. It is often part of a machinist's combination set.

9. _millimeter_____ The basic increment used in industry for both linear and angular metric measurements.

10. _International System of Units_ A measurement system used throughout most of the world, based on the use of the numeral 10 or multiples of 10.

A. angle
B. degree
C. dimension
D. dual-dimensioning system
E. International System of Units
F. meter
G. metrology
H. micrometer
I. millimeter
J. *p*
K. protractor
L. radian
M. steel rule
N. vernier

©1995 West Publishing Company

32   Section 2: *Competency Quiz*

PART C TECHNICAL MATHEMATICS

1. Convert the following into feet. (10 pts)

    a. 27 yards   = _____81_____   **b.** 47.8 yards   = _____15.9_____

    c. 8 yards   = _____24_____   **d.** 254 inches   = _____21.16_____

    e. 84 inches   = _____7_____   **f.** 1 mile   = _____5280_____

    g. 12 inches   = _____1_____   **h.** 3 miles   = _____15840_____

    i. 121 inches   = _____10.08_____   **j.** 4.7 miles   = _____24816.0_____

2. Convert the following into yards. (6 pts)

    a. 15 feet   = _____5_____   **b.** 1 mile   = _____1760_____

    c. 37 feet   = _____12.3_____   **d.** 4 miles   = _____7040_____

    e. 72.3 feet   = _____24.1_____   **f.** 13 miles   = _____22880_____

3. Convert the following into inches. (10 pts)

    a. 2 feet   = _____24_____   **b.** 29 feet   = _____348_____

    c. 11 feet   = _____132_____   **d.** 3 yards   = _____108_____

    e. 5 yards   = _____180_____   **f.** 1 mile   = _____63360_____

    g. 14.5 yards   = _____522_____   **h.** 39.2 feet   = _____3.26_____

    i. 19.5 feet   = _____234_____   **j.** 2.3 miles   = _____145728_____

4. If a section of pipe is 12 feet 6 inches in length, how many pieces 6 inches long can be cut from it? Do not consider any waste. (10 pts)

    900

Section 2: *Measuring and Measurement Systems* **33**

Student _____   Date _____

5. How many 4-inch-long parts can be made from a piece of bar stock that is 5 feet 3 inches long? Allow $^1/_{16}$ inch between parts for cutoff waste. (12 pts)

6. Convert the following into inches. (10 pts)

a. 2 millimeters = _.07874_ **b.** 3 meters = _11.811_

c. 7 millimeters = _.27559_ **d.** 6.3 meters = _24.8021_

e. 59 millimeters = _2.32283_ **f.** 10 centimeters = _3.937_

g. 135 millimeters = _5.3495_ **h.** 23 centimeters = _9.0551_

i. 374 millimeters = _14.72438_ **j.** 149 centimeters = _58.663_

7. Convert the following into millimeters. (10 pts)

a. 3 inches = _76.2_ **b.** .25 yards = _228.60_

c. 6.5 inches = _165.1_ **d.** 1.11 centimeters = _11.1_

e. 8.3 inches = _210.82_ **f.** 6.75 centimeters = _67.5_

g. 11.75 inches = _298.45_ **h.** 31.25 centimeters = _312.5_

i. 24.625 inches = _625.475_ **j.** .68 yards = _621.792_

8. Complete the following. (10 pts)

a. 50.8 millimeters = _1.99996_ inches

b. 7 inches = _177.8_ millimeters

c. 11.375 inches = _288.925_ millimeters

d. 4 centimeters = _1.574803_ inches

$1'' = 25.4 \, mm$

$.03937'' = 1 \, mm$

©1995 West Publishing Company

34   **Section 2:** *Competency Quiz*

e. 13 centimeters   =   _5.1181102_ inches

f. 79.25 inches   =   _2012.95_ meters

g. 9 meters   =   _9000_ millimeters

h. 33 centimeters   =   _330_ millimeters

i. 3 inches + 25.4 millimeters = _4_ inches

j. 123 millimeters + 9.625 inches = _367.4496_ millimeters

9. What is the total length of brass rod required to make 18 pieces 75 millimeters long? Allow 4 millimeters of material as cutoff waste between pieces. Answer in millimeters. (12 pts)

_1422 mm._

10. A flat metal plate is rectangular in shape, 93 millimeters by 176 millimeters. What are the measurements if converted to centimeters? (10 pts)

_9.3 × 17.6_

SECTION 3

Drawing Media and Formats

LEARNER OUTCOMES

You will be able to:

- Define *drawing medium*.
- Identify three drawing media and state their differences.
- List the most common drawing sizes for both the inch and metric formats.
- Discuss the four major parts of a drawing, where they are located, and the purpose of each.
- Explain the purpose of the ECN.
- Define ten technical terms.
- Explain how parts are identified on an assembly drawing.
- Demonstrate your learning through the successful completion of competency exercises.

3.1 INTRODUCTION

Several types of drawings are required to satisfy the needs of business and industry. Because there are small parts and large parts, details, subassemblies, and assemblies to be drawn or copied in different ways, a large number of very different drafting and tracing papers are used. As a result, there is a wide selection of paper and film media from which to choose to meet the specific drawing requirements of any given industry. This section will cover the different options available as well as various drawing formats and their details.

3.2 DRAWING MEDIA

Materials for producing manual and computer-generated drawings include paper, cloth, and film-based items. Each of these materials is called a **drawing medium**. Drafting paper has various characteristics; the one characteristic that is most critical to producing quality reproductions is translucency. If a paper is **translucent**, a person cannot see through it, but it does allow light to pass through. Therefore, the more translucent a drawing medium, the better copies it will produce. Other qualities to look for in a paper medium are its strength, permanency, and erasability. *Strength* is a paper's ability to withstand constant use, abuse, and handling. **Permanency** refers to a paper's ability to retain its qualities over a period of years. For example, will it become brittle, turn yellow, or fade over time? **Erasability** is a drawing medium's ability to withstand erasures, whether due to drawing errors or required changes to a drawing because of part revisions.

The most common medium used in a typical engineering department is *vellum*, which is a white, translucent, rag-content paper. It is a fine general-purpose medium that can be used in several different printing processes and pieces of equipment. It is available in all standard sizes for both the decimal-inch and metric systems.

Another paper product is plain, lined, or unlined bond paper (with a reproducible or nonreproducible grid). It is usually found in sizes $8^1/_2$ by 11 inches or $8^1/_2$ by 14 inches and is available in single sheets or pads of 250 sheets. The advent of the office copier using a toner has made this product easy to use.

The second most common type of medium used is one that has a film base. It is durable, dimensionally stable, and can be stored for many years with no degradation of drawing quality. This film-type medium is available under several commercial trademark names and is really a *polyester film*. Special leads and erasers need to be used when drawing on this type of medium. It is available in various sizes, including a roll size for design layouts. This product costs more than paper, but its permanency and quality over the long term makes it very desirable.

The third most common type of medium used in drafting rooms is *cloth*. It is not used with great frequency today. It is made from a finely woven linen sandwiched between either a starch or plastic material. It can be used to produce ink or pencil drawings or tracings. Cloth is more permanent than paper but less permanent than polyester film.

3.2.1 Standard Drawing Sizes

Drawing media are available in standard sizes, either plain or with the company's format printed on one side. Most industries use both the inch and metric sizes. The inch system is based on the dimensional size of commercial letter-size paper, $8^1/_2$ by 11 inches, and standard paper roll sizes of 36 and 42 inches. The metric drawing size is based on the A0 size (see Table 3.1), which has an area of 1 square meter (m^2) and a length-to-width ratio of 1:$\sqrt{2}$.

Figure 3.1 illustrates a standard medium layout and the difference between the paper width and the border width, while Table 3.1 identifies the most frequently

TABLE 3.1
Frequently used drawing sizes

DRAWING SIZE (INCHES)	BORDER SIZE	OVERALL PAPER SIZE
A	8.00 x 10.50	8.50 x 11.00
B	10.50 x 16.50	11.00 x 17.00
C	16.25 x 21.25	17.00 x 22.00
D	21.00 x 33.00	22.00 x 34.00
E	33.00 x 43.00	34.00 x 44.00

DRAWING SIZE (MILLIMETERS)	BORDER SIZE	OVERALL PAPER SIZE
A4	190 X 267	210 X 297
A3	277 X 390	297 X 420
A2	400 X 564	420 X 594
A1	574 X 811	594 X 841
A0	821 X 1159	841 X 1189

FIGURE 3.1
Standard medium layout

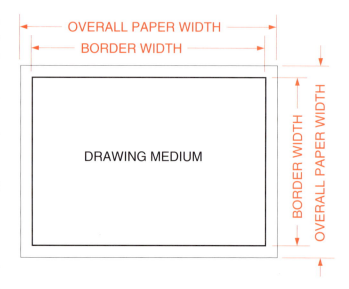

used standard sizes available in both inch and metric dimensions, in accordance with **MIL-STD-100A**, Engineering Drawing Practices, a military standard for industries under contract to the U.S. Government.

3.3 PARTS OF A DRAWING

It is very difficult to interpret, read, and understand an engineering drawing or print without first having

Section 3: *Drawing Media and Formats*   37

knowledge of the parts of such a drawing, their relative importance and relationship to each other. Every drawing contains sufficient general information for the user, such as the title of the part, the part number, the scale of the drawing, the design and drafting approval data, the date the drawing was produced, the drawing field, and so on. The method of presenting and locating this information on a drawing medium varies from one industry to another, but for the purpose of this book, the American National Standards Institute specification **ANSI Y 14.1**, Drawing Sheet Size and Format, will be used.

3.3.1 Drawing Field

Every drawing has a **drawing field**. The drawing field is the main body of the medium and is the area used for producing all the necessary views, including dimensions and text, to complete the drawing. An example of the drawing field area is given in Figure 3.2. Note that, depending on the complexity of the object to be drawn, it may appear off center on the medium because of other data that may conflict with the drawing field area.

3.3.2 Title Block

The **title block** is that area of a drawing containing information not provided on the drawing field itself. It is located in the lower-right-hand area of the drawing format and includes individual areas for specific information, as seen in Figure 3.2. A more detailed example of a typical title block (per ANSI Y 14.1) is pictured in Figure 3.3. Note that all the relevant information in this case is complete. All companies, large and small, prefer to use preprinted stock with the company's name, address, and logo as well as other pertinent information related to its particular business. Title block sizes and the type of information required will vary depending on the size of the company and the types of products it manufactures. Some companies require basic information in the block, while others, especially those under government contract, need to provide more data. Most title blocks are more basic than the example shown.

The various regions, or blocks, that comprise the title block shown in Figure 3.3 are identified here:

- *Contract number block*—The contract number is entered here only when appropriate.

FIGURE 3.2
The drawing field

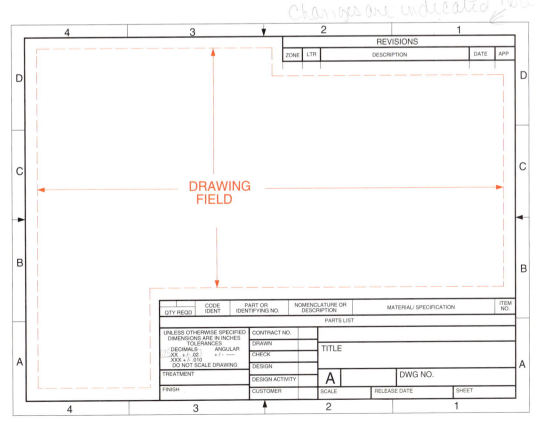

FIGURE 3.3
Typical title block

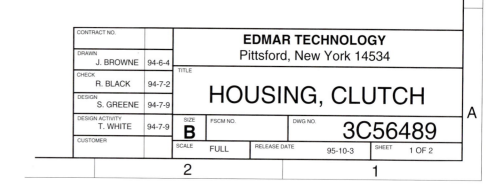

- *Drawn block*—The drafter's name appears here in lettered form. In addition, the year, month, and day the drawing was completed is usually shown (e.g., 94-6-4). Remember that this method of dating the drawing is for those under government contract or doing business with the military. Most companies prefer the standard month, day, year arrangement (6/4/94). In Canada, the preference is day, month, year (4/6/94).
- *Approval block*—Some or all of the individual blocks that comprise the approval block are used, depending on the complexity of the project, the structure of the organization, and whether the company is under contract to the U.S. government or with another company. The Check block is for the name of the engineering checker who is assigned to the project. The Design block is for the name of the responsible engineer, and the Design Activity block is for the name of the program manager or the chief engineer. Each block is signed and dated by authorized project or program personnel.
- *Design activity name and address block*—This block is for the company name, address, and trademark or logo. In our case, this block reads "Edmar Technology, Pittsford, New York 14534."
- *Drawing title block*—This area is for the basic name or title of the part or assembly shown in the field of the drawing. For our purpose, this block reads "Housing, Clutch." This drawing title seems to be listed in an unusual way, that is, backwards. However, all detail parts or assembly titles throughout this worktext are listed this way; it is the approved military method.
- *Size block*—This block contains a letter indicating the appropriate size for the drawing (see Table 3.1).
- *Drawing number block*—The drawing number for the drawing is inserted here. In our case, it is "3C56489."
- *Scale block*—This block identifies to what scale the part is drawn (i.e., full, half, 1:1, 1:2, etc.).
- *Release date block*—When the drawing is complete and has received all the necessary approvals, the date that the drawing is released to manufacturing is inserted here.
- *Sheet block*—Some drawings, because of their complexity and size, require several pages or sheets. When this occurs, the sheet number (e.g., 1 of 2) is inserted in this block. If the drawing consists of only one sheet, the sheet block is left blank.
- *FSCM number block*—This block is completed only for those companies under contract to the U.S. government. The government assigns each company its own identifying number. (FSCM stands for Federal Supply Code for Manufacturers.)

In addition to the information inserted in the title block, more data is often required. Figure 3.4 identifies this **supplemental data**, which is positioned to the immediate left of, and bordering on, the title block (such data is also shown in Figure 3.2). Only the more important of these items are detailed here:

- *Standard tolerance block*—This area is devoted to standard tolerances that apply to dimensions on the drawing. These tolerances, at times, are uniformly stated for an entire project.

Section 3: *Drawing Media and Formats* **39**

FIGURE 3.4
Supplemental data block

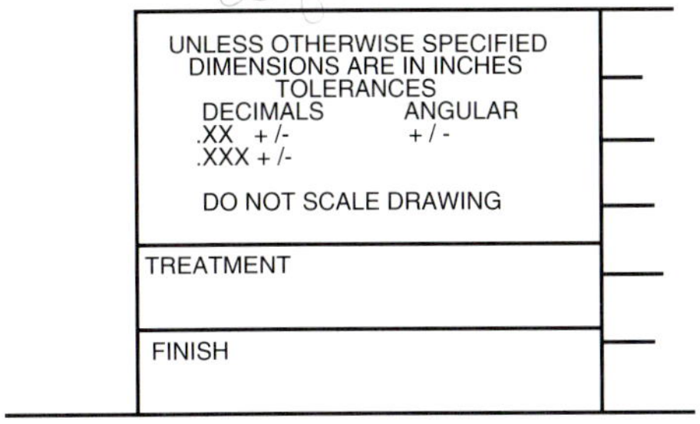

default tolerances

```
UNLESS OTHERWISE SPECIFIED
DIMENSIONS ARE IN INCHES
        TOLERANCES
DECIMALS         ANGULAR
.XX  + /-           + / -
.XXX + /-

DO NOT SCALE DRAWING

TREATMENT

FINISH
```

- *Finish block*—The finish block is used to specify the type of coating that is to be applied to the part or item. It may be a specification for a protective coating, a decorative finish, or both.
- *Treatment block*—Information relative to heat treatment, temper, or hardness, usually in the form of a specification number, is inserted here.

The preceding information may seem complex to the beginning tradesperson; this is because working under a government contract or for governmental agencies can be very complex for an industrial organization.

■ 3.3.3 Revision Area

Most drawings require one or more revisions during their lifetime—revisions in dimensions, shape, text, or form due to changes in design, changes in a customer's requirements, or to correct errors in design or drafting. To perform an engineering change to a drawing, an engineer, designer, or drafter must write a formal document to make certain that the revision is recorded so that the drawing user will know the reason and time for the change. Most industries call this process writing an **ECN (Engineering Change Notice)**. To accommodate engineering changes on drawings, a **revision area**, or *revision block*, has been established. It is normally preprinted as part of the drawing format. According to ANSI recommendations, the revision area should be located in the upper-

right-hand corner of the drawing and should provide space for a description of the change, a revision symbol zone location, and approval for the change. An example of a revision block is provided in Figure 3.5, showing the various areas completed for a given revision to a drawing.

■ 3.3.4 Parts List

The **parts list** is actually a list of materials that identifies those items required to complete an assembly or all the necessary parts of some functional grouping. It is sometimes referred to as a *bill of material*. The various items normally found on the parts list include the required quantity of each part, the part or identifying number, the title or description of the part, the material of the part (when required), and an item number for each part. The parts list is located in the lower-right-hand corner of the drawing medium, directly above the title block (see Figure 3.2). When the parts list is an integral part of the drawing, the lettering is done by the drafter. If the parts list is not part of the drawing, it can either be lettered by the drafter or take the form of a computer printout. Figure 3.6 illustrates a parts list for an assembly entitled "Assy, Tap Wrench."

■ 3.3.5 Part Identification

Note that on the parts list shown in Figure 3.6, an item number is required for each part shown on the assembly. In this case, item numbers 1, 2, and 3 identify the handle, body, and collet, respectively. This ties in the part to the parts list. The primary method for keying the two is through the use of **balloon numbers**, which consist of identification numbers placed inside circles called balloons. Each balloon is attached to the related part with a leader. The **leader** is a thin line that usually ends in an arrowhead (or solid dot) that points to and touches the related part in the field of the drawing. The leader is never shown as either a horizontal or a vertical line, to avoid confu-

FIGURE 3.5
Completed revision block

REVISIONS					
ZONE	LTR	DESCRIPTION		DATE	APPROVED
C - 3	A	.750 " WAS .687 ", R .437 WAS R .375		94-10-6	W.E. STOKES

FIGURE 3.6
Assembly drawing with parts list

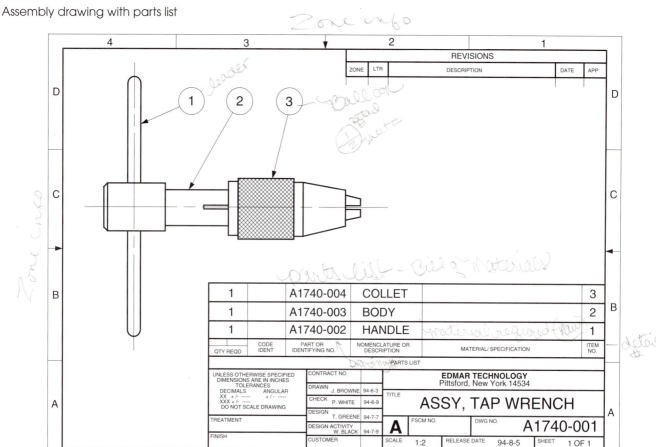

sion with other lines on the drawing. Note in Figure 3.6 that the balloon numbers are not shown at random locations but have some order—either horizontally or vertically aligned or both. Balloons vary in size from .31 to .75 inch in diameter, depending on the size and scale of the assembly drawing.

■ 3.3.6 Part-Numbering System

Every company has some form of *part-numbering system* for identifying and recording drawings for documentation purposes. There is no single best system to use. Most industries use a system in which drawing numbers are assigned sequentially as each drawing is completed. Other companies assign an entire block of numbers to a project, especially if it is a large project. At times, prefix or suffix letters are added to the part number assigned to a drawing. Some companies use as few as four digits in their numbering system, while others use as many as fifteen. In most cases, the part-numbering system includes a base number followed by a dash number. An example of a part-numbering system is presented in Figure 3.7.

3.4 SUMMARY

There is a large selection of paper and film media from which to choose in industry. Each of the three major types—film, paper, and cloth—was examined in this section. Strength, translucency, permanency, and erasability are four qualities that are important to any drawing medium. Drawing media are available in standard sizes in both decimal-inch and metric format. Several standard sizes were identified, with both border and overall sizes included. Details of the various parts of a drawing format were discussed, including the drawing field, title block, drawn block, approval block, design activity name and address block, size block, drawing number block, release date

FIGURE 3.7

Example of a part numbering system

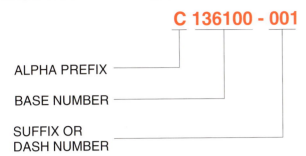

WHEN MORE THAN ONE ITEM IS SHOWN ON A DRAWING, EACH IS IDENTIFIED BY ITS OWN UNIQUE AND CONSECUTIVE PART NUMBER, AS FOLLOWS:

A128090 - 015
A128090 - 016
A128090 - 017
A128090 - 018
A128090 - 019

block, and sheet block. Covered also were supplementary data on tolerances, finish, and treatment. Engineering changes, the revision area, part information, and the use of balloon numbers on an assembly drawing were discussed. Finally, because every company uses some form of part-numbering system, an example of such a system was provided.

TECHNICAL TERMS FOR STUDY

ANSI Y 14.1 A specification (produced by the American National Standards Institute) that covers drafting standards, including drawing sheet size and format.

balloon number An identification number placed inside a circle. Used to relate items on the parts list with those shown graphically on the drawing. Each balloon is attached to a leader.

drawing field The main body of a drawing; the area for producing all the necessary views, including dimensions and text, to complete the drawing.

drawing medium The material on which a manual or computer-generated drawing is produced. Usually a paper, film, or cloth product.

ECN Engineering Change Notice. A formal written document for recording a revision to a drawing.

erasability A drawing medium's ability to withstand erasures, whether due to drawing errors or required changes to a drawing because of part revisions.

leader A thin line attached to a balloon at one end and ending with an arrowhead or solid dot that points to and touches an object or part at the other end. Used for item number identification purposes on an assembly drawing.

MIL-STD-100A A military standard for engineering drawing practices. Used primarily by those companies under contract with the U.S. government or one of its agencies.

parts list A list of materials identifying those items required to complete an assembly. Sometimes referred to as a bill of material. Items shown on the parts list may include the required quantity of each part, the part identification number, the title or description of the part, the material of the part, and an item number for each part.

permanency The ability of a drawing medium to retain its quality over a period of years.

revision area An area located in the upper-right-hand corner of a drawing format to accommodate the documentation for a required revision to a drawing, so that the drawing user will know the reason for the change. Part of the ECN process.

supplemental data block Additional information sometimes needed on a drawing format, such as a tolerance block, a finish block, or a material treatment block.

title block The area of a drawing that contains information not provided in the drawing field itself. It is located in the lower-right-hand corner of the drawing format and includes individual areas for specific information.

translucent The quality of a drawing medium that allows light to pass through, although one cannot see through the medium.

Section 3: *Drawing Media and Formats*   **43**

Student _____   Date _____

Section 3: Competency Quiz

PART A COMPREHENSION

1. Name three common types of drawing media. (6 pts)

   vellum
   mylar
   cloth

2. What are the qualities of the polyester film medium? (6 pts)

   erasable, permanency storable

3. What are the standard drawing sizes for the following? (12 pts)

   A = _____   A4 = _____

   C = _____   A2 = _____

   E = _____   A0 = _____

4. What is the purpose of the drawing field? (10 pts)

   To place all views dimensions of text

5. Identify seven pieces of information normally found in a title block. (14 pts)

   Title Drawing #, release date, sheet
   scale, size

6. Name three items that are usually included as supplemental data on a drawing. (6 pts)

   Tolerance, finish treatment

©1995 West Publishing Company

44   **Section 3:** *Competency Quiz*

7. On an assembly drawing, what is the purpose of the parts list? (10 pts)

to know what the bill of materials are to have items to make the part.

8. What purpose does the ECN serve? (8 pts)

to know what revision + engineering # you are working with - list current.

9. Why is it important to document a change to a drawing? (9 pts)

to know if it is current -

10. Why is a parts list important to an assembly drawing? (9 pts)

to have needed items at time of build.

11. What is a balloon number? Explain its relationship to a leader line on an assembly drawing. (10 pts)

used to relate parts in list to drawing

Section 3: *Drawing Media and Formats*   45

Student _____   Date _____

PART B TECHNICAL TERMS

For each definition, select the correct technical term from the list on the bottom of the page. (10 pts each)

1. ___*BALLOON NUMBER*___ An identification number placed inside a circle to relate items on the parts list with those on the drawing.

2. ___*MIL-STD-100A*___ A military standard for engineering drawing practices, used primarily by those companies under contract with the government.

3. ___*ECN*___ A formal written document for recording a revision to a drawing.

4. ___*ANSI Y 14.1*___ A specification that covers drafting standards, including drawing sheet sizes and format.

5. ___*PARTS LIST*___ A list of materials, identifying those items required to complete an assembly.

6. ___*DRAWING FIELD*___ The main body of a drawing; the area for producing all the necessary views, including dimensions and text, to complete a drawing.

7. ___*PERMANENCY*___ The ability of a drawing to retain its quality over a period of years.

8. ___*DRAWING MEDIUM*___ Materials on which manual and computer-generated drawings are produced.

9. ___*TITLE BLOCK*___ The area of a drawing that contains information not provided in the drawing field of the drawing.

10. ___*LEADER*___ A thin line attached to a balloon at one end and ending with an arrowhead that points to and touches a part on the other end.

A. ANSI Y 14.1
B. balloon number
C. drawing field
D. drawing medium
E. ECN
F. erasability
G. leader
H. MIL-STD-100A
I. parts list
J. permanency
K. revision area
L. supplemental data block
M. title block
N. translucency

©1995 West Publishing Company

Section 3: Competency Quiz

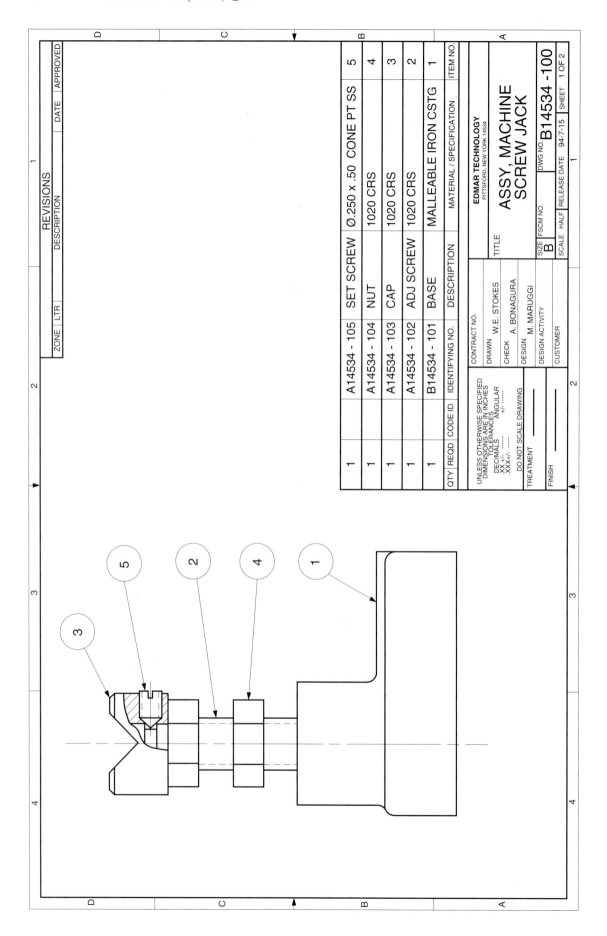

Section 3: *Drawing Media and Formats*   **47**

Student _____   Date _____

PART C DRAWING INTERPRETATION

Refer to Assembly Drawing B14534-100 (on page 46) and the approved abbreviations in Appendix D (pages 389–392) for responding to the following questions.

1. What is the drawing title? (6 pts)

2. How many different parts are there on the assembly? (10 pts)

3. How many revisions have been made to the drawing? (6 pts)

4. What is the part number for the Adjusting Screw? (6 pts)

5. What part number does balloon 5 refer to? (6 pts)

6. What does CRS mean in reference to the Cap, the Adjusting Screw, and the Nut? (12 pts)

7. In what zone is item 2? (6 pts)

8. How many Set Screws are required on the assembly? (6 pts)

9. What is the material for the base? (6 pts)

10. What does SS mean in relation to the Set Screw? (6 pts)

©1995 West Publishing Company

48   **Section 3:** *Competency Quiz*

11. Who was the drafter of the drawing? (6 pts)

12. What is the drawing title for A14534-004? (6 pts)

13. How many sheets are in this set of drawings? (6 pts)

14. What is the drawing size? (6 pts)

15. Balloon 1 identifies which part number? (6 pts)

SECTION 4

Line Conventions and Text Presentation

LEARNER OUTCOMES

You will be able to:

- Define *line convention*.
- List three methods for producing text on engineering drawings.
- Name the style of lettering commonly used on engineering drawings.
- Explain the six basic steps for producing alphanumeric characters.
- Produce acceptable lettering.
- Define ten technical terms.
- Identify two types of mechanical lettering aids.
- Demonstrate your learning through the successful completion of competency exercises.

4.1 INTRODUCTION

The **line** is the most fundamental and important aspect of any drawing or print. It is part of every mechanical or technical drawing. Lines come in many forms, from straight lines and curved lines used to show the size and shape and outline of an object, to lines made freehand to perform lettering tasks. Lines are a permanent part of a drawing, and eventually they translate into actual parts on the manufacturing floor. Several different lines are produced on each drawing. They are used in combination with each other on all prints and drawings in this worktext. There are thick lines and thin lines, each with its distinct intent and use. Lines can be produced freehand, by mechanical means, or with a computer. In any case, lines should be dark, sharp, and clearly visible, so that the drawing user can make no mistake as to their interpretation.

4.2 LINE CONVENTION

A normally accepted use of lines on a drawing is referred to as a **line convention**. According to the ANSI Y 14.2 M-1979 specification entitled *Line Conventions and Lettering*, there are two acceptable weights, or widths, of lines for common types of technical drawings required by industry. These weights are thin (.015 to .022 inch) and thick (.030 to .038 inch). All lines on drawings are clean-cut, opaque, and of uniform width throughout their lengths. There are fourteen different lines in the ANSI specification; their

FIGURE 4.1
Examples of line conventions

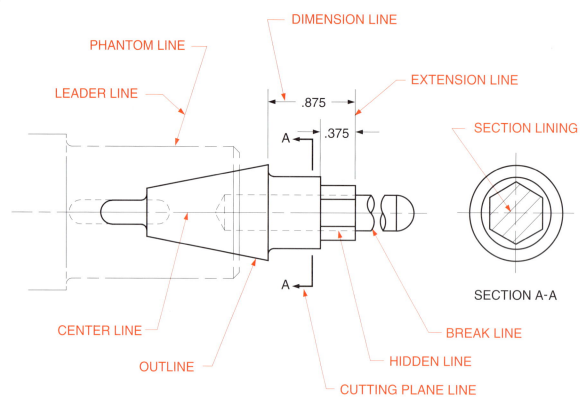

uses are as follows, and an example on how they are used in combination with each other is pictured in Figure 4.1.

- *Center line*—A thin line made up of long and short dashes, alternately spaced and consistent in length, beginning and ending with a long dash. Center lines cross each other without voids. Very short center lines are unbroken if there is no possibility of confusion with other lines. A center line is used to indicate the axis of a part feature, a path or motion, or a theoretical line about which a part or feature is symmetrical.
- *Dimension line*—A thin line the same width as the center line, terminating with arrowheads at both ends and unbroken except where the dimension line is located. This line is used to indicate the extent and direction of a dimension.
- *Leader line*—A thin line the same width as the center line, terminating with an arrowhead or a dot. The arrowhead always terminates at a line, usually an object line. The leader is inclined (slanted) so as to not interfere with other horizontal or vertical lines on the drawing. Also part of the leader line is a short horizontal line that extends to the mid-height of the first or last letter of a note. A leader line is used to indicate a part or the portion of a drawing to which a number, note, or reference applies.
- *Break line*—A thin line with freehand zigzags for long breaks or a thick line for short breaks. Break lines are used to show an area or portion of a part that has been removed to reveal internal detail, to limit a partial section or view, or to eliminate repeated detail.
- *Extension line*—A thin, continuous line that does not touch the outline of the object. Extension lines are used to extend points or planes to indicate dimensional limits.
- *Phantom line*—A thin line consisting of one long and two short dashes alternately and evenly spaced, with a long dash at each end. Phantom lines are used to show alternate positions of

parts, relative positions of adjacent parts shown for reference, or to eliminate repeated detail.

- *Section lining*—A series of thin, continuous lines usually shown at 45-degree angles. (Section lining is used to indicate surfaces that have been cut, to show the inside of a part.)
- *Hidden line*—A thin line of short dashes (approximately $1/8$ inch long), closely and evenly spaced. A hidden line begins and ends with dashes in contact with the lines that define its end points. It is not as outstanding a line as an object or visible line. A hidden line is used to identify hidden features of a part.
- *Outline* or *visible line*—A thick line, usually the most prominent one on a drawing. Outlines are used for all lines representing visible edges or lines of an object.
- *Cutting plane* or *viewing plane line*—A thick line with arrowheads drawn at 90 degrees to the cutting plane line or viewing plane line to indicate the viewing direction. A cutting plane line indicates the plane in which the section is taken; a viewing plane line indicates the plane from which a surface is viewed.

4.3 TEXT PRESENTATION

The words **printing** and **lettering** have been used for many years to describe alphanumeric characters (letters and numerals) placed on drawings. A more appropriate term used today is **text presentation**. The word **text** came into prominence with the advent of the computer, specifically in the field of computer-aided drafting (CAD). Text on engineering drawings consists of dimensions, notes, charts, instructions, specifications, legends, parts lists, and any other information that is best communicated through the use of alphanumeric characters. Text and how it is presented is an important part of any drawing.

There are three generally accepted types of text for engineering drawings. They include the **freehand**, **mechanical**, and **computer-generated** types.

■ 4.3.1 Freehand Lettering Techniques

Good freehand lettering techniques produce acceptable text material on engineering drawings or on freehand sketches. The ability to produce text that is neat, clear, legible, uniform, and of proper density is a necessary skill for persons in the machine trades industry. All text materials for drawings must be of high quality regardless of the final reduction size of the drawing. Some people have a natural talent for lettering, while others need to work hard to become proficient. Skill in lettering can be acquired only by continued practice. During idle moments or free time, one should, with pencil in hand, practice lettering. Practice is the key.

■ 4.3.2 Styles of Lettering

Of the several styles of lettering that have been developed over the years, the acceptable one for use on engineering drawings is the *single-stroke*, *uppercase*, **commercial Gothic font**, also referred to as block lettering. *Single-stroke* means that the required thickness or weight of each letter is formed using one stroke. *Uppercase* indicates that all letters are capitalized, and a *Gothic* font is one in which all strokes of each letter are even or uniform. *Font* means "style" and is a word that is used frequently today because of its relationship to computers and CAD language. A font is a complete assortment of alphanumeric characters of a particular size and style, which can be used for several different applications by an artist, drafter, or illustrator. For example, the artwork and drawings for this worktext were produced by a Hewlett-Packard HP IIIP LaserJet printer, which has thirty-five common fonts available. Figure 4.2 identifies several of these fonts.

In a Gothic font, the alphanumeric characters are produced using six basic strokes, as illustrated in Figure 4.3. *Stroke 1* is a single stroke downward from upper left to lower right at an angle of approximately 30 to 60 degrees. *Stroke 2* is drawn downward from upper right to lower left at an angle of approximately 30 to 60 degrees. *Stroke 3* is a vertical line drawn from top to bottom. *Stroke 4* is a horizontal line drawn from left to right. *Stroke 5* is a semicircular stroke that moves downward in a curved, counterclockwise movement. *Stroke 6* is a semicircular stroke that moves downward in a curved, clockwise movement (the mirror image of stroke 5). All numerals and letters can be produced by using combina-

FIGURE 4.2
Common font names and styles

```
AVANT GARDE
ABCDEFGHIJKLMNOPQRSTUVWXYZ
1234567890

CHICAGO
ABCDEFGHIJKLMNOPQRSTUVWXYZ
1234567890

COURIER
ABCDEFGHIJKLMNOPQRSTUVWXYZ
1234567890

GENEVA
ABCDEFGHIJKLMNOPQRSTUVWXYZ
1234567890

HELVETICA
ABCDEFGHIJKLMNOPQRSTUVWXYZ
1234567890

MONACO
ABCDEFGHIJKLMNOPQRSTUVWXYZ
1234567890

NEW YORK
ABCDEFGHIJKLMNOPQRSTUVWZYZ
1234567890

TIMES
ABCDEFGHIJKLMNOPQRSTUVWXYZ
1234567890

VENICE
ABCDEFGHIJKLMNOPQRSTUVWXYZ
1234567890
```

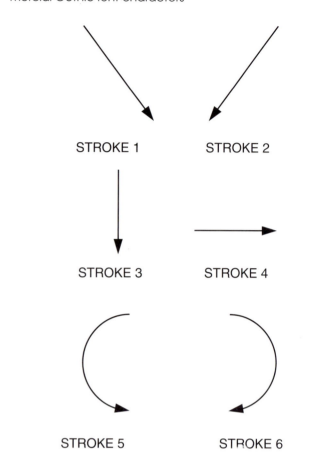

FIGURE 4.3
Basic strokes for producing single-stroke, uppercase, commercial Gothic font characters

FIGURE 4.4
Acceptable and unacceptable lettering practices

tions of these six basic strokes. Tools required for practicing lettering are a lined pad of paper, a 2H or H pencil, and an eraser. Acceptable and unacceptable lettering practices are illustrated in Figure 4.4.

In the machine trades, one is likely to encounter both vertical and slant lettering; however, the most

Section 4: *Line Conventions and Text Presentation* **53**

FIGURE 4.5
Vertical Gothic-style lettering

ABCDEFGHIJKL
MNOPQRSTUVWX
YZ1234567890

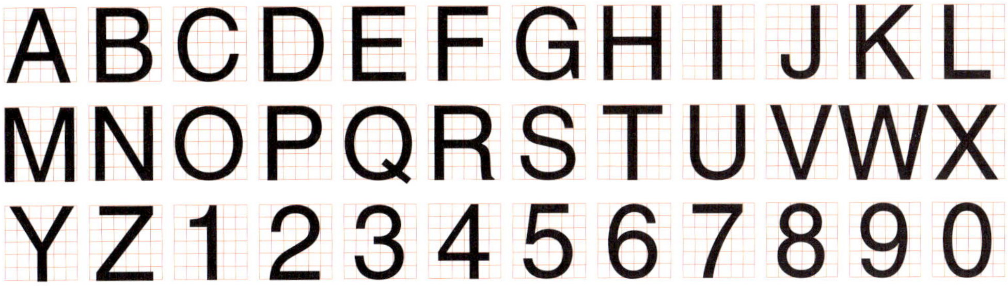

used and accepted text is vertical. Only one type of lettering, either vertical or slant, appears on a drawing. Figure 4.5 illustrates the form and character style of vertical Gothic-style lettering.

The preferred method of identifying fractions and mixed numbers is to use a horizontal bar to separate the fractional numerals. An older method allows the use of an angled bar to provide this separation. Figure 4.6 illustrates both methods.

FIGURE 4.6
Preferred practices for producing fractions and mixed numbers

$$2\frac{3}{4} \quad 7\frac{5}{8} \quad \frac{13}{16} \quad \frac{15}{32}$$

$$1.562 \quad .3125 \quad 1.625$$

CURRENT METHOD

2 3/4   7 5/8   13/16   15/32

OLDER METHOD

Two types of notes are usually found on drawings. These types are a **local note**, which is information regarding hole sizes, diameters, radii, and tap and drill sizes, and a **general note**, which includes data about finish, additional specifications, or other instructions. Text material for general or local notes on drawings is positioned so that the notes align parallel with the bottom of the drawing format, as pictured in Figure 4.7.

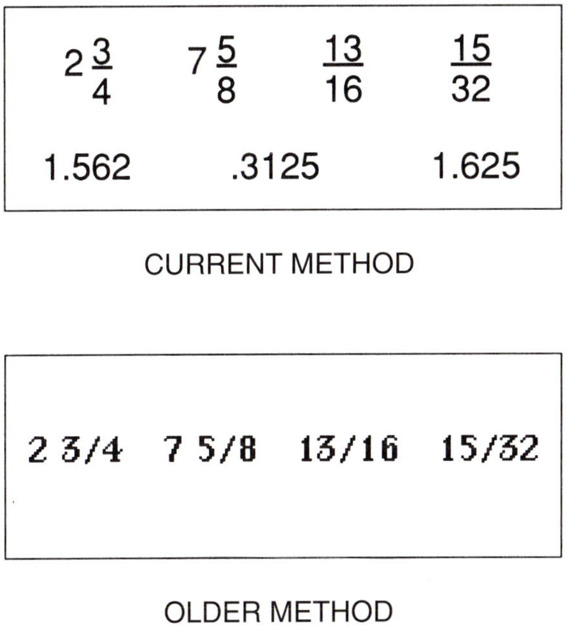

■ 4.3.3 Mechanical Lettering Aids

Several manufacturers produce mechanical lettering aids called **lettering templates**. Lettering templates are used to produce alphanumeric text. They are available in various thicknesses of plastic material and are made in many character sizes and fonts to meet the engineering department's needs. They are capable of producing either vertical or angled lettering. A sample of available templates for vertical mechanical lettering is shown in Figure 4.8.

Another type of aid is **transfer-type lettering**. Transfer-type lettering is usually found on translucent paper sheets with either an adhesive backing or letters that are rubbed off by providing a firm and even pressure to the desired area. **Burnishing** (rubbing the letter with a blunt tool) is required to ensure that the transferred characters remain permanently attached. Transfer-type lettering ensures a uniform shape to characters, but accuracy in the spacing between characters depends on the person performing the transfer operation. Figure 4.9 provides examples of available transfer-type lettering.

54 **Section 4:** *Line Conventions and Text Presentation*

FIGURE 4.7
Alignment of notes

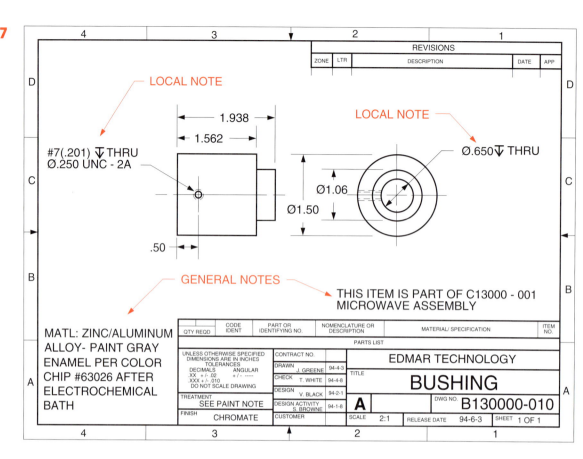

FIGURE 4.8
Lettering templates

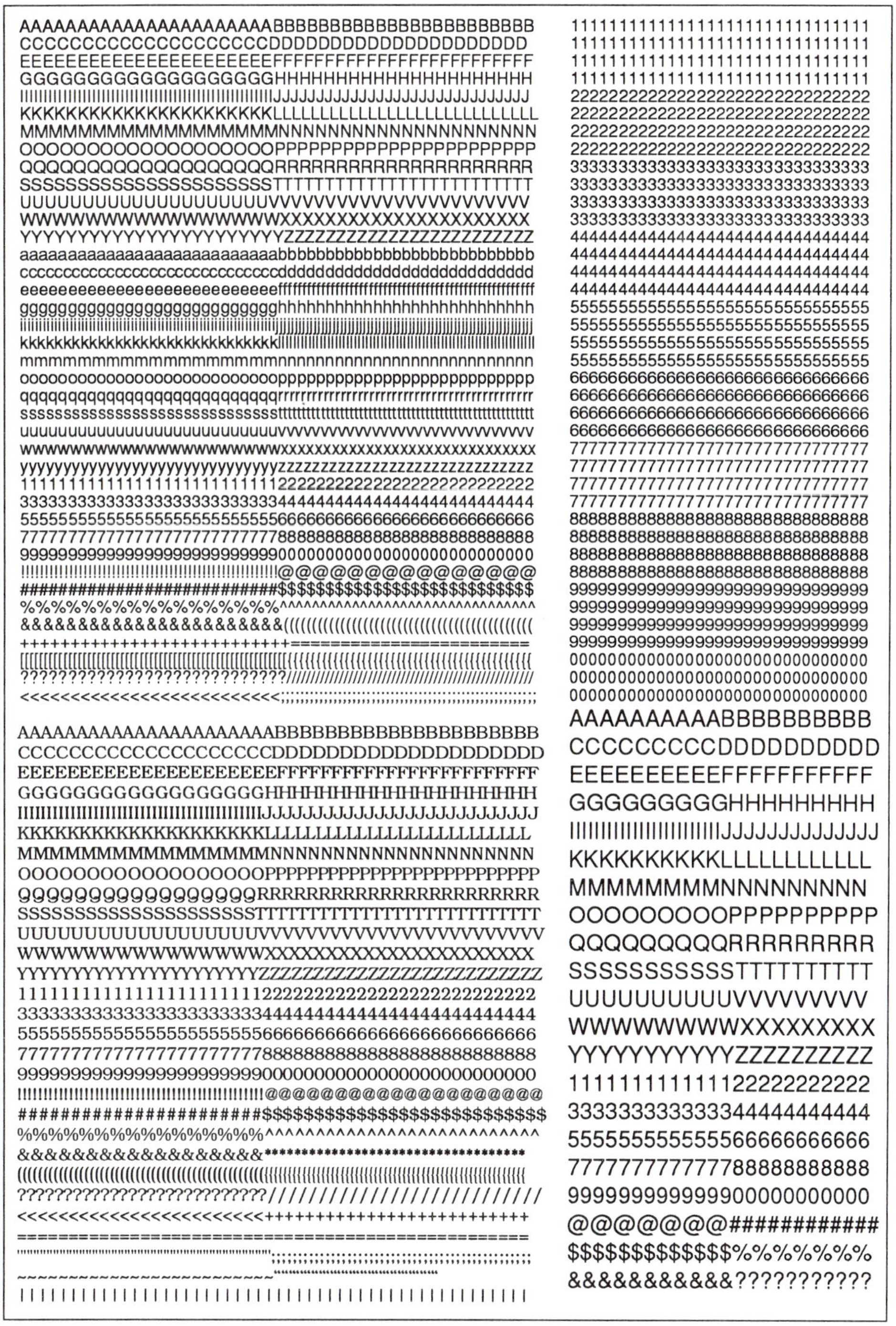

FIGURE 4.9
Transfer-type lettering aids

■ 4.3.4 Computer-Generated Text

Any mainframe, mini-, micro-, or laptop computer that is manufactured today—especially computer-aided drafting equipment—is capable of producing computer-generated text material. Each of these systems can perform word-processing functions. Adding text to engineering drawings is a simple task of inputting alphanumeric characters, using the keyboard. On a computer, text can be inserted and justified to either margin, or it can be centered. It also can be placed at any desired angle. Several different scalable fonts of various heights can be produced in the text mode of a typical computer.

56 **Section 4:** *Line Conventions and Text Presentation*

4.4 SUMMARY

In this section, emphasis was placed on line conventions and text presentation. The line is the most fundamental and important aspect of any engineering drawing. ANSI specification Y 14.2 M-1979 identifies two weights of lines as being appropriate for use on engineering drawings: thick and thin. This section covered all the major types of lines in depth.

Terms connected with the placement of alphanumeric characters on engineering drawings were introduced. The most appropriate term used today to describe lettering and notes on drawings is *text*. In an engineering sense, *text* means the dimensions, notes, charts, instructions, specifications, legends, and parts lists added to drawings. Three methods of producing text material were discussed: the freehand, mechanical, and computer-generated methods. The predominate style of lettering used by engineering departments is the single-stroke, uppercase, commercial Gothic font. Basic strokes were illustrated, and acceptable and unacceptable methods for producing alphanumeric characters were shown. In addition, aids to freehand lettering and mechanical lettering were discussed.

TECHNICAL TERMS FOR STUDY

burnishing   A method of permanently transferring (by rubbing) characters from a sheet of transfer-type lettering to an engineering drawing.

commercial Gothic font lettering   The accepted lettering style for engineering drawings throughout industry. It can be produced either vertically or angled (slanted).

computer-generated text   Text material added to a drawing through the use of the keyboard of a computer.

font   A series of alphanumeric characters of a particular size and style that can be used for several different applications by an artist, drafter, or illustrator.

freehand lettering   Alphanumeric characters produced by hand; such lettering should be neat, clear, legible, uniform, and of proper density.

general note   A note on a drawing, which may include one or more of the following: specifications, general instructions, or special instructions regarding finish, heat treatment, and so forth.

lettering   The form that letters and numbers take on an engineering drawing.

lettering templates   A method of producing lettering and text material on a drawing by using a mechanical device called a template, which is available in various thicknesses of various plastic materials.

line   The most fundamental aspect of every engineering drawing. Lines are drawn according to certain line conventions and take many different forms.

line convention   The normally accepted uses of lines on a drawing. Two line weights (thick and thin) are recommended according to ANSI specification Y 14.2 M-1979. When drawn, lines should be clean-cut, opaque, and of uniform width throughout their length.

local note   Information regarding hole sizes, diameters, radii, tap and drill sizes, and so forth on a drawing.

mechanical lettering aids   Same as **lettering templates**

printing   Same as **lettering** on an engineering drawing.

text   A term used recently to refer to alphanumeric characters on a drawing; same as **printing** and **lettering**

text presentation   The form that alphanumeric characters take on an engineering drawing.

transfer-type lettering   A repetitive series of alphanumeric characters usually found on translucent sheets of paper with either an adhesive backing or with letters that can be rubbed off.

Section 4: *Line Conventions and Text Presentation*   **57**

Student _____   Date _____

Section 4: Competency Quiz

PART A **COMPREHENSION**

1. What is the normally accepted use of lines on a drawing called? (6 pts)

2. List three methods of producing text on engineering drawings. (9 pts)

3. According to ANSI specification Y 14.2 M-1979, how many acceptable line weights are used for common types of technical drawings? What are they? (6 pts)

4. Explain the characteristics of the following lines and their use. (20 pts)

   center line:

   extension line:

   visible line:

   section lining:

   break line:

©1995 West Publishing Company

58   **Section 4:** *Competency Quiz*

5. Name five items that may be considered as text material on an engineering drawing. (10 pts)

6. What style of lettering is commonly used throughout industry? (6 pts)

7. Draw the six basic strokes used in the production of Gothic-style lettering. (12 pts)

8. Provide three examples of a local note and two examples of a general note. (15 pts)

9. Identify two types of mechanical lettering aids. (8 pts)

10. Explain how transfer-type lettering is applied to a drawing. (8 pts)

Section 4: *Line Conventions and Text Presentation*   **59**

Student _____   Date _____

PART B **TECHNICAL TERMS**

For each definition, select the correct technical term from the list on the bottom of the page. (10 pts each)

1. _____ The most fundamental aspect of every engineering drawing; takes many forms according to line conventions.

2. _____ Adding text material to a drawing by using the keyboard on a computer.

3. _____ A method of permanently transferring characters from a sheet of transfer-type lettering to an engineering drawing.

4. _____ A term used recently to refer to alphanumeric characters on a drawing; same as *printing* and *lettering*.

5. _____ The generally accepted freehand lettering style for engineering drawings throughout industry.

6. _____ A method of producing lettering and text material on a drawing by using a mechanical device.

7. _____ On a drawing, information that may include one or more of the following: specifications, general instructions, or special instructions regarding finish, heat treatment, and so on.

8. _____ The normally accepted uses of lines on a drawing.

9. _____ A repetitive series of alphanumeric characters usually found on translucent sheets of paper with either an adhesive backing or letters that can be rubbed off.

10. _____ On a drawing, information pertaining to hole sizes, diameters, tap and drill sizes, and so forth.

A. burnishing
B. commercial Gothic font
C. computer-generated text
D. freehand lettering
E. general note
F. lettering
G. lettering templates
H. line
I. line convention
J. local note
K. printing
L. text presentation
M. text
N. transfer-type lettering

©1995 West Publishing Company

PART C LETTERING ACTIVITY

1. Letter the alphanumeric characters shown in the following example to the required heights using vertical, uppercase, Gothic-style characters. Produce the characters freehand, using the guidelines provided. Letter each example twice.

EXAMPLE: ABCDEFGHIJKLMNOPQRSTUVWXYZ 1234567890

.125" HIGH

.190" HIGH

.250" HIGH

2. Produce, freehand, the following lettering and text presentation using .190-inch-high vertical, uppercase, Gothic-style letters.

QUALITY FREEHAND LETTERING AND TEXT PRESENTATION IS A SKILL THAT EVERY TECHNICAL PERSON SHOULD MASTER. TO ACHIEVE THIS GOAL, PRACTICE IS NECESSARY. ALWAYS USE LIGHTWEIGHT GUIDELINES WHEN LETTERING.

SECTION 5

Technical Sketching

LEARNER OUTCOMES

You will be able to:

- State the intent of the technical sketch.
- Identify materials required for producing technical sketches.
- Identify four types of technical sketches.
- Describe the differences between the various axonometric drawings.
- Explain the steps required for producing circles and arcs.
- List the differences between multiview, isometric, oblique, and perspective sketches.
- List five suggestions for producing quality technical sketches.
- Demonstrate your learning through the successful completion of competency exercises.

5.1 INTRODUCTION

The intent of the **technical sketch**, which is usually drawn **freehand**, is to communicate a concept, idea, or plan in a rapid but effective manner. Knowing how to interpret and draw technical sketches is a key skill for machine trades personnel. Very often, the technical sketch is the primary, and sometimes the only, means of communication between various groups in the workplace. Sketches are drawn by engineers during the design and development of new products. Model shop personnel, instrument makers, and machine builders draw sketches as part of their work. Production and industrial engineers use sketches to show initial or conceptual thoughts on improving production processes. In short, anyone involved with the engineering and production of goods and services can use and benefit from the technical sketch because all communication needs cannot be met by the written or spoken word.

Many preliminary concepts and ideas take the form of a technical sketch with descriptive data, which may later become a finished drawing. Technical sketches, although not usually drawn to scale, are drawn to correct proportions.

The only materials required to produce a technical sketch include one of the softer grades of **leads**, preferably an H or 2H; a **mechanical pencil** or **lead holder**; an *eraser*; and a **drawing medium**. It is a good idea to have two pencils available; one with a very sharp point for sketching thin lines (center lines and invisible lines), and the second with a dull or rounded point for thick lines (cutting plane lines and visible lines). The most suitable medium to use is

FIGURE 5.1
Tools needed to produce a freehand technical sketch

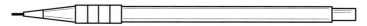

MECHANICAL PENCIL, OR LEAD HOLDER

ERASER

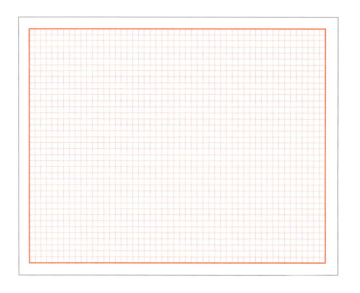

GRID PAPER

bond paper, ruled with a reproducible or nonreproducible grid of 4, 8, 10, or 12 squares to the inch. A plain bond (gridless) translucent paper can also be used if a grid master is placed underneath it to provide grid lines. Examples of the tools necessary for producing a freehand technical sketch are illustrated in Figure 5.1.

5.2 TYPES OF TECHNICAL SKETCHES

Several types of technical sketches are used in the manufacture of industrial mechanical equipment. The most frequently used freehand sketches are the **multiview sketch**, and the **isometric sketch**. Others used to a lesser degree are the **oblique sketch** and the **perspective sketch**. Each will be covered in this section.

■ 5.2.1 Multiview Sketch

A **multiview sketch** is a drawing that shows more than one view and whose lines are drawn using the principles of *orthographic projection* (which is covered in detail in Section 6). In multiview sketching, all lines are produced using the line conventions discussed in Section 4.2. It is important that views be placed in their proper location; that is, the top view should be placed above the front view because they

FIGURE 5.2
Step bracket—multiview freehand sketch

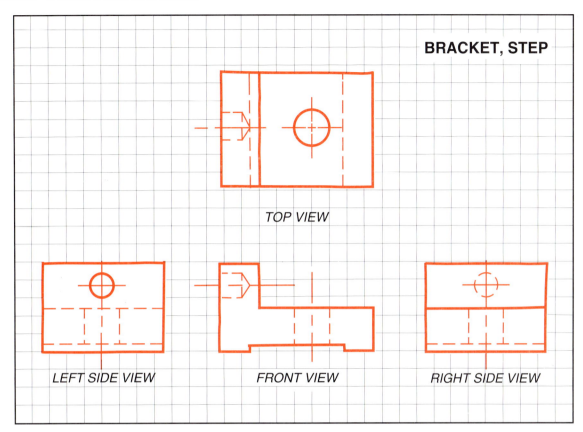

both share the same width dimension. The side view should be placed to the right or left of the front view because they both share the same height dimension. The number of views needed depends on the complexity of the object to be drawn. As many as six views may be required: front, top, right side, left side, bottom, rear, and possibly auxiliary views (see Section 7). Figure 5.2 shows a freehand multiview sketch on grid paper of a simple Step Bracket. The sketch requires four views. Note the location of the views and their relationship to each other.

■ 5.2.2 Isometric Sketch

An **isometric sketch** is a form of **axonometric drawing**. The term *axonometric* means that an object is presented on a surface so that its perpendicular projections appear as inclined surfaces in which three faces are shown. There are two other types of axonometric drawings: the **dimetric drawing**, in which two of the three principal faces are equally inclined to the plane of projection; and the **trimetric drawing**, in which all three faces and axes of the object make different angles with the plane of projection. Figure 5.3 provides illustrations of the three types of axonometric drawings.

Figure 5.3a shows the basic principles of the isometric sketch, including lines that are located 120 degrees apart, or equidistant from each other. In other words, they are drawn vertically and at 30 degrees from the horizontal. When sketching an isometric drawing for the first time, it is wise to begin with three lightweight lines, as illustrated in Figure 5.4, and then use the principles just discussed. Note also that all isometric lines are parallel to the 30-degree isometric lines. Object lines for an isometric sketch are direct reading lines; that is, if a line measures 3 inches at a 30-degree angle, it really is 3 inches long. The same is true for vertical lines.

■ 5.2.3 Oblique Sketch

The principles for drawing a freehand oblique sketch are very similar to those used for drawing an isomet-

FIGURE 5.3
Types of axonometric drawings

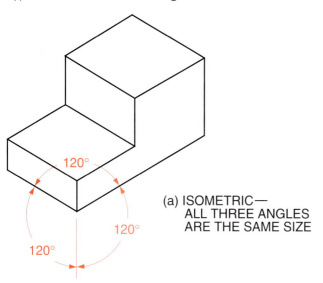
(a) ISOMETRIC—ALL THREE ANGLES ARE THE SAME SIZE

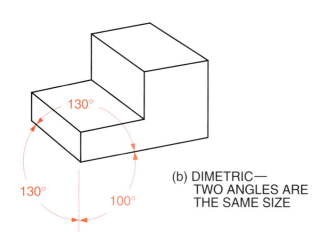
(b) DIMETRIC—TWO ANGLES ARE THE SAME SIZE

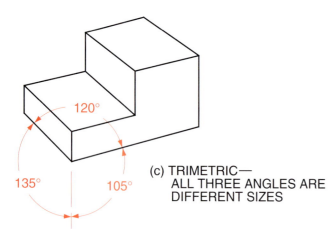
(c) TRIMETRIC—ALL THREE ANGLES ARE DIFFERENT SIZES

FIGURE 5.4
Starting an isometric sketch

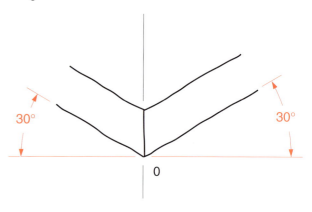

ric sketch. In an oblique sketch, the receding (depth) lines are usually drawn at 30, 45, or 60 degrees to the left or to the right of the front view. This type of sketch involves the combination of a flat front surface and the depth lines receding at the selected angle. The front surface always shows the true size and shape of the object. Figure 5.5 provides two examples of oblique freehand sketches, each with different but acceptable receding lines. In Figure 5.5a, the receding lines are at a 45-degree angle; Figure 5.5b, shows the same object but with the receding lines drawn at a 30-degree angle.

■ **5.2.4 Perspective Sketch**

A perspective sketch is more difficult to draw than either the isometric or the oblique sketches. The advantage of this drawing is that it truly represents what the human eye actually sees when viewing an object. For example, if one were to see a tall building at some distance, it would appear to the human eye as shown in Figure 5.6. Notice that the angled lines are not parallel. This is true because the lines tend to be foreshortened when viewed by the naked eye and vanish into the distance.

A perspective sketch is, therefore, a form of projection in which the projecting or extended lines tend to radiate from a line of sight called a **vanishing point (VP)**. A **horizon line (HL)** is also important because it represents the eye level of the observer. Figure 5.7 shows the same building as in Figure 5.6 except that it identifies the vanishing points and the horizon line and thus demonstrates how a perspective sketch is formed.

Section 5: *Technical Sketching*   **65**

FIGURE 5.5
Freehand oblique sketches

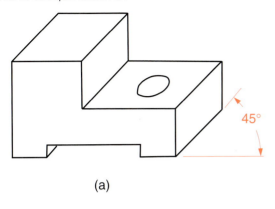

(a)

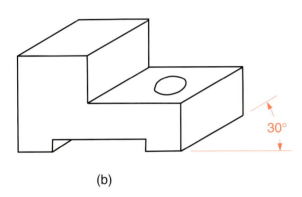

(b)

FIGURE 5.6
How the human eye sees a tall building

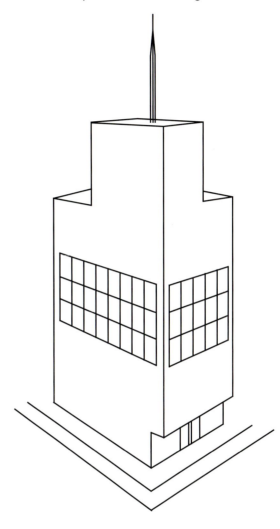

FIGURE 5.7
How a perspective sketch is formed

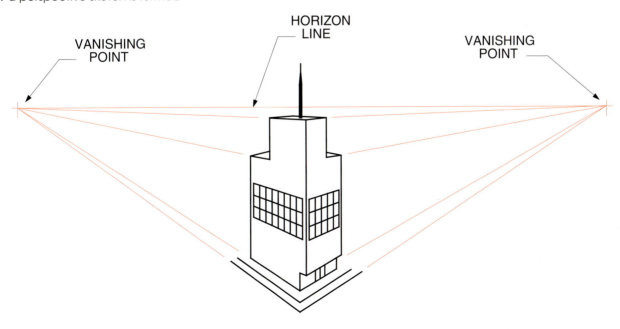

FIGURE 5.8
One-, two-, and three-vanishing-point freehand perspective sketches

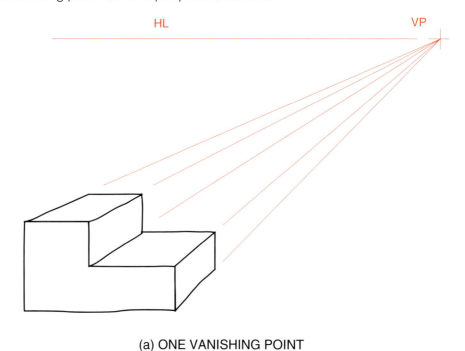

(a) ONE VANISHING POINT

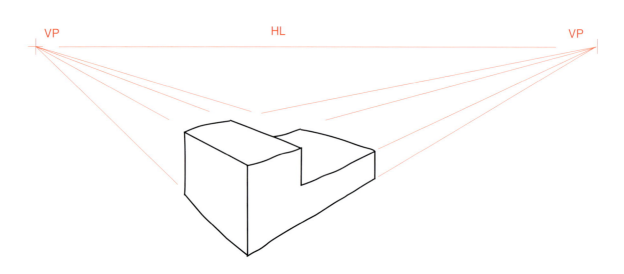

(b) TWO VANISHING POINTS

When laying out a freehand perspective sketch, first determine the observation point, which is the horizon line, and then locate the two vanishing points. The position of the observer can either be above or below the object. Perspective sketches can have one, two, or three vanishing points. The two-vanishing-point sketch is the most common in industry. Examples of the one-, two-, and three-vanishing point perspectives are illustrated in Figure 5.8. Note the differences in the three types. In each case, the observation point is above the object.

5.3 SKETCHING SQUARES, RECTANGLES, AND TRIANGLES

Sketching a **square**, a **rectangle**, or a **triangle** freehand is as simple as drawing horizontal lines. For most people, sketching horizontal lines is easier than

Section 5: *Technical Sketching*   **67**

FIGURE 5.8 CONTINUED
One-, two-, and three-vanishing-point freehand perspective sketches

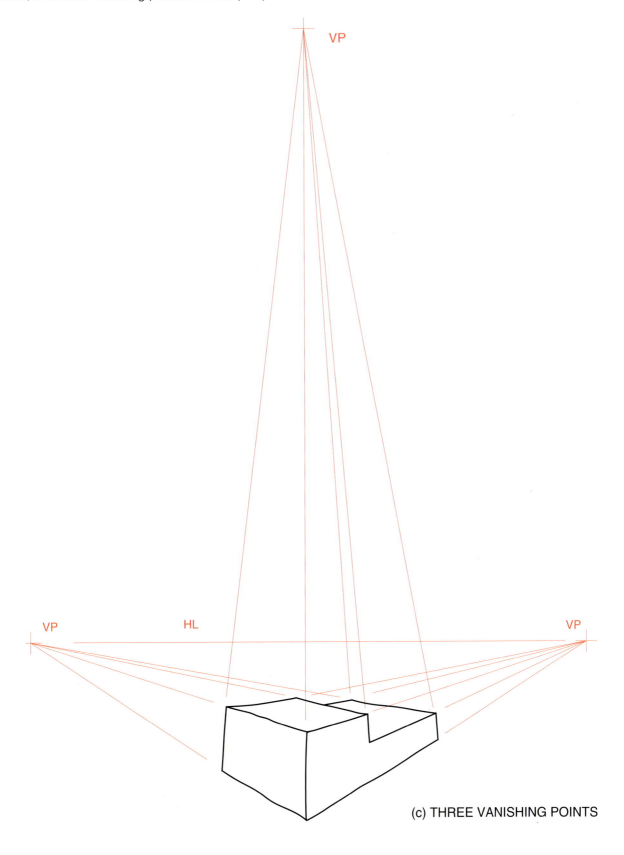

(c) THREE VANISHING POINTS

drawing vertical or angled ones. Horizontal lines are drawn with a firm but steady movement of the arm on a supported surface. The arm should not be allowed to dangle. Movement can be from left to right or from right to left. It is recommended that the starting point and the end point of the line be marked so the sketched line will be as straight as possible. To ensure lines of uniform weight and thickness, try to rotate the pencil or lead holder so that the line being drawn will not become fuzzy as the lead wears down. Avoid too much pressure between the lead and the medium, which could result in grooves in the medium that would be difficult to erase in the event of a mistake. To avoid serious errors in drawing a freehand sketch, draw all lines lightly and then darken them when satisfied with the sketch. Figure 5.9 illustrates how to sketch squares, rectangles, and triangles by trying to draw most lines with a horizontal stroke. To accomplish this, it is necessary to rotate the medium several times until the sketch is complete.

FIGURE 5.9
Sketching squares, rectangles, and triangles

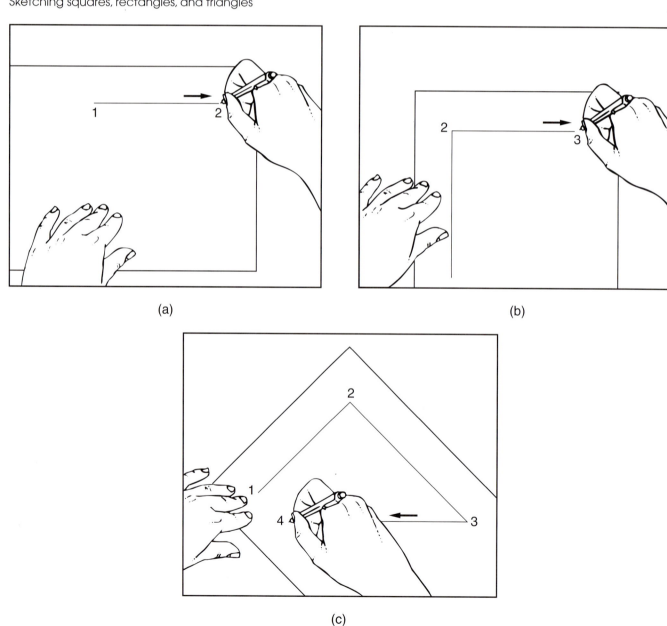

5.4 SKETCHING CIRCLES AND ARCS

Although sketching a **circle** or an **arc** can be particularly difficult, the two methods illustrated in Figure 5.10 will assist in the process. Both methods require that a vertical and a horizontal line be sketched first, intersecting at a point where the center of the circle will be located (step A). In the first method, 5.10a, a lightly sketched square, whose distance across is approximately equal to the diameter of the expected circle, is drawn (step B). Short dashed lines are then sketched to the desired diameter (step C), followed by a heavy continuous stroke to complete the circle (steps D and E).

In the method shown in 5.10b, one must try to visualize the diameter of the circle to be drawn. Translate this mental picture into equally spaced radial points or short dashed lines (steps A and B). The sequence illustrates that as few as eight or as many as sixteen points may be plotted (step C). A larger number of points used will produce a smoother, more perfect circle. Finally, connect the radial points using an even, constant pressure of the lead on the medium (step D). Completing an arc equivalent to one quadrant (one-fourth) of the circle and then rotating the drawing until all four quadrants are complete may result in a more even circle.

5.5 SUGGESTIONS FOR PRODUCING TECHNICAL SKETCHES

The quality of a technical sketch depends on several factors, such as drafting style, experience of the drafter, steadiness of hand, and the use of proper sketching materials. The following suggestions are offered to help you produce quality freehand technical sketches:

1. Select the proper drawing medium and lead for sketching based on the size, complexity, and reproducibility requirements of the drawing.

FIGURE 5.10
Sketching circles and arcs

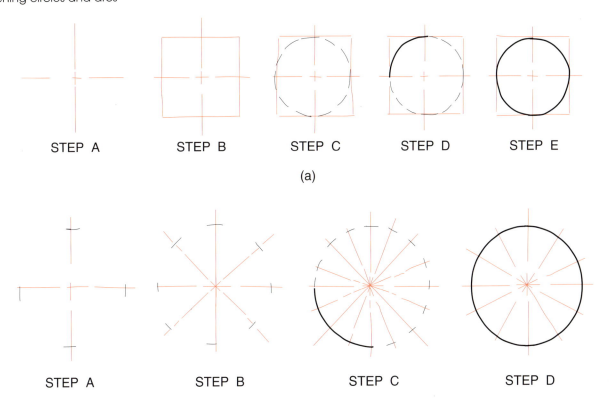

70 **Section 5:** *Technical Sketching*

2. Determine the type of sketch required. Will it be a multiview, isometric, oblique, or perspective drawing? Will it require dimensions, text material, or both?

3. Do not attach the drawing medium to the drafting surface, so that the medium can be placed at any comfortable angle for sketching. Lines can be drawn straighter, more smoothly, and more accurately if as much as the arm as possible is supported by a table or a drafting board surface. Do not allow the unsupported arm to hang off the edge of the drawing board or table when sketching.

4. Grip the pencil or lead holder firmly and maintain a comfortable angle between the drawing surface and the pencil or lead holder. Sketch the complete outline of the drawing using lightweight lines. View the sketch. When satisfied, heavy up all lines.

5. Regardless of the type of sketch required, try to produce it using as many horizontal lines as possible, rotating the medium as often as necessary until completion.

5.6 SUMMARY

The freehand technical sketch provides an easy method for communicating an idea, concept, or plan in a rapid but effective manner. The types of technical sketches most used by industry include the multiview, isometric, and perspective sketch. Technical sketches can be produced by anyone whose needs for communication in an industrial setting cannot be met by the written or spoken word. Materials required to produce a freehand technical sketch include a soft grade of lead, a mechanical pencil or lead holder, an eraser, and a drawing medium. The most suitable medium is bond paper, ruled with a reproducible or nonreproducible grid.

This section described various methods for producing a freehand technical sketch. In addition, details on how to sketch squares, rectangles, triangles, circles, and arcs were covered. Finally, a step-by-step process for producing quality technical sketches was outlined.

TECHNICAL TERMS FOR STUDY

arc   A small segment of a circle or a curve.

axonometric drawing   A drawing whose object is presented on a surface so that three faces appear as inclined faces.

circle   A closed plane curve on which all points are equidistant from the center.

dimetric drawing   A drawing in which two of the three principal faces are equally inclined to the plane of projection (i.e., two angles of projection are identical).

drawing medium   A surface on which to produce a drawing, such as paper, vellum, film, or tracing cloth.

freehand   A term that describes a sketch or drawing produced without the aid of drafting tools or equipment.

horizon line (HL)   A line used when producing a perspective sketch; it represents the observation point of the viewer.

isometric sketch   An axonometric drawing whose lines are located 120 degrees apart starting at 30 degrees from the horizontal.

lead holder   A device used for holding leads for the purpose of drawing or sketching.

lead   A writing material used to sketch or draw. Graphite leads are used on all paper surfaces, while plastic leads are normally used on polyester film media. They are available in very soft to very hard material.

mechanical pencil   A device for holding leads for the purpose of creating a drawing.

multiview sketch   A drawing showing more than one view of an object. A drawing whose lines are drawn using the principles of orthographic projection.

oblique sketch   A drawing that involves the combination of a flat surface and depth lines that recede at a selected angle, and whose front view always displays the true shape and form of the object.

Section 5: *Technical Sketching* **71**

perspective sketch   A sketch that represents how the human eye actually views an object; a form of projection in which the projecting or extended lines tend to radiate from a line of sight called a **vanishing point.** A **horizon line** represents the eye level of the observer.

rectangle   A quadrilateral whose opposite sides are parallel and of different lengths.

square   A regular, four-sided polygon whose sides are all equal in length.

technical sketch   A drawing used for communicating a concept, idea, or plan in a rapid but effective manner. It is usually produced freehand.

triangle   A closed plane figure having three sides.

trimetric drawing   A form of axonometric drawing in which all three principal faces and axes of the object make different angles with the plane of projection.

vanishing point (VP)   In a perspective sketch, the position where projecting or extended lines tend to originate.

Student _____   Date _____

Section 5: Competency Quiz

PART A COMPREHENSION

1. What is the purpose of a technical sketch? (10 pts)

2. What basic tools and materials are required to produce a technical sketch? (9 pts)

3. What are the four types of technical sketches used by industry? (8 pts)

4. Name three types of axonometric sketches. (9 pts)

5. At what angles are isometric lines drawn? (5 pts)

6. At what angles are the receding lines of an oblique sketch normally drawn? (9 pts)

7. Describe the perspective sketch. (10 pts)

8. What do vanishing points represent on a perspective sketch? (10 pts)

9. What is the viewing location of the observer called on a perspective sketch? (5 pts)

74   **Section 5:** *Competency Quiz*

10. Identify the steps required in producing circles and arcs. (10 pts)

11. List five suggestions for producing a quality technical sketch. (15 pts)

Section 5: *Technical Sketching*   **75**

Student _____   Date _____

PART B TECHNICAL TERMS

For each definition, select the correct technical term from the list on the bottom of the page. (10 pts each)

1. _____ A sketch that represents how the human eye actually views an object.

2. _____ A drawing whose object is presented on a surface so that all three faces appear as inclined surfaces.

3. _____ A drawing in which two of the three principal faces are equally inclined to the plane of projection (two angles of the projection are identical).

4. _____ An axonometric drawing whose lines are located 120 degrees apart starting at 30 degrees from the horizontal.

5. _____ A device for holding leads for the purpose of drawing or sketching.

6. _____ A sketch showing more than one view of an object.

7. _____ A writing material used to sketch or draw. The graphite form of this material is used on all paper surfaces, while the plastic form is used on all polyester film media.

8. _____ In a perspective sketch, the position where projecting or extended lines tend to originate.

9. _____ A drawing used for communicating a concept, idea, or plan in a rapid but effective manner. It is usually produced freehand.

10. _____ A form of axonometric drawing in which all three of the principal faces and axes of the object make different angles with the plane of projection.

A. axonometric drawing
B. circle
C. dimetric drawing
D. isometric sketch
E. lead holder
F. lead
G. multiview sketch
H. oblique sketch
I. perspective sketch
J. square
K. technical sketch
L. triangle
M. trimetric drawing
N. vanishing point (VP)

©1995 West Publishing Company

PART C SKETCHING

1. Produce a freehand sketch of each figure shown below. Draw each at twice its present size.

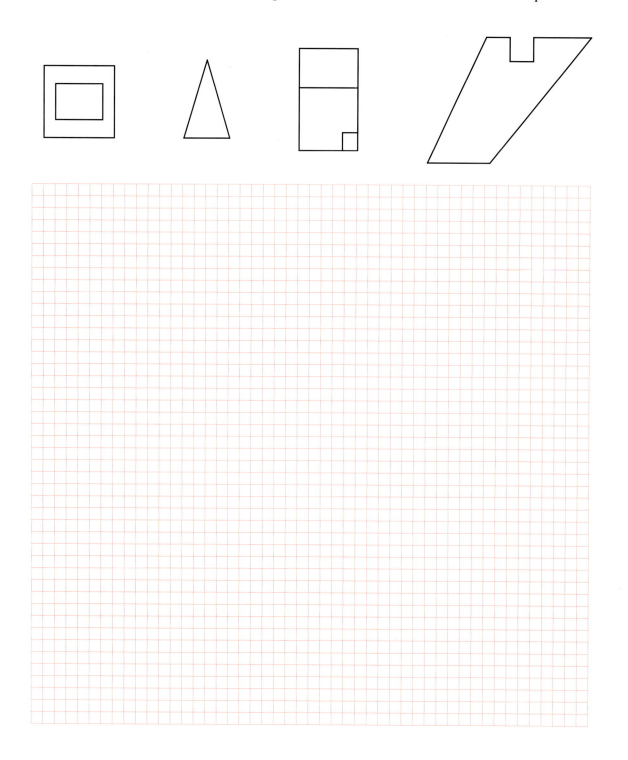

Section 5: *Technical Sketching*   77

Student _____   Date _____

2. Given the circles A, B, and C, duplicate them using freehand sketching methods in the space provided.

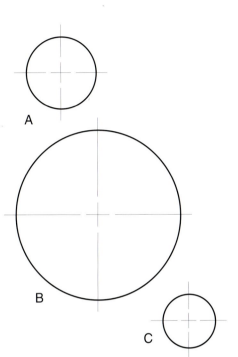
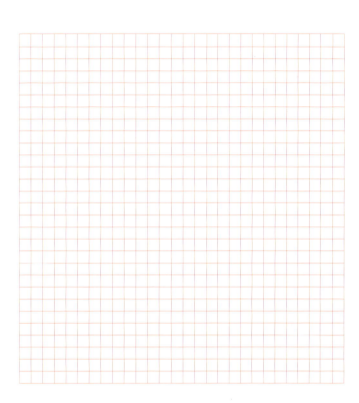

3. From the two-view object shown, produce an isometric drawing. Draw freehand at full scale. Use the starting point as a guide.

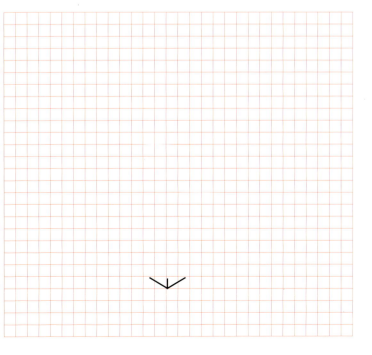
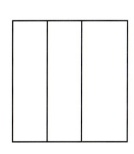
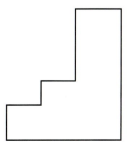

©1995 West Publishing Company

78 Section 5: *Competency Quiz*

4. Draw two views (front and top) for each object shown below.

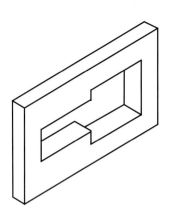
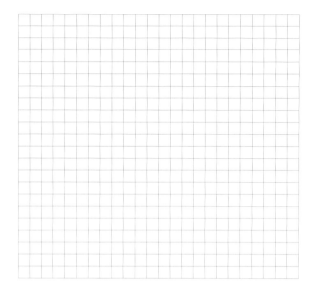
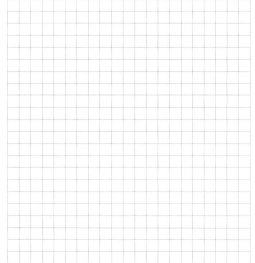
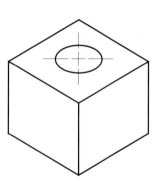
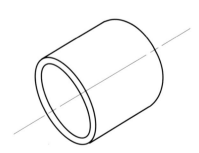

Section 5: *Technical Sketching* 79

Student _____ Date _____

5. Redraw the following two-view drawing as a freehand isometric sketch. Use the 30-degree grid provided. Draw full size with H or 2H lead. Viewing may be from the left or right.

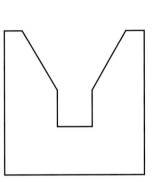  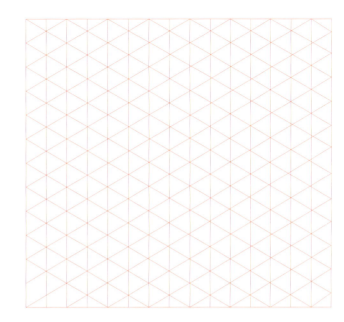

6. Scale the drawing below. Using H or 2H lead, draw a freehand isometric sketch of the two-view drawing. Viewing may be from the left or right.

©1995 West Publishing Company

7. Redraw the following oblique drawing at full size, freehand.

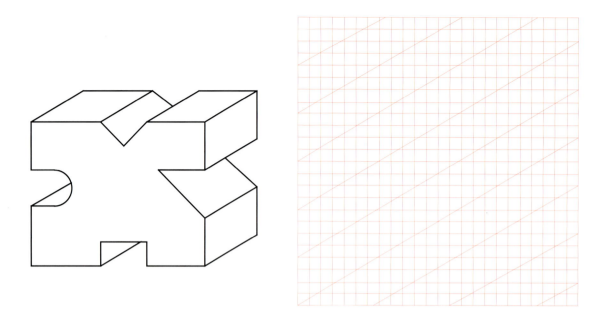

8. Redraw the object below as a freehand perspective drawing. Use the vanishing points, horizon line, and locating dimensions as a guide. Draw the object at twice its present size. Use B- or C-size vellum or another suitable medium.

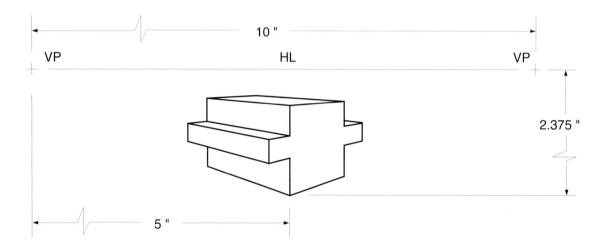

SECTION

Principles of Orthographic Projection

LEARNER OUTCOMES

You will be able to:

- Explain the systematic placement of views for orthographic projection.
- Describe the differences between third-angle and first-angle projection.
- Identify the purpose and function of the multiview drawing.
- State the importance of the selection and placement of views.
- List six view placement principles.
- Describe the difference between lines and surfaces.
- Explain the precedence-of-lines concept.
- Define several technical terms.
- Demonstrate your learning through the successful completion of competency exercises.

6.1 INTRODUCTION

The predominant method of producing technical drawings throughout the world is through the principles of **orthographic projection**. In orthographic projection, different views of an object (front, top, and side, for example) are systematically arranged on a drawing medium to convey information to the reader. Features, surfaces, lines, and edges are projected from one view to another. The term *orthographic* has its origin in the Greek words *orthos*, which means "straight, correct, at right angles to," and *graphikos*, which means "to write or to describe by drawing lines." Two types of orthographic projection are used throughout business and industry: third-angle and first-angle projection.

6.1.1 Third-Angle Projection

Third-angle projection is used primarily in the United States and Canada. To understand the theory behind this system, visualize the transparent box illustrated in Figure 6.1, with the object to be drawn placed inside the box so that its axes are parallel to the axes of the box. The projections on the sides of the box are the views one sees by looking straight at the object through each side. If each view is drawn as seen on the side of the box and the box is unfolded and laid flat, as shown in Figure 6.2a, the result is a six-view orthographic projection of the object. Any of the views of the object could be regarded as the **principal view**, and the box could be folded out from one view as easily as from another. When the box is unfolded from the front view of the object, as pictured in Figure 6.2b, it is called **third-angle projection.**

FIGURE 6.1
Transparent box encompassing an object to be drawn

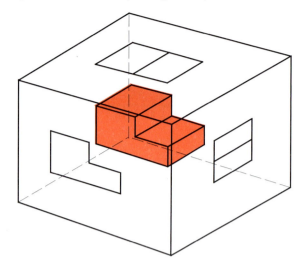

FIGURE 6.2
(a) Unfolding box; (b) six-view third-angle projection

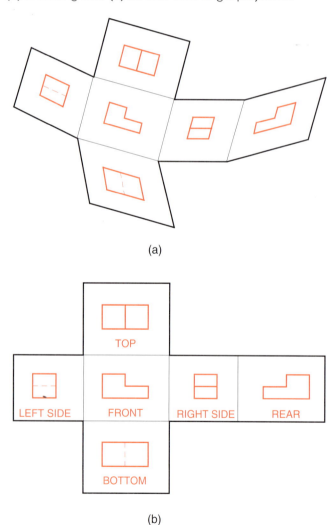

6.1.2 First-Angle Projection

Although **first-angle projection** is not generally used in the United States, it is being presented here because people working for multinational companies in the machining industries are likely to encounter this kind of drawing at some point during their working life.

First-angle projection is the preferred method for drafting in most European and Eastern countries. The major difference between the third-angle and first-angle methods is how the object is projected and the position of the views on a drawing. In third-angle projection, the projection plane is perceived to be between the observer and the various views of the object, with the views being forward to that plane. In first-angle projection, the projection plane is perceived to be on the far side of the object from the viewer. Figure 6.3 illustrates first-angle projection. Note how the views are projected to the rear and onto the projection plane instead of being projected forward. Compare Figures 6.2a and 6.2b with Figure 6.3 to see the differences between first-angle and third-angle orthographic projections.

6.2 THE MULTIVIEW DRAWING

Of all the different types of drawings produced in engineering departments around the world, none is produced more often than the **multiview drawing**. It is the major type of technical drawing used by industry. The purpose of the multiview drawing is to represent the various faces of an object in two or more views on planes at right angles to each other by extending perpendicular **projection lines** from the object to these planes. The multiview drawing allows the observer to view an object from more than one orientation. The *top view*, for example, illustrates how the object appears when viewed from above; the *front view* shows how the object looks when viewed from the front; and the *side view* represents the object's features when viewed from the side. Depending on the complexity of the object to be drawn, one view or as many as six or more views may be required to display all the object's features. Each view is in direct relationship with the next, but each view portrays the object from a different orientation. Figure 6.4 provides

Section 6: *Principles of Orthographic Projection* 83

FIGURE 6.3
First-angle projection

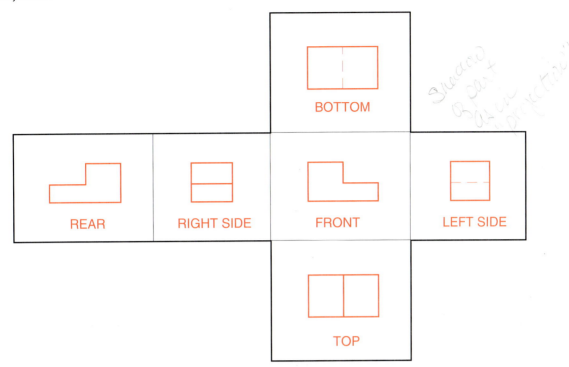

FIGURE 6.4
Multiview drawing

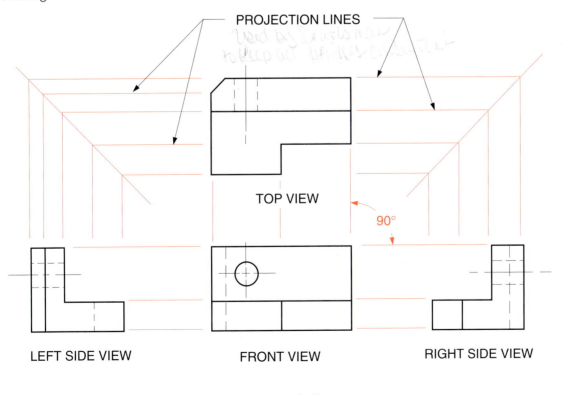

CLIP, ANGLE

an example of a multiview drawing with the projection lines shown as light lines. The particular object shown requires four views. It should be noted here that projection lines are shown only when the drawing is in process. They do not normally appear on a finished drawing.

6.3 THE SELECTION AND PLACEMENT OF VIEWS

In a multiview drawing, there is a certain order involved in the selection and placement of the several views of an object. **View selection** is determined by examining the object to be drawn. If it is a complex part with several surfaces, holes, notches, and reliefs, three or more views may be required. If the part is a thin, flat piece or a round, spherical part, only one or two views may be needed to clearly show its features.

In determining view selection, remember that (1) only those views that clearly show the size and shape of the part are drawn, (2) views are selected to ensure the fewest number of hidden lines, and (3) the front view is the one on which most of the dimensions are placed. The front view is normally called the **principal view** for this reason.

When interpreting a drawing, a machine trades worker is in a position to determine if **view placement** principles have been used by the drafter. The following view placement principles are used throughout industry and are in accordance with the American National Standards Institute (ANSI) guidelines. See Figure 6.4 for an example of the application of these guidelines.

1. The front and top views are aligned vertically, with the top view above the front view.
2. The front view and side view or views are always aligned horizontally. The right side view is shown to the right of the front view, and the left side view appears to the left of the front view.
3. The depth of the top view is the same as the depth of the side view or views.
4. The length of the top view is the same as the length of the front view.
5. The height of the side view or views is the same as the height of the front view.
6. The bottom view, if required, is shown directly below and vertically aligned with the front view. It is also the same width as the front view.

6.4 THE ONE-VIEW DRAWING

At times, shop personnel will encounter a *one-view drawing*. These drawings are acceptable for such items as cylinders, spheres, and square parts if important dimensions are indicated. Thin objects, such as shims, gaskets, plates, and parts made of thin-gage sheet metal, may also be found as one-view drawings, but key descriptors, such as DIA, SQ, and THK (for diameter, square, and thick or thickness), will also be indicated on the drawing. Two examples of one-view drawings are shown in Figure 6.5.

FIGURE 6.5
One-view drawing

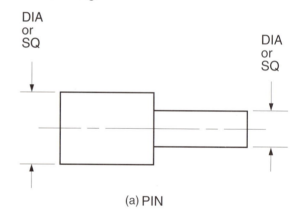

(a) PIN

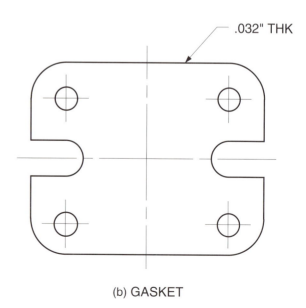

(b) GASKET

6.5 THE TWO-VIEW DRAWING

Certain objects are visually described using two views. A **two-view drawing** can be arranged as any two adjacent views (see Figure 6.2b to see which views are next to each other). Some drawings, such as those of a cylindrical shape, may consist of a top and a front view or a front and a side view. Adding a third view would be an unnecessary duplication of time, energy, paper, and information. In a two-view drawing, view selection is made by determining which views will show the fewest number of hidden lines, as pictured in Figure 6.6.

FIGURE 6.6
Two-view drawing

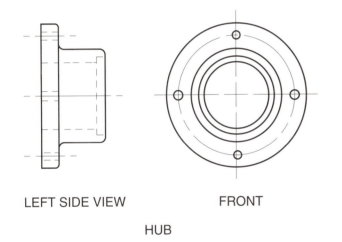

LEFT SIDE VIEW   FRONT

HUB

6.6 THE THREE-VIEW DRAWING

A **three-view drawing** consists of three adjacent views. The most common views selected for the three-view drawing, as indicated earlier, are the front, top, and right side views; however, other arrangements are acceptable. If Figure 6.2b were divided into enough segments showing the object as a multiview drawing, the possibilities shown in Figure 6.7 would result.

In Figure 6.7, some of the three-view drawings are acceptable, but others are not because they show unnecessary hidden lines. Figure 6.7 only serves to illustrate the several different combinations of views that may be selected for three-view drawings. Figure 6.8 offers examples of three-view drawings with views that clearly outline the shape and size of the object.

6.7 THE PROJECTION OF POINTS, LINES, AND SURFACES

Producing the several views of an object is accomplished by taking measurements of certain **points**, **lines**, and **surfaces** from one view and projecting them to one or more views. The projection of points, lines, and surfaces provides greater accuracy in the alignment of views than does measuring each view separately with a scale or transferring them with dividers.

Figure 6.9 shows how a **miter line** provides a rapid and accurate method of projecting point, line, or surface measurements from one view to another. Note that the miter line is a single 45-degree angled line that originates at point O. The distance, D is established for the necessary space between views for dimensions, dimension and extension lines, and notes. It is a variable distance depending on space requirements.

In the production of a multiview drawing, the person making the drawing does not actually see the surface of an object. Visible lines are drawn to indicate where one surface meets another surface. A line may indicate a surface in one view, but it may take the form of a point in another view. A line or a surface that cannot be seen because it is hidden or behind a surface is identified by a hidden or invisible (dashed) line. Figure 6.10 shows a simple object that can be studied to learn the relationships between points, lines, and surfaces. A pictorial view of the slide block is also shown for ease of point, line, and surface visualization.

■ 6.7.1 Points

A **point** is defined as something that has a position, but no extension. It is usually indicated with a small crosshair, sometimes with a small solid circle. On a drawing, a point may be found at the intersection of two or more lines, at the termination point of a line, or at the corner of an object. When lines meet, they form a point. Figure 6.10 shows that (1) lines 3 and 9

86 Section 6: *Principles of Orthographic Projection*

FIGURE 6.7
Examples of three-view drawings

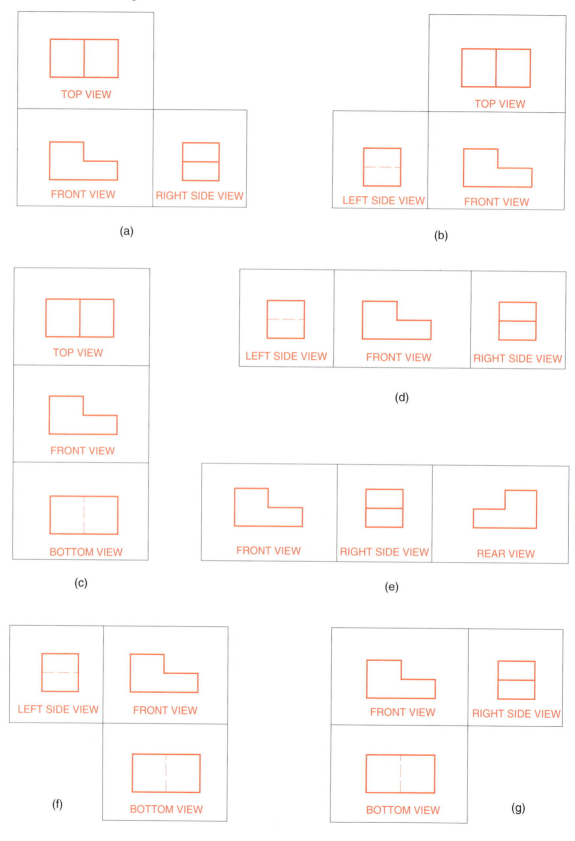

FIGURE 6.8
Three-view drawings clearly showing the shape and size of the object

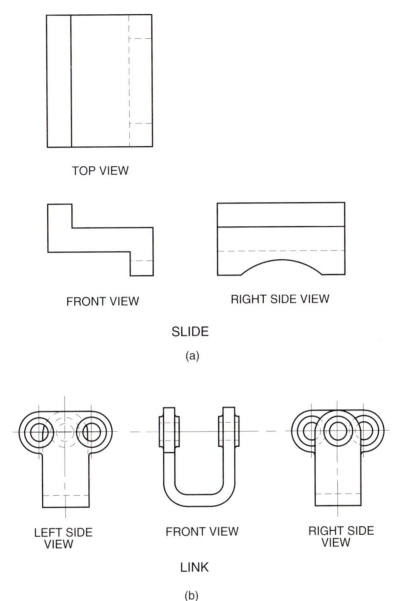

in the front view and line 7 in the side view form point X, as indicated in all three views as well as in the pictorial view; and that (2) lines 5 and 10 in the front view, 1 and 12 in the top view, and 6 in the side view form point Y.

6.7.2 Lines

A **line** is a mark drawn using a pen or pencil. It can extend in any direction. A **straight line** is the shortest distance between two points. A *curved line* may be a variety of curved forms or arcs. In a multiview drawing, a surface in one view may become a line in an adjacent view. Figure 6.10 demonstrates that (1) surface A in the front view of the object becomes line 1 in the top view; (2) surfaces B, C, and D in the top view become lines 3, 5, and 4 in the front view and lines 7, 6, and 8 in the right side view (line 8 is shown as a hidden line corresponding to surface D); and (3) surfaces E and F in the right side view are shown as lines 9 and 10 in the front view and as lines 11 and 12 in the top view.

Section 6: *Principles of Orthographic Projection*

FIGURE 6.9
Projection of points, lines, and surfaces

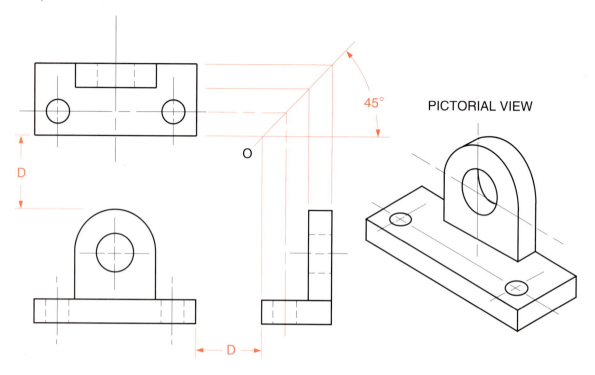

SUPPORT, BEARING

FIGURE 6.10
Relationships between points, lines, and surfaces

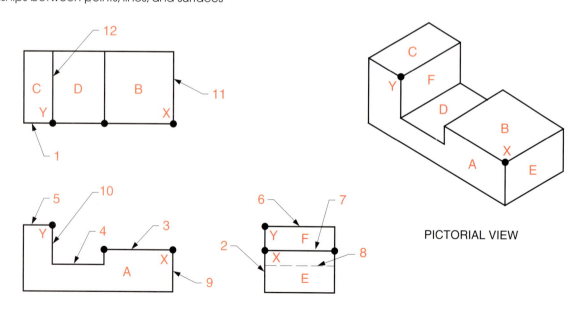

SLIDE BLOCK

Section 6: *Principles of Orthographic Projection*

■ 6.7.3 Surfaces

Most of the **surface** features on machine parts represent curved or plane areas. Examples of curved surfaces may be found on cylindrical shapes, such as pistons, shafts, and holes. Plane surfaces are found on flat, cubic, and rectangular parts. Surfaces may be horizontal, vertical, inclined (or slanted), curved, irregular, or oblique in shape. Figure 6.10 illustrates that (1) surface A in the front view of the object is also surface A in the pictorial view, but is line 2 in the right side view and line 1 in the top view; and (2) vertical surfaces E and F in the right side view are shown as vertical lines 9 and 10 in the front view, vertical lines 11 and 12 in the top view, and surfaces E and F in the pictorial view.

6.8 THE PRECEDENCE OF LINES

To fully interpret an object, the viewer should realize that a multiview drawing contains lines that represent edges, surfaces, intersections, size, limits, and points. In any given view, there may be portions of an object that are not clearly visible because they are covered by parts of the object that are closer to the eye of the observer. The areas of an object that are not clearly visible are, therefore, invisible and are represented by hidden lines, as illustrated in Figure 4.1, Section 4.2.

The first lines normally drawn on a multiview drawing are the center lines. Center lines are always light in weight and form the horizontal and vertical axes of circular features, such as cylinders, circles, spheres, and cones. Every circle or part of a circle shown has its center represented by the intersection of two mutually perpendicular center lines, as described and shown in Figure 4.1, Section 4.2.

In any given view of a multiview drawing, there may be a conflict, or a **coincidence of lines**. For example, a coincidence of lines occurs when an object line (visible line) projects to coincide with (or requires the same precise location as) a center line or a hidden line, or when a center line coincides with a cutting plane line. Dimension and extension lines should always be positioned so as not to interfere or coincide with other lines. To make certain that there is a standard "pecking order" for lines on a drawing— that is, a system for determining which line is drawn

if two or more lines occupy the same position, the **precedence of lines** was established. The following list defines the order of precedence of lines, which can also be referred to as the *importance of lines*:

1. Object, or visible, line
2. Hidden, or invisible, line
3. Cutting plane line
4. Center line
5. Break line
6. Dimension and extension lines
7. Sectioning line

In Figure 6.11, the drawing of an adjusting lever demonstrates a coincidence of lines that results from separate features having identical positions of identical size or behind (hidden from) one another. Examine Figure 6.11 carefully to determine which lines are in conflict and which lines hold precedence over others.

6.9 SUMMARY

The topic of orthographic projection was introduced as being the predominant method for producing technical drawings throughout the world. It is a method of drawing that calls for the placement of the views of an object in a certain order. Third-angle projection, which is primarily used in the United States and Canada, and first-angle projection, used in most European and Eastern countries, were detailed, and their differences were noted. The purpose of a multiview drawing, it was stated, is to represent the various faces of an object in two or more views on planes at right angles to each other. The importance of the selection and placement of views was stressed. Suggestions for determining view selection were made. Principles for view placement, in accordance with ANSI guidelines, were outlined. Reference was made to one-view, two-view, and three-view drawings, showing examples of each and outlining their similarities and differences. Instructions and illustrations were provided for the projection of points, lines, and surfaces. Definitions of these terms were presented. Rules for establishing the precedence of lines were identified, and the order of importance for various types of lines was detailed.

FIGURE 6.11
Precedence of lines

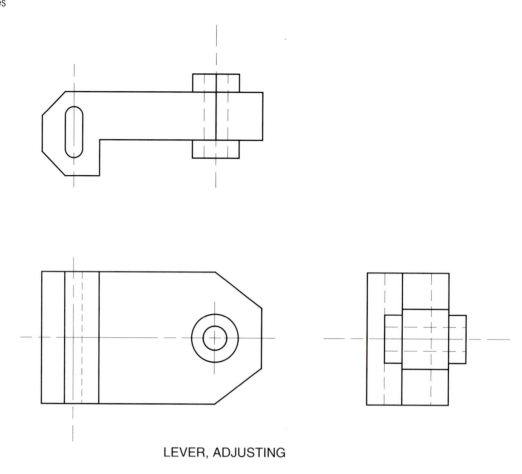

LEVER, ADJUSTING

TECHNICAL TERMS FOR STUDY

coincidence of lines What occurs when a line requires the same precise location or position as another line. This is when a conflict, or a coincidence of lines, occurs.

first-angle projection A method of projection in which the projection plane is observed to be on the far side of the object from the viewer.

line A drawn mark that can extend in any direction. A straight line is the shortest distance between two points.

miter line A single 45-degree angled line used for projecting points, lines, or surfaces from one view to another.

multiview drawing A drawing whose purpose is to represent the various faces of an object in two or more views on planes at right angles to each other by extending projection lines from the object to these planes.

orthographic projection A method of producing technical drawings in which different views of an object are systematically arranged on a drawing medium in a certain order.

point A geometrical element that has a position, but no extension. It is usually indicated by a small crosshair or a small solid circle and is found at the intersection of two or more lines.

precedence of lines A system for determining which line is drawn if two or more lines occupy the same position.

principal view Usually the front view on a drawing, the view on which most of the dimensions are placed.

Section 6: *Principles of Orthographic Projection* **91**

projection lines Perpendicular lines drawn from an object to align the various views (top, front, and side, for example). Used in conjunction with a miter line.

surface An area on an object that may be horizontal, vertical, inclined, curved, irregular, or oblique in shape.

third-angle projection A method of projection in which the projection plane is perceived to be on the near side of the object from the viewer.

three-view drawing A drawing with three views of an object, usually the top, front, and side, projected orthographically.

two-view drawing A drawing containing two views, usually the top and front or the front and side views.

view placement The orientation of the views of an object in relation to other views.

view selection The determination of which views of an object should be drawn, after careful examination of the object.

Section 6: *Principles of Orthographic Projection* 93

Student _____ Date _____

Section 6: Competency Quiz

PART A COMPREHENSION

1. Explain the concept of orthographic projection. (10 pts)

2. Name the six views that are normally shown in an orthographic projection. (6 pts)

3. Describe the differences between first-angle and third-angle projection. (8 pts)

4. What is the purpose of a multiview drawing? (6 pts)

5. Why is the selection and placement of views important? (8 pts)

6. Identify six view placement principles. (6 pts)

7. What is a miter line? Why is it useful? (6 pts)

8. Define *view placement*. (4 pts)

©1995 West Publishing Company

94   **Section 6:** *Competency Quiz*

9. What are the most common views on a three-view drawing? (3 pts)

10. Define *point*. (6 pts)

11. List three objects with curved surfaces. (3 pts)

12. Identify three objects that require only two views on a drawing. (3 pts)

13. List three objects with a plane surface. (3 pts)

14. What types of objects require only one view on a drawing ? (3 pts)

15. Explain the precedence-of-lines concept. (10 pts)

16. Identify seven types of lines in order of precedence. (7 pts)

Section 6: *Principles of Orthographic Projection*   **95**

Student_____   Date _____

17. What type of line is used to represent a surface or line that is not clearly visible to the observer? (4 pts)

18. What is the purpose of a center line? (4 pts)

96   Section 6: *Competency Quiz*

PART B TECHNICAL TERMS

For each definition, select the correct technical term from the list on the bottom of the page. (10 pts each)

1. _____ A drawn mark that can extend in any direction; the shortest distance between two points.

2. _____ Usually the front view on a drawing; the view on which most of the dimensions are placed.

3. _____ A single 45-degree angled line used for projecting points, lines, and surfaces from one view to another.

4. _____ What occurs when a line requires the same location or position as another line.

5. _____ A drawing whose purpose is to represent the various faces of an object in two or more planes at right angles to each other by extending projection lines from the object to these planes.

6. _____ The determination of which views of an object should be drawn, after careful examination of the object.

7. _____ A method of producing technical drawings in which different views are systematically arranged in a certain order on a drawing medium.

8. _____ A geometrical element that has a position, but no extension. It is usually indicated by a crosshair or a small solid circle and is found at the intersection of two or more lines.

9. _____ Perpendicular lines starting from the object to align the various views. Used in conjunction with a miter line.

10. _____ The orientation of the views of an object in relation to other views.

A. coincidence of lines
B. line
C. miter line
D. multiview drawing
E. orthographic projection
F. point
G. principal view
H. projection lines
I. surface
J. third-angle projection
K. three-view drawing
L. two-view drawing
M. view placement
N. view selection

Section 6: *Principles of Orthographic Projection*   97

Student _____   Date _____

PART C VIEW AND SURFACE IDENTIFICATION

1. Match the numbers on the three-view drawing below with the letters on the pictorial isometric drawing. Place your answers on the chart.

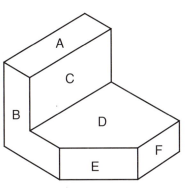

PICTORIAL

	TOP	FRONT	SIDE
A	7	6	13
B	9	1	18
C	11	5	14
D	8	2	17
E	10	2	15
F	12	3	

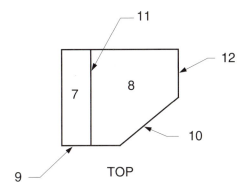

TOP

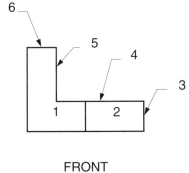

FRONT

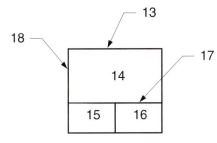

RIGHT SIDE

©1995 West Publishing Company

98 Section 6: *Competency Quiz*

2. Match the numbers on the three-view drawing below with the letters on the pictorial isometric drawing. Place your answers on the chart.

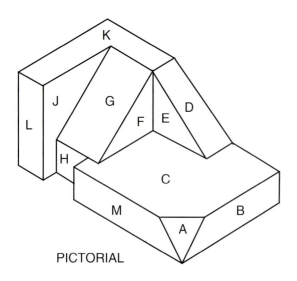

PICTORIAL

	TOP	FRONT	SIDE
A	13	2	24
B	18	7	23
C	19	8	28
D	17	10	27
E	14	1	29
F	20	9	26
G	15	6	30
H	21	4	31
J	22	11	25
K	16	12	32
L	33	5	34
M	35	3	36

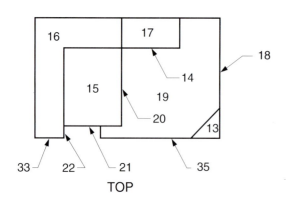

TOP

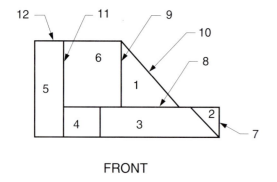

FRONT

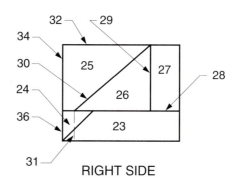

RIGHT SIDE

Section 6: *Principles of Orthographic Projection* **99**

Student _____ Date _____

PART D POINT, LINE, AND SURFACE IDENTIFICATION

1. Add all missing points, lines, and surfaces to the three-view drawings shown below. Identify all views. Suggestion: Light projection lines would be very helpful.

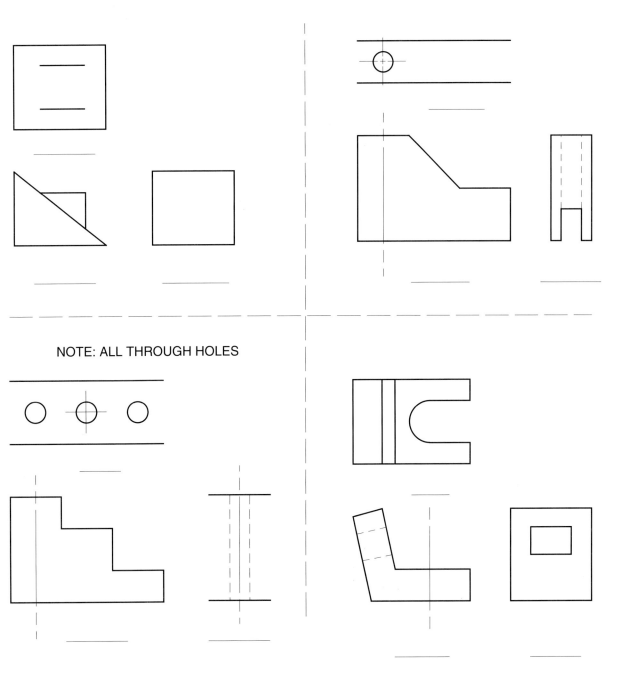

100 Section 6: *Competency Quiz*

2. Complete the three views (top, front, right side) of the shoe plate using the dimensions shown on the chart. Add all dimensions to the incomplete drawing. The drawing may be completed freehand.

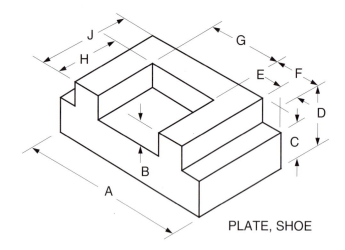

CHART	
A	1.75"
B	.38"
C	.44"
D	.88"
E	.18"
F	.50"
G	.75"
H	.68"
J	1.06"

PLATE, SHOE

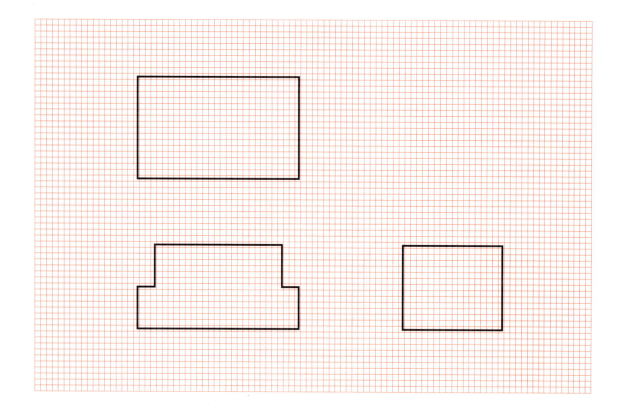

Section 6: *Principles of Orthographic Projection* **101**

Student _____ Date _____

PART E TECHNICAL MATHEMATICS

1. For an exercise in the addition and subtraction of numbers, examine the guide bracket below and fill in the dimensions on the chart. (10 pts each)

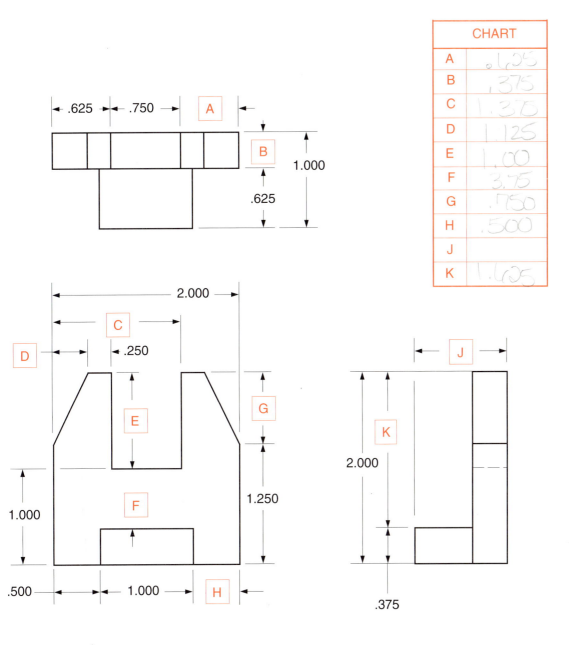

CHART	
A	.625
B	.375
C	1.375
D	1.125
E	1.00
F	3.75
G	.750
H	.500
J	
K	1.625

BRACKET, GUIDE

SECTION 7

Sectional and Auxiliary Views

LEARNER OUTCOMES

You will be able to:

- State the purpose of sectional and auxiliary views.
- Identify seven different types of sectional views.
- Explain section-lining procedures for single objects as well as for those in assembly.
- Describe a cutting plane and demonstrate its use.
- Complete several different sectional views given partial or incomplete information.
- List those items that do not require section lining.
- Name three types of basic auxiliary views.
- Identify three kinds of primary auxiliary views.
- Sketch several auxiliary views given basic orthographic information.
- Demonstrate your learning through the successful completion of competency exercises.

7.1 INTRODUCTION

The **sectional view** and the **auxiliary view** are parts of a drawing required when an object has either a complex internal shape or has a sloped, slanted, or inclined surface that is not parallel to any of the principal planes of projection. These views are used to convey, more accurately and more readily, detailed information about the object.

7.2 SECTIONAL VIEWS

The purpose of a sectional view is to show the internal form and detail of an object. It is a cutaway section that usually eliminates the need for hidden lines, while retaining the important outlines of an object. A sectional view is obtained by passing an imaginary plane through some specific part of an object and presenting the object as though the area cut off by this plane has been removed. An example of the removal of an area of a pulley is illustrated in Figure 7.1.

In Figure 7.1, the front view of the pulley is shown. To complete the drawing, the right side view is also identified so that the internal form and detail of the object can be revealed. The completed drawing is shown in Figure 7.2. The example illustrates that one-fourth of the object is removed. An important note: the sectional view should be drawn using correct orthographic principles in relation to other views on a drawing.

104 Section 7: *Sectional and Auxiliary Views*

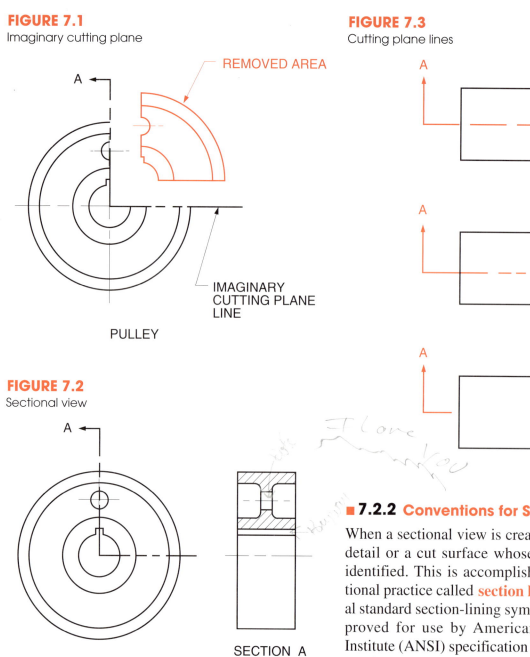

FIGURE 7.1
Imaginary cutting plane

FIGURE 7.2
Sectional view

FIGURE 7.3
Cutting plane lines

■ 7.2.1 The Cutting Plane Line

The cutting plane is determined by a **cutting plane line**, which is used to indicate the location or position of the cutting plane for sectional views. The line is represented by a dashed or broken line in one of three forms identified in Figure 7.3. It is likely that a machine trades worker will encounter one or all of these forms in the shop during his or her work experience. More about the cutting plane line can be found in Section 4.2.

■ 7.2.2 Conventions for Sectioning

When a sectional view is created, it exposes internal detail or a cut surface whose material needs to be identified. This is accomplished through a conventional practice called **section lining**. There are several standard section-lining symbols that have been approved for use by American National Standards Institute (ANSI) specification Y 14.2. These symbols are shown in Figure 7.4 and should be used when interpreting engineering drawings. Note that each code symbol represents a different material.

Section lining is also referred to as **cross-hatching** or just plain **hatching**. Section lining consists of sharp, uniformly spaced lines usually shown at a 45-degree angle to the principal lines of the view. Section lines are parallel to each other and spaced from $1/16$ to $1/4$ inch apart, depending on the size of the object and the scale to which the drawing is produced.

It is likely that shop personnel will see several of these code symbols on older drawings; however, the symbol used predominately today for all materials is the one shown in Figure 7.4a, which is the symbol

FIGURE 7.4
Standard code symbols for section lining (ANSI Y 14.2)

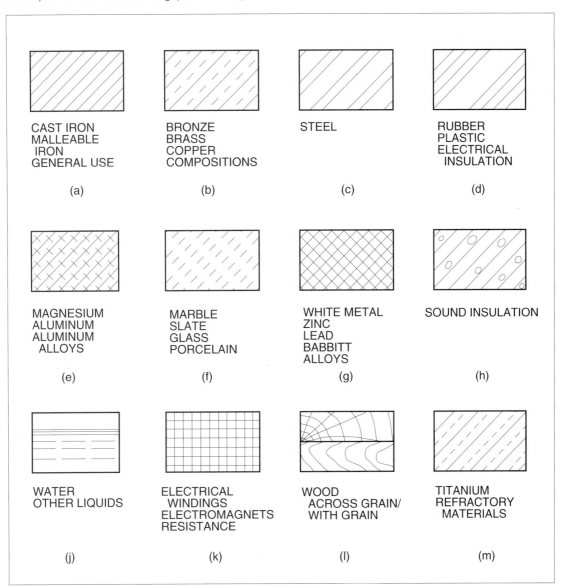

for cast or malleable iron but is also for general-purpose use. Figure 7.5 is an enlarged view of the general-purpose symbol that should be used for all exercises requiring section lining at the end of this section.

When section lining is needed for two or more mating parts, such as for the Hub and Bushing Assembly, the 45-degree section lines are drawn in opposite directions, as shown in Figure 7.6.

Figure 7.7 pictures an assembly and a thin component in section. Shims, gaskets, or thin sheet metal parts appear in a solid color because they are too thin to be shown otherwise.

FIGURE 7.5
General-purpose section lining

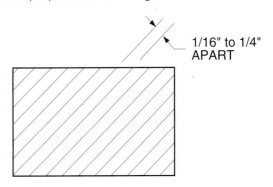

FIGURE 7.6
Section lining for mating or adjacent parts

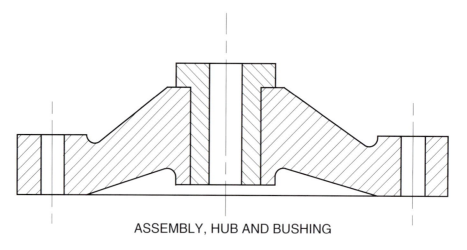

ASSEMBLY, HUB AND BUSHING

FIGURE 7.7
Section lining for a gasket

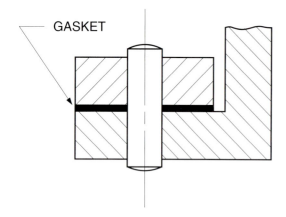

7.3 TYPES OF SECTIONAL VIEWS

Several different types of sectional views are used by industry to better clarify the interiors of objects and to assist in the production of mechanical parts, subassemblies, and assemblies. The most common sectional views include the full section, half section, **broken-out section**, revolved section, offset section, and removed section as well as sections through webs, spokes, ribs, and breaks in elongated objects.

Each of these will be covered in depth in the following paragraphs.

■ 7.3.1 The Full Section

A **full section** is formed when the cutting plane passes completely through an object, leaving one half of the object exposed to the observer for viewing its internal details. Figure 7.8a shows a full section of the top view of a Housing Bracket Support. Figure 7.8b portrays a full section of the front view and identifies the internal details of the object where the full sectional view A-A is taken. Note that the cutting plane line passes through the entire length of the object. The arrows for A-A represent the direction for viewing the section. Section lining is shown only where the cutting plane cuts through the object. If sectional view B-B is desired, it would be represented correctly as in Figure 7.8c. The cutting plane line B-B cuts through the width of the object and indicates the viewing direction.

Because the full section is used so frequently in industry, an additional example, a Bearing Housing, is pictured in Figure 7.9. Note how the internal details are exposed to the viewer.

■ 7.3.2 The Half Section

A **half section** is usually a view of a symmetrical object, showing both internal and external features. This is accomplished by passing two cutting planes at right angles to each other along the center lines of symmetrical axes so that one-quarter of the object is removed and the interior detail is exposed to the viewer. Figure 7.10 is an example of a half section of a jet pump housing. Note the internal and external features that can be seen.

Section 7: *Sectional and Auxiliary Views* **107**

FIGURE 7.8
Sectional views

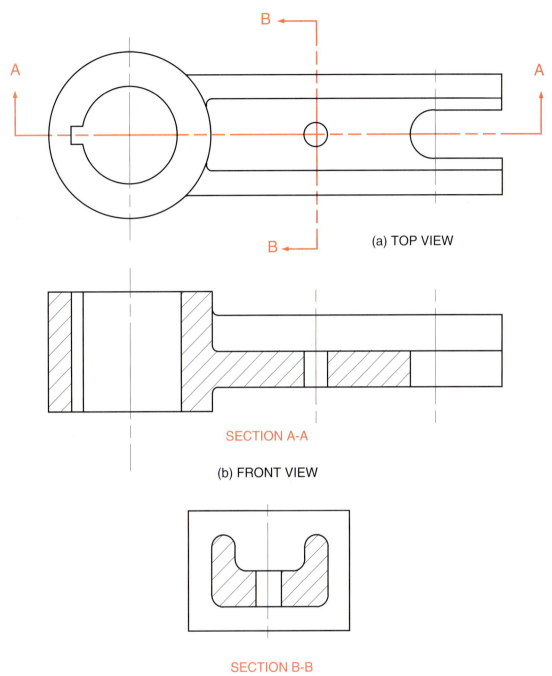

Section 7: *Sectional and Auxiliary Views*

FIGURE 7.9
Full section

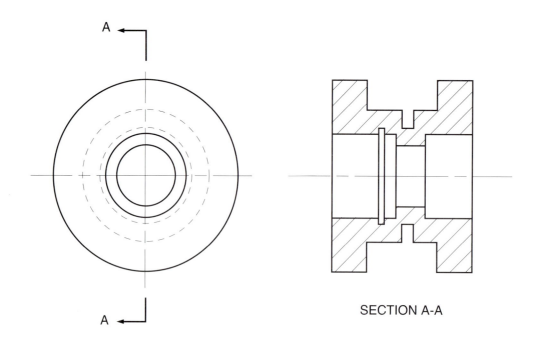

HOUSING, BEARING

FIGURE 7.10
Half section of a symmetrical object

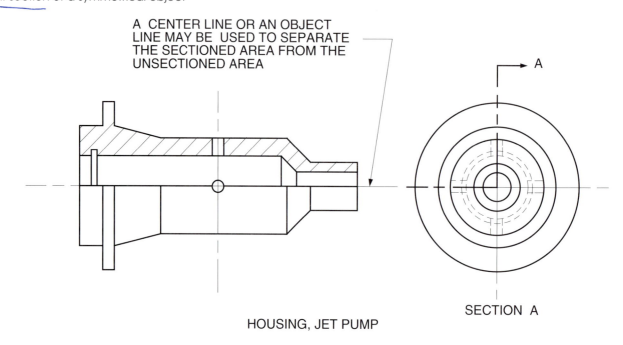

HOUSING, JET PUMP

7.3.3 The Broken-Out Section

When a sectional view of only a portion of an object is required, a **broken-out section** can be used. An irregular break line is normally used to determine the extent of the section. In this situation, neither a cutting plane line nor letters (A-A or B-B) to identify the section are required. Two examples of broken-out sections are provided in Figure 7.11.

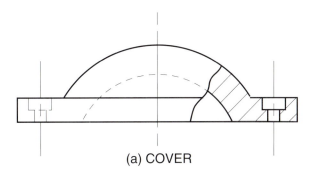

(a) COVER

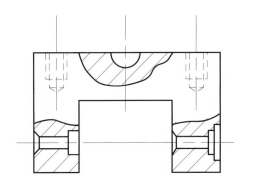

(b) YOKE

FIGURE 7.11
Broken-out sections

7.3.4 The Revolved Section

A **revolved section** provides a means of illustrating the shape of the cross section of an object and is shown directly on the exterior, as shown in Figure 7.12. This drawing is a partial view showing three different revolved sections. Note that the cutting plane is perpendicular to the center line or axis of the part to be sectioned, and the resulting section is rotated in place. When the object's visible lines interfere with the section, the view is broken away by an irregular line to allow a clear space for the sectional views, and the cutting plane indications are omitted.

7.3.5 The Offset Section

At times, the cutting plane appears in a direction other than that of the main axis to show internal features not positioned in a straight line. When the cutting plane is not a single continuous plane, the resulting section is called an **offset section**. Figure 7.13 identifies offset section A-A of a Filter Housing; note that the cutting plane is not a straight line but shows the interior of the object very clearly because of its

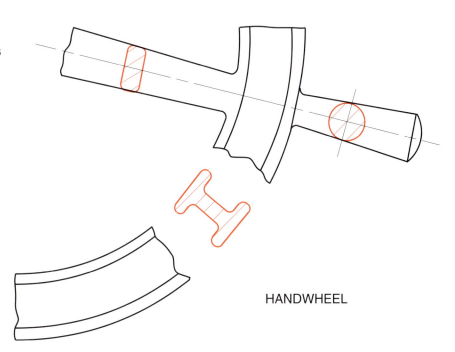

HANDWHEEL

FIGURE 7.12
Revolved sections

FIGURE 7.13
Offset section

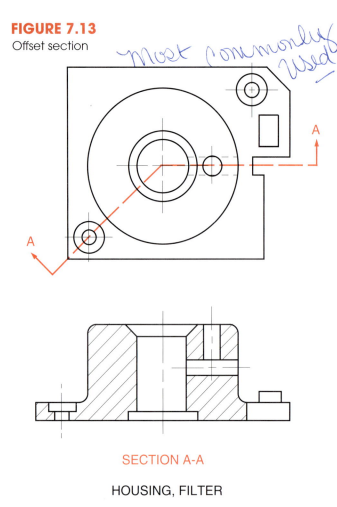

SECTION A-A

HOUSING, FILTER

offset nature. The section lining of the object is shown as if the offset were not present. Therefore, the offset section cutting plane is always identified on the drawing because the offset cannot be detected in the sectional view.

7.3.6 The Removed Section

A **removed section** may be used to illustrate specific areas or parts of an object. It is much like a revolved section except that the section is placed outside the object, as shown in Figure 7.14. Note that the object has different shapes along its length, and these shapes are illustrated through the use of the removed section. Note also that the cutting plane lines are not shown. This type of section is often drawn larger than it appears on the object itself to show important data in greater detail. When this occurs, the scale to which the section is drawn is noted directly below the section.

7.3.7 Sections through Webs, Spokes, and Ribs

Sometimes, the true projection for sections through webs, spokes, and ribs creates false impressions about the actual shape of an object and may mislead the observer into thinking the object is shaped in a certain manner when, in fact, it is not. For this reason, sectional views through webs, spokes, and ribs are treated in a special way.

7.3.7.1 Webs and Spokes

A **web** is an interior piece of material used to connect heavier sections of an object. A **spoke** is a rod, finger, arm, or brace that connects the hub with the rim of a wheel. Figure 7.15a shows a pulley whose interior connecting section (web) is attached to the hub (center) and the rim (outer section) of the pulley. The solid web illustrated in Figure 7.15b is correctly shown as part of full section A-A.

However, if the pulley had four spokes instead of a solid web, its full section would be as shown in Figure 7.16. Even though the cutting plane passes directly through two of the spokes, the sectional view of the figure is drawn without section lining to avoid the appearance of a solid web.

7.3.7.2 Ribs

A **rib** is a part or a piece that serves to shape, support, or strengthen an object. A bearing cap with ribs is illustrated in Figure 7.17a. The conventional, acceptable full section through the object is shown in Figure 7.17b.

7.3.8 The Unlined Section

There are parts that flow through the production line that do not require section linings on drawings. The term **unlined section** refers to those parts that are considered standard, "off-the-shelf," purchased items of hardware. Shafts, nuts, bolts, rods, rivets, screws, keys, pins, ball bearings, set screws, and other types of fasteners whose axes lie in the cutting plane will not be found with section lines. The reason for this is that application of section lining to such objects,

Section 7: *Sectional and Auxiliary Views*   **111**

FIGURE 7.14
Removed sections

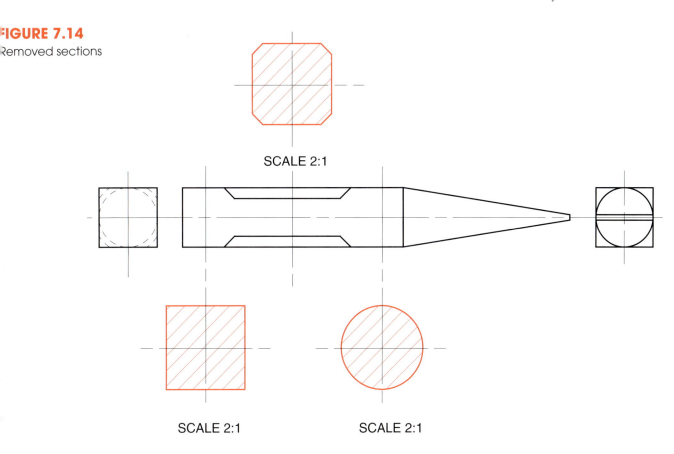

CHISEL, STONE

FIGURE 7.15
Section through webs and spokes

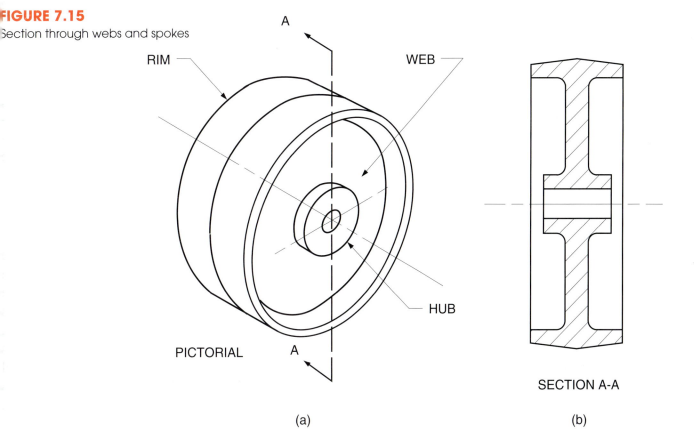

PULLEY

which have no internal detail, would serve no useful purpose and, in fact, would tend to confuse the observer. Figure 7.18 provides several illustrations of items that do not require section lining or crosshatching.

FIGURE 7.16
Spokes in section

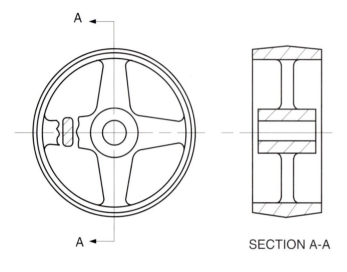

PULLEY

FIGURE 7.17
Section through ribs

7.4 BREAKS IN ELONGATED OBJECTS

Normally, elongated objects such as shafts, bars, tubing, and wood products cannot be shown to their entire lengths on a drawing medium. Therefore, objects of this type require a conventional **break** at a convenient position, and their true length is indicated with a dimension. Another reason for the break is that an object can be drawn to a larger scale, sometimes, if a break is used. Very often, breaks are drawn freehand by the drafter, as seen in some of the examples shown in Figure 7.19.

7.5 AUXILIARY VIEWS

When an object has a sloped, slanted, or inclined surface, it is treated in a special way because it is difficult to distinguish the true lengths of these surfaces. Figure 7.20 is an illustration of such an object. When the block is shown in orthographic projection, as in Figure 7.21, in the top and side view the inclined surface A appears to be *foreshortened*, and the true shape of this surface is not apparent to the observer. To truly

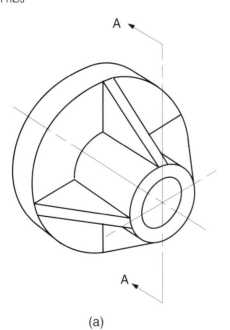

(a)

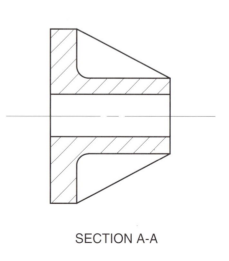

(b)

CAP, BEARING

FIGURE 7.18
Items not requiring section lining

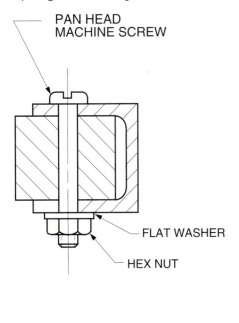
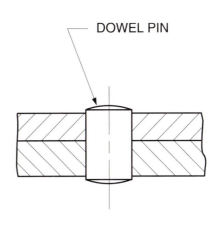
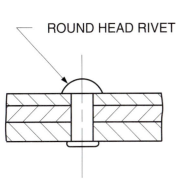
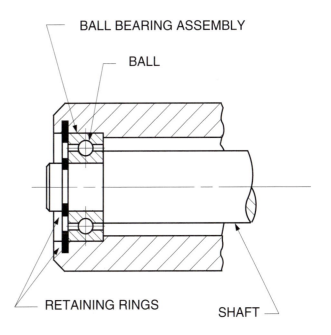

represent the inclined surface and correctly show its true shape, an **auxiliary view** is used.

Figure 7.22 not only shows the three orthographic views but also identifies the true shape of surface A through the addition of an auxiliary view. An auxiliary view is developed as though the *auxiliary plane* were hinged to the plane of the object to which it is perpendicular and then revolved into the plane of the paper. An observer is not likely to see hidden lines on an auxiliary view unless they are required for clarity.

■ 7.5.1 Types of Auxiliary Views

In interpreting engineering drawings, the tradesperson is likely to encounter several types of auxiliary views. Three basic types—the primary, secondary, and partial auxiliary views—will be covered in this section.

7.5.1.1 The Primary Auxiliary View

Auxiliary views can assume many different positions in relation to the three principal planes of projection.

FIGURE 7.19
Conventional breaks on elongated parts

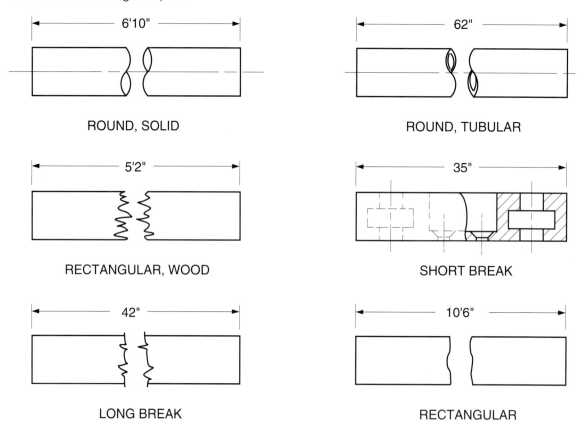

FIGURE 7.20
An object with an inclined surface

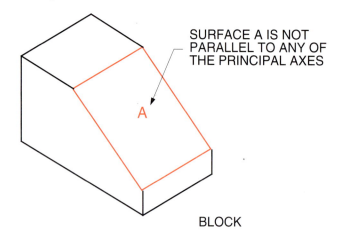

The **primary auxiliary view** can be categorized into three general types, the *front auxiliary view*, the *top auxiliary view*, and the *side auxiliary view*, depending on which orthographic view is used to create the auxiliary view. For example, the front auxiliary view is drawn as a surface of the front view, the top auxiliary view is taken from the top view, and the side auxiliary view is related to the side view. Illustrations of the three types of auxiliary views are shown in Figures 7.23, 7.24, and 7.25. Carefully note which view has been used to create the auxiliary view.

7.5.1.2 The Secondary Auxiliary View

At times, a primary auxiliary view will not sufficiently illustrate the complete detail of an object and, therefore, a **secondary auxiliary view** is necessary. The secondary auxiliary view is a supplementary view projected directly from the primary auxiliary

Section 7: *Sectional and Auxiliary Views*   **115**

FIGURE 7.21
Orthographic projection of a block

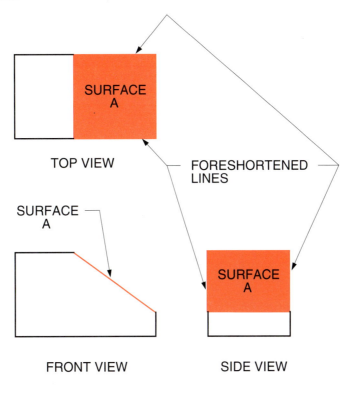

FIGURE 7.22
Orthographic projection with auxiliary view

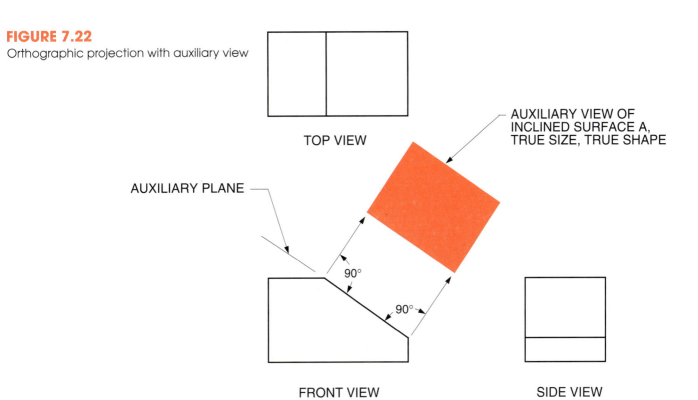

FIGURE 7.23
Front auxiliary view

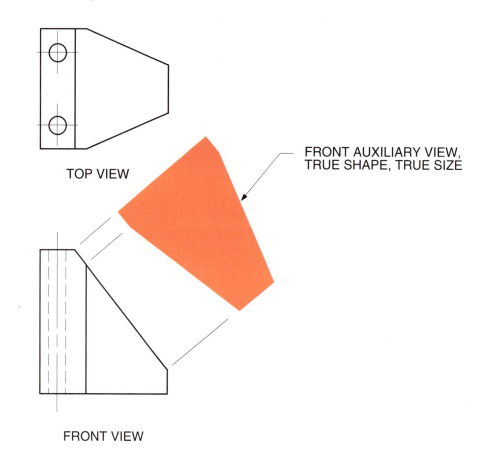

FIGURE 7.24
Top auxiliary view

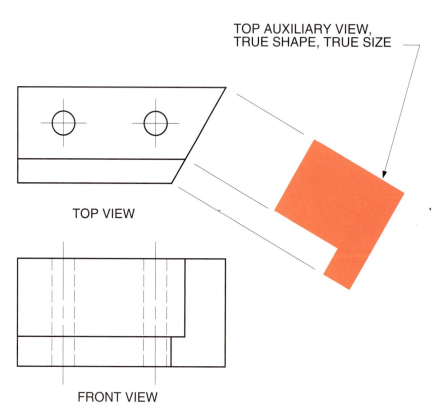

FIGURE 7.25
Side auxiliary view

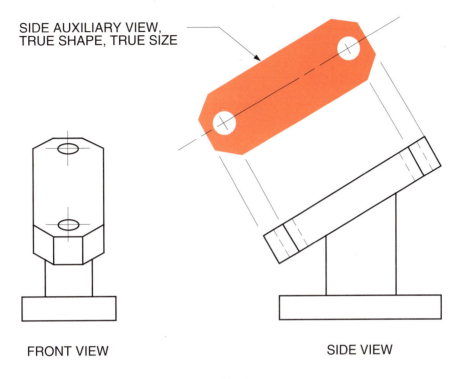

view. Figure 7.26 shows a secondary auxiliary view and its relationship to the front, top, and primary auxiliary views.

7.5.1.3 The Partial Auxiliary View

Another frequently used supplementary view is called the **partial auxiliary view**. It is a view that allows for easier reading of a drawing. It also allows the drawing to be more functional without sacrificing clarity. The adjustable bracket in Figure 7.27 is illustrated with a complete side view and two partial auxiliary front views. Break lines are seen at convenient locations in the partial views.

7.6 SUMMARY

This section dealt with sectional views and auxiliary views and how to interpret them. The purpose of a sectional view is to illustrate the internal form and detail of an object whose interior construction cannot be easily seen in an exterior view. The sectional view is obtained by passing an imaginary plane, called a cutting plane line, through some specific part of an object and presenting the object as though the part cut off by this plane had been removed. Three different cutting plane lines were introduced. Conventions for sectioning were outlined, and several approved symbols for section lining were identified.

The most commonly used sectional views were described and covered in detail, including the full section, half section, broken-out section, revolved section, offset section, and removed section as well as sections through webs, spokes, ribs, and breaks in elongated parts. The various sectional views were defined, and examples of each were shown in detail. The use of unlined sections and breaks in elongated parts were discussed, and examples of each were presented.

The purpose of an auxiliary view is to clearly show the true shape and true size of an object that has an inclined surface. The auxiliary view is drawn as if it were hinged to the surface to which it is perpendicular. Three basic auxiliary views were addressed—the primary, secondary, and partial auxiliary views. As subsets of the primary auxiliary views, the front, top, and side auxiliary views were detailed.

118 Section 7: *Sectional and Auxiliary Views*

FIGURE 7.26
Secondary auxiliary view

FIGURE 7.27
Partial auxiliary views

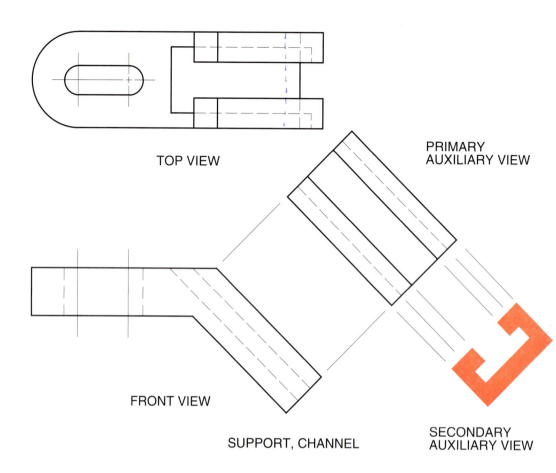

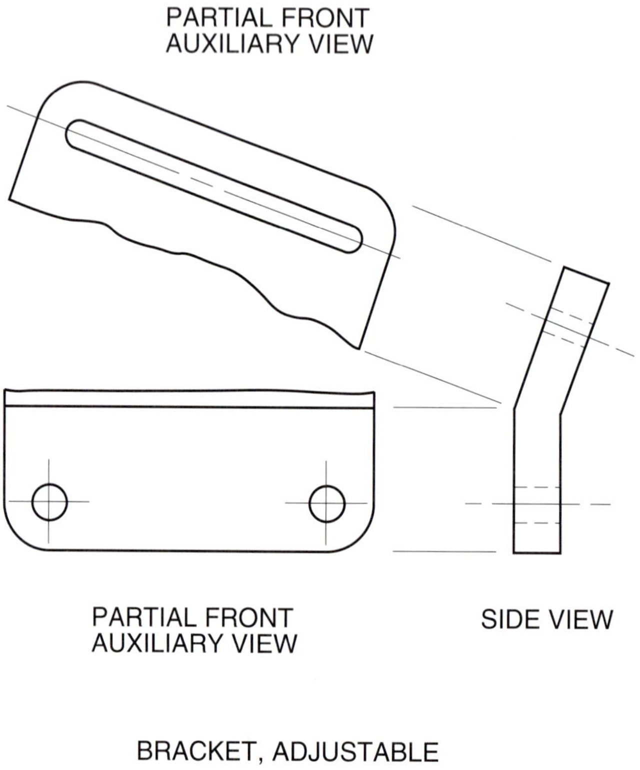

TECHNICAL TERMS FOR STUDY

auxiliary view   The view or views of an object that truly represent the true shape and size of an inclined surface.

break   An interruption of a view due to its size or length.

broken-out section   The sectional view created when only a portion of the internal part of an object is seen. An irregular line is used to determine the extent of the section.

cross-hatching (hatching)   Same as **section lining**.

cutting plane line   A line used to indicate the location or position of the cutting plane for sectional views.

full section   The sectional view formed when the cutting plane passes completely through an object, leaving one-half of the internal details of the object exposed to the observer.

half section   The sectional view formed by passing two cutting planes at right angles to each other along the center lines so that one-quarter of the internal details of an object are seen.

offset section   The sectional view created when the cutting plane is not in a single, continuous, straight line. A simplified method of showing the internal features of an object.

partial auxiliary view   A supplementary view that allows for easier reading of a drawing. It allows the drawing to be more functional without sacrificing quality.

primary auxiliary view   A view that can assume one of three forms—the front auxiliary, top auxiliary, or side auxiliary view. These are considered primary because they are directly related to the orthographic view from which they are taken.

removed section   Similar to a revolved section except that this section is placed outside the object; normally drawn shown larger than the object itself to show important details.

revolved section   A sectional view that is rotated or turned in place so the viewer can see the interior section of an object at that point; like a slice of a specific area. This type of section is located within the part itself.

rib   A part or a piece that serves to shape, support, or strengthen an object.

secondary auxiliary view   A supplementary view of an object, projected directly from a primary auxiliary view.

sectional view   A view used to describe the internal form and detail of an object by passing an imaginary line through some specific part of it.

section lining   Lines used to describe the internal detail of an object. Consists of a series of sharp, uniformly spaced parallel lines, usually shown at a 45-degree angle to the principal lines of a view. Same as cross-hatching, or hatching.

spoke   A rod, finger, arm, or brace that connects the hub with the rim of a wheel.

unlined section   Objects whose axes lie in the cutting plane but that do not require section lining, such as shafts, nuts, bolts, rods, rivets, screws, keys, pins, ball bearings, set screws, and so forth.

web   An interior piece of material used to connect heavier sections of an object.

Section 7: *Sectional and Auxiliary Views* **121**

Student _____   Date _____

Section 7: Competency Quiz

PART A COMPREHENSION

1. What is the purpose of a sectional view? (5 pts)

2. Of what use is a cutting plane? (5 pts)

3. Sketch three examples of cutting plane lines. (3 pts)

4. List six different kinds of sectional views. (6 pts)

5. In a full section, how much of the object is removed? (5 pts)

6. In a half section, how much of the object is removed? (5 pts)

7. What kind of line is used to identify a broken-out section? (5 pts)

8. What kind of sectional view is recommended when the cutting plane lies in a direction other than that of the main axis? (5 pts)

©1995 West Publishing Company

122   Section 7: *Competency Quiz*

9. Explain why webs, spokes, and ribs are not crosshatched. (5 pts)

10. List eight items whose axes lie in the cutting plane but that do not require section lining. (8 pts)

11. Why are breaks necessary in some long parts? (5 pts)

12. What do the arrows indicate on a cutting plane line? (5 pts)

13. What is the purpose of an auxiliary view? (5 pts)

14. List three basic types of auxiliary views. (3 pts)

15. Name three kinds of primary auxiliary views. (3 pts)

16. What must be done before a secondary auxiliary view can be drawn? (6 pts)

17. Complete the following. (10 pts)

    a. The top auxiliary view is one in which the inclined surface is perpendicular to and projected from the _____ view.

    b. The _____ auxiliary view is one in which the inclined surface is perpendicular to and projected from the front view.

Section 7: *Sectional and Auxiliary Views* **123**

Student Name _____ Date _____

18. At what angle do projection lines appear from the inclined surface of an object? (5 pts)

19. What type of auxiliary view simplifies the drawing, shortens the drawing time, and makes a drawing easier to read? (6 pts)

124   Section 7: *Competency Quiz*

PART B TECHNICAL TERMS

For each definition, select the correct technical term from the list on the bottom of the page. (10 pts each)

1. _Cutting plane line_ A line used to indicate the location or position of the cutting plane for sectional views.

2. _break_ An interruption of a view because of its length.

3. _auxiliary view_ Views of an object that truly represent the true shape and size of an object.

4. _rib_ A part or a piece that serves to shape, support, or strengthen an object.

5. _removed section_ A section that is placed outside the object and that is normally shown larger than the object itself.

6. _broken out_ A view used to describe the internal form and detail of an object by passing an imaginary line through some specific part of the object.

7. _offset_ The sectional view created when the cutting plane is not in a single, continuous, straight line.

8. _web_ An interior piece of material used to connect heavier sections of an object.

9. _partial auxiliary_ A supplementary view that allows for easier reading of a drawing. It helps the drawing to be more functional without sacrificing quality.

10. _revolved section_ A sectional view that is rotated or turned in place, so the viewer can see the interior section of an object at that point.

A. auxiliary view
B. break
C. broken-out section
D. cross-hatching
E. cutting plane line
F. full section
G. half section
H. offset section
I. partial auxiliary view
J. removed section
K. revolved section
L. rib
M. sectional view
N. web

Section 7: *Sectional and Auxiliary Views* **125**

Student Name _____ Date _____

PART C SECTIONAL VIEW IDENTIFICATION

1. In the grid area, produce a freehand sketch of full section A-A of the flanged housing.

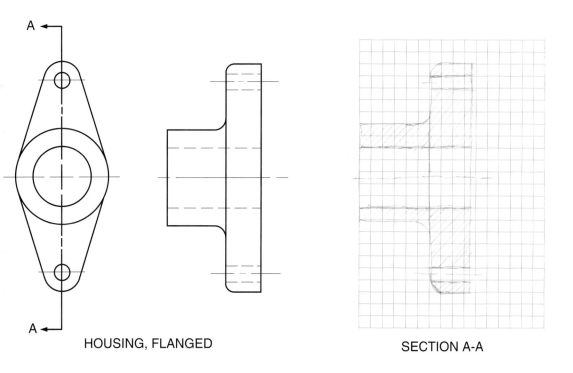

HOUSING, FLANGED   SECTION A-A

2. In the grid area, produce a freehand sketch of half section A of the shaft coupling.

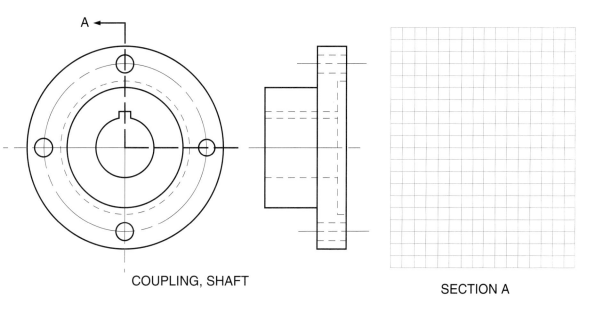

COUPLING, SHAFT   SECTION A

©1995 West Publishing Company

3. Select a highly detailed area of the retainer bushing and produce a broken-out section of the selected area.

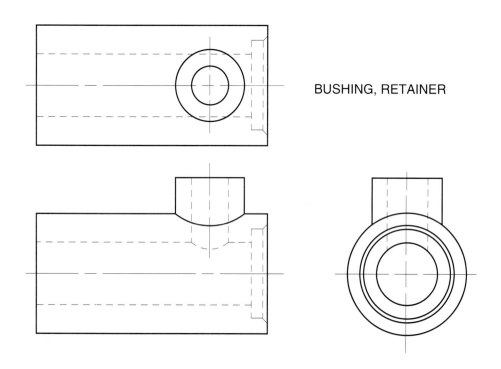

4. Show a revolved section of the pivot arm at section A-A.

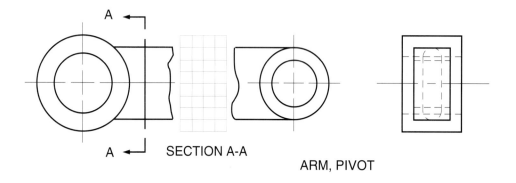

5. Draw a removed section A-A of the adjusting tool below.

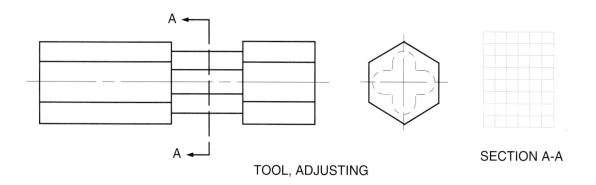

Section 7: *Sectional and Auxiliary Views* **127**

Student Name _____ Date _____

6. In the space provided, sketch offset section A-A of the bearing housing.

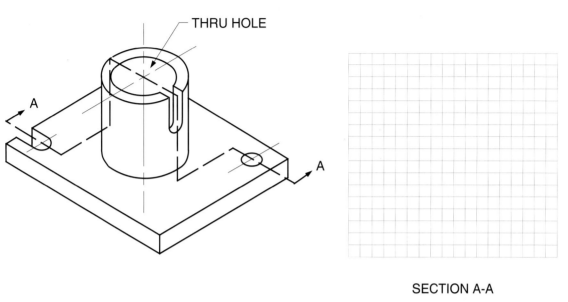

HOUSING, BEARING

SECTION A-A

7. In the space provided, sketch full section of A-A of the drive pulley.

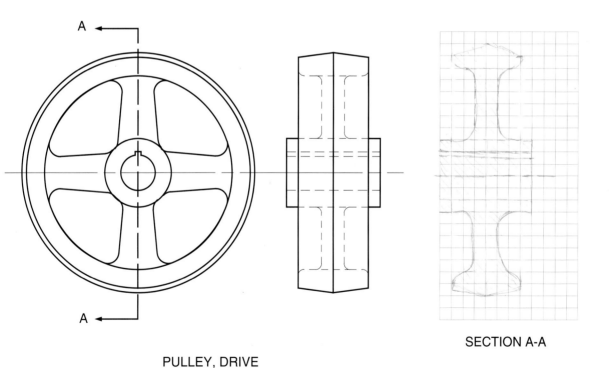

PULLEY, DRIVE

SECTION A-A

©1995 West Publishing Company

128 Section 7: *Competency Quiz*

8. In the grid area, sketch full section A-A of the face plate. In the front view, show the location of the cutting plane line.

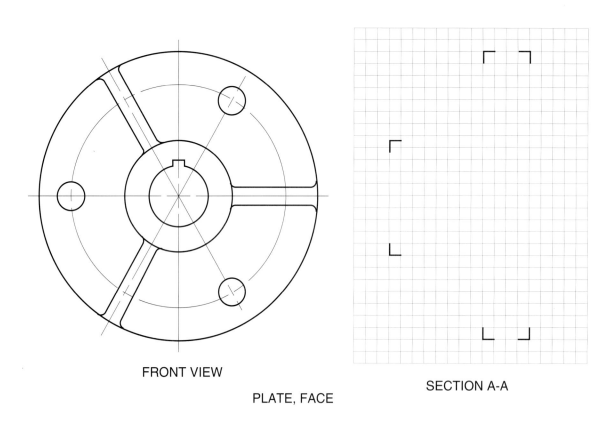

FRONT VIEW

PLATE, FACE

SECTION A-A

9. Shown is a full section of a shaft coupling assembly. Add conventional, general-purpose section lining to the appropriate parts.

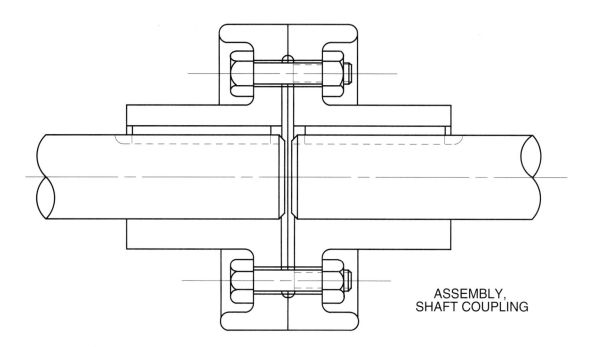

ASSEMBLY, SHAFT COUPLING

Section 7: *Sectional and Auxiliary Views*   **129**

Student Name _____ Date _____

PART D AUXILIARY VIEW IDENTIFICATION

1. Sketch the front auxiliary view of surface A of the inclined block. Use the grid for the sketch.

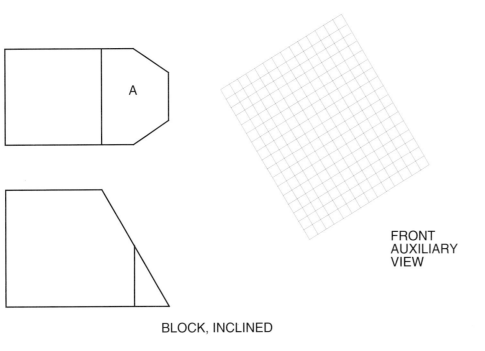

BLOCK, INCLINED

FRONT AUXILIARY VIEW

2. Construct a top auxiliary view of the shaft mount. Use the grid for your drawing.

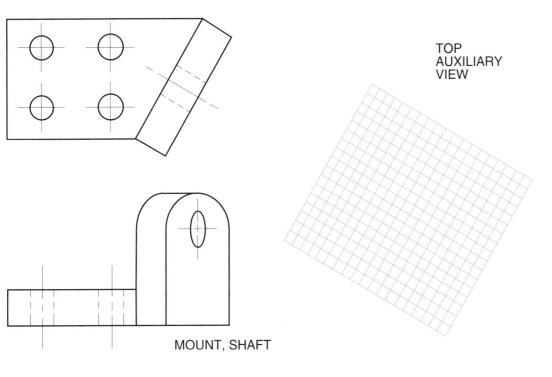

MOUNT, SHAFT

TOP AUXILIARY VIEW

©1995 West Publishing Company

3. Given the front and side views of a wooden model of a slide block, construct a side auxiliary view of surface A.

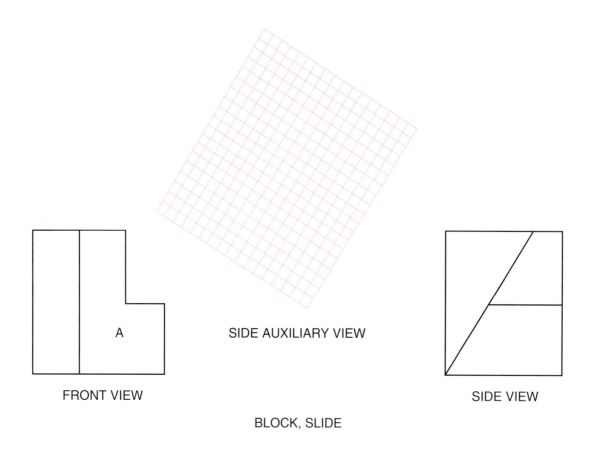

4. Sketch a partial auxiliary view of surface A of the strap.

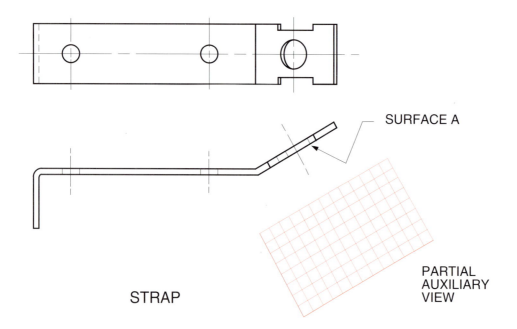

SECTION 8

Dimensioning

LEARNER OUTCOMES

You will be able to:

- Explain the purpose of dimensions.
- Identify the two predominant dimensioning systems.
- List the fundamental rules for dimensioning.
- Identify several special symbols used in the application of dimensions.
- Define ten technical terms.
- Follow appropriate practices for the dimensioning of part features.
- List the differences between countersunk, counterbored, and counterdrilled holes.
- Demonstrate your learning through the successful completion of competency exercises.

8.1 THE PURPOSE OF DIMENSIONS

In producing a drawing, the features of the object are drawn first, and then the dimensions are added. Dimensions on a drawing provide the producer of the object (the machine tradesperson) with the necessary information regarding the exact size and location of each feature of the part to be made. Without dimensions, an object cannot be produced accurately. For our purposes, a **dimension** may be defined as a numerical value expressed in appropriate units of measure (inches, millimeters, feet, yards, etc.). Dimensions are placed on a drawing, along with lines, symbols, and text, to define the geometrical characteristics of an object or part. Dimensions are used to complete the description of an object, and they are placed on a drawing with the ultimate user of the drawing kept in mind. For example, when placing dimensions on a drawing, the drafter should ask, Will dimensions that are added facilitate the assembly of a group of parts? Will they aid the mold maker? Are they to be used for machining purposes, or for some other manufacturing or production purpose?

The dimensions that are seen on a drawing are placed there by one who has the ability to determine, locate, and apply them in an orderly, acceptable manner. The most widely accepted standard for dimensioning is the American National Standards Institute (ANSI) specification Y 14.5 M-1982, Dimensioning and Tolerancing for Engineering Drawings. Over the years, most medium- and large-size industries, including governmental agencies, have changed their drafting practices to this revised standard. The 1982 ANSI specification was the first effort toward the standardization of production drawings with the thought of ensuring the interchangeability of the parts produced by one manufacturer with the same parts produced by another manufacturer. The dimensioning practices outlined in this section reflect the various terms and symbols used in the 1982 specification.

8.2 FUNDAMENTAL RULES OF DIMENSIONING

For dimensions to define geometrical characteristics clearly and concisely, the drafter must follow a few basic rules to produce a quality engineering drawing. The tradesperson should be aware of these rules in an effort to seek out correct as well as incorrect methods used by the drafter in producing the drawing. The shop worker is in a position to examine the drawing to see if the basic rules have been followed and to raise questions such as these:

- Are sufficient dimensions shown so that intended sizes and shapes can be determined without performing calculations or assuming distances?
- Are dimensions stated clearly so that they can be interpreted in only one way?
- Have dimensions been selected so that the accumulation of tolerances is minimized or eliminated?
- To avoid confusion, has each dimension been identified only once?
- Have dimensions to hidden lines been avoided, where possible?
- Have dimensions been placed so that they are outside the outline of the part?

The *principal view* is the view that most completely shows the object's shape, and therefore, it is the location of most of the object's dimensions. Figure 8.1 clearly identifies the front view as the principal view of the slide bar because most of the dimensions are taken from this view.

8.3 TYPES OF DIMENSIONING SYSTEMS

Several types of dimensioning systems are used in industry. They include the **unidirectional**, **aligned**, *tabular*, *arrowless*, and *chart-drawing* types. The two

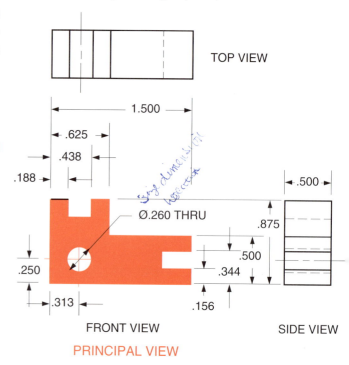

FIGURE 8.1
Multiview drawing showing its principal view

BAR, SLIDE

most common are the unidirectional and the aligned systems. Both will be covered in this section.

■ 8.3.1 The Unidirectional System

The unidirectional system of dimensioning is probably the most commonly used system throughout industry because it is the easiest to read. The term *unidirectional* implies that the dimensions are placed in one way or direction. This means that all numerals, figures, and text are lettered horizontally and are placed parallel to the bottom edge of the drawing medium. It allows the reader to view the drawing without having to turn it. It is read directly. The clevis pin illustrated in Figure 8.2 is dimensioned using this method.

■ 8.3.2 The Aligned System

The aligned system requires that numerals be in alignment (in line) with dimension lines so that they are read parallel to the bottom of the drawing for horizontal dimensions and parallel to the right side for

FIGURE 8.2
Unidirectional dimensioning practices

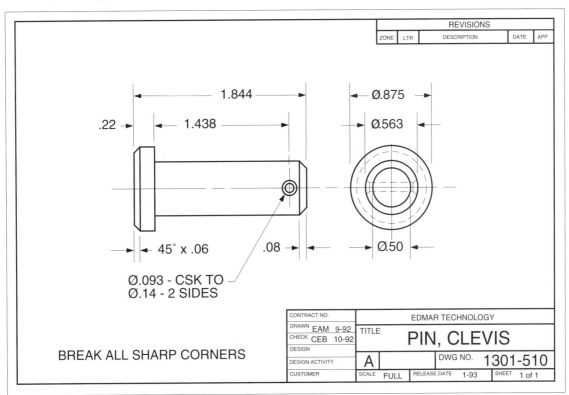

8.4 BASIC DIMENSIONING TERMS

vertical dimensions. Figure 8.3 demonstrates acceptable and unacceptable practices for this system. The tradesperson should be able to identify these differences.

There are a few basic dimensioning terms that are important to this topic. They are terms used repeatedly. Therefore, the manufacturing technician should become familiar with their meanings and usages.

- **Nominal size**—The approximate size of a part or object, normally described in fractional form. The nominal size, for example, for the pin shown in Figure 8.4 is ⌀ ½ inch because both decimal dimensions are close to ⌀ ½ inch.
- **Basic size, or basic dimension**—A numerical value used to describe the theoretical exact size of a part or object, presented in decimal form. It is the starting point for working out the tolerances for a dimension. The basic size for the pin in Figure 8.4 is ⌀.5000 inch. Tolerances will be covered, in depth, in Section 9.
- **Actual size**—The actual measured size or dimension of a part that a measuring tool would provide, shown as ⌀.4950 in Figure 8.4.
- **True position**—The exact location of a feature established by basic dimensions. The location of the hole in the pin in Figure 8.4 is a true position (1.531 inch).
- **Limit dimensions**—The largest and smallest acceptable dimensions. For the pin in Figure 8.4, these dimensions are as follows:

⌀.4955 is the high limit (maximum value)
⌀.4945 is the low limit (minimum value)

- **Datum**—A theoretically exact point. A datum is the origin from which the location of the features of a part are established. An illustration of a datum appears in Figure 8.5, where the datum

FIGURE 8.3
Aligned system practices

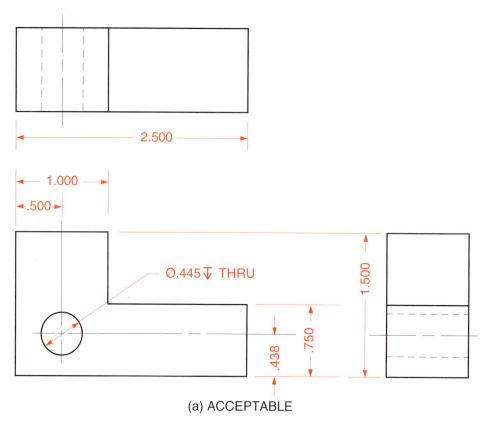

(a) ACCEPTABLE

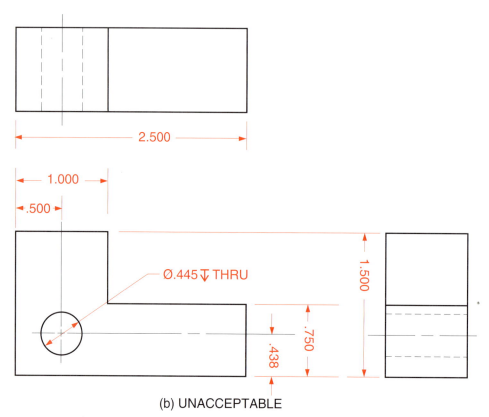

(b) UNACCEPTABLE

FIGURE 8.4
Basic dimensioning terms

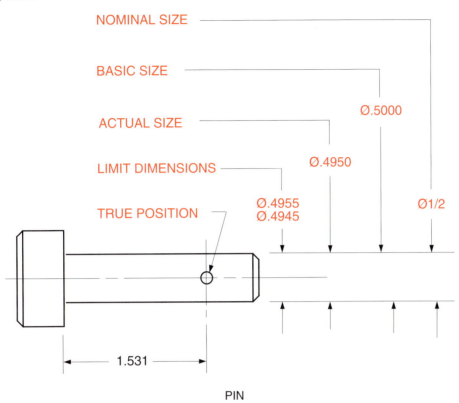

FIGURE 8.5
Datum illustration

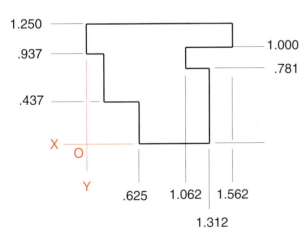

is at point O, at the intersection of the X and Y axes. A datum is the basis for arrowless or datum dimensioning practices.

- **Feature**—A general term at refers to a physical attribute of an object or part. It may also include an object's shape or form.

- **Reference dimension**—A dimension, usually without a tolerance, used for informational purposes only. It is considered auxiliary information and is not normally used in inspection procedures. Methods for identification include enclosing the reference dimension in parentheses, as shown in the two separate examples in Figure 8.6.

8.5 APPLICATION OF DIMENSIONS

Up to this point, general information has been provided relative to the topic of dimensioning. The sections that follow deal with specific techniques for the *application of dimensions* and should benefit the shop worker in the interpretation of engineering drawings.

8.5.1 The Dimension Line

The **dimension line**, with terminations called arrowheads, shows the direction and extent of a dimension. Numerals indicate the number of units of a measure-

FIGURE 8.6
Reference dimensions

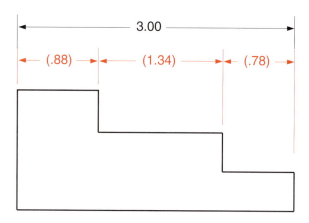

(a) INTERMEDIATE REFERENCE DIMENSIONS

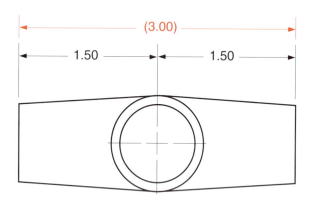

(b) OVERALL REFERENCE DIMENSIONS

FIGURE 8.7
Conventional dimension lines

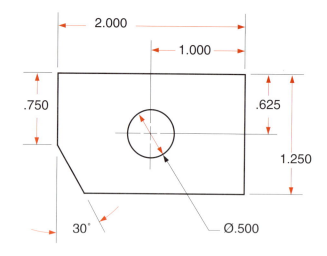

FIGURE 8.8
Alternate dimension lines

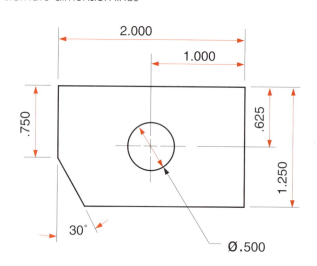

ment, and normally the dimension line is broken to allow for the insertion of the numerals, as illustrated in Figure 8.7. An alternate, but acceptable, method has the dimensions appear above and parallel to an unbroken dimension line, as shown in Figure 8.8.

■ 8.5.2 The Extension Line

An **extension line,** which is also referred to as a *projection line*, is used to show the extension of a surface to a location outside the object (part) outline, where it may be clearly dimensioned. An extension line normally begins with a short, visible gap from the outline of the part and extends beyond the outermost related dimension line. Extension lines are normally shown perpendicular to dimension lines, but under certain circumstances they may appear at an oblique angle to clearly show the proper application. When oblique extension lines are used, the dimension line appears in the direction in which the dimension applies. Typical extension lines are represented in Figure 8.9. Breaks in extension lines may be seen when these lines cross arrowheads, as illustrated in Figure 8.10.

■ 8.5.3 The Leader

A **leader** is a line used to direct a dimension, note, or symbol to the intended location on the drawing. It takes the form of an inclined straight line except for a short horizontal portion extending to mid-height of

Section 8: *Dimensioning*   **137**

FIGURE 8.9
Extension lines

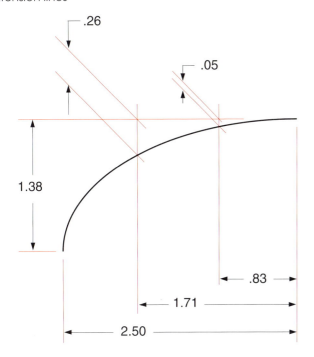

FIGURE 8.10
Breaks in extension lines

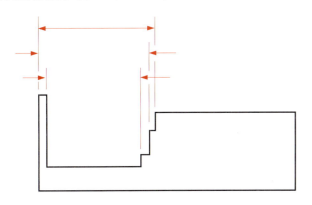

FIGURE 8.11
Leaders

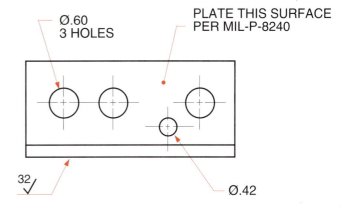

FIGURE 8.12
Leader-directed dimensions

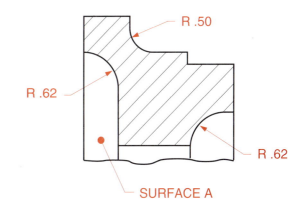

the first or last letter or digit of a note or dimension (see Figure 8.11). A leader normally ends with an arrowhead touching a line. When a leader makes reference to a surface within the surface area, the leader ends with a dot approximately .06 inch in diameter. Examples of leader-directed dimensions are shown in Figure 8.12.

■ 8.5.4 Dimensions Not to Scale

When a dimension is not to scale (the dimension and the actual distance drawn are not equal), the dimension in question is underlined with a straight, thick line so that the person interpreting the drawing will know of this out-of-scale condition. On older drawings, one is likely to find a line with curves to represent this condition. Both these ways of indicating dimensions not to scale are depicted in Figure 8.13.

8.6 DIMENSIONING OF FEATURES

Various characteristics and features of objects require special or unique methods of dimensioning. The practices discussed in the following sections are considered important enough for the tradesperson to become familiar with them.

■ 8.6.1 Diameters

The definition of the term *diameter* states that it is the length of a straight line passing through the center of a circle and ending at the circumference on each end. When a dimension refers to a round object, or more specifically, to a **diameter**, the diameter symbol, ⌀, will be seen preceding all diametral values, as required by the ANSI 1982 specification.

When the diameter of a spherical feature is needed, the diametral value is preceded by the spherical diameter symbol, S, in addition to the diameter symbol, ⌀. Figure 8.14 presents samples of each of these conditions.

8.6.2 Radii

A **radius** may be defined as the length of a straight line that originates at the center of a circle and ends at the perimeter (circumference). When a radius value is specified on a drawing, it is preceded by the radius symbol, R. The radius dimension ends with an arrowhead at the arc end. Figure 8.15 shows several examples of how radii are dimensioned for both internal and external dimensions. The major difference in identifying the internal versus the external types is the **foreshortened radius lines** for external radii, as illustrated in Figure 8.15b. Note the letter, R preceding each of the radii in the figures. One of the key factors in the dimensioning of radii is space availability on a given drawing.

For radii whose centers have a known location, Figure 8.16a illustrates how extension/center lines locate the center of the radius. Examples of dimensioning radii whose center location is unknown are provided in Figure 8.16b.

When a spherical radius symbol is needed to show a spherical radius, the letters *SR* precede the radius dimension, as illustrated in Figure 8.17.

FIGURE 8.13
Dimensions not to scale

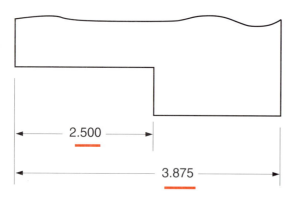

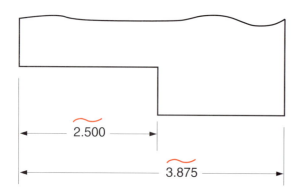

FIGURE 8.14
Dimensioning of diameters

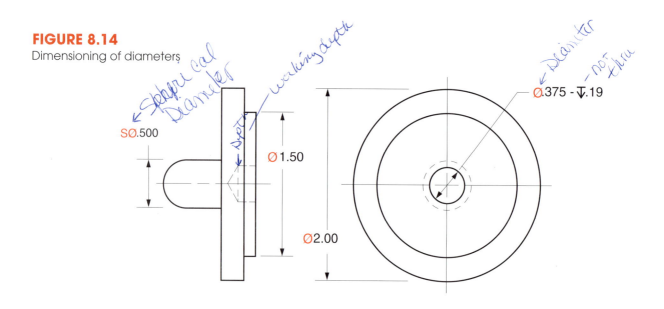

Section 8: *Dimensioning* 139

FIGURE 8.15
Dimensioning internal and external radii

FIGURE 8.16
Dimensioning radii with located and unlocated centers

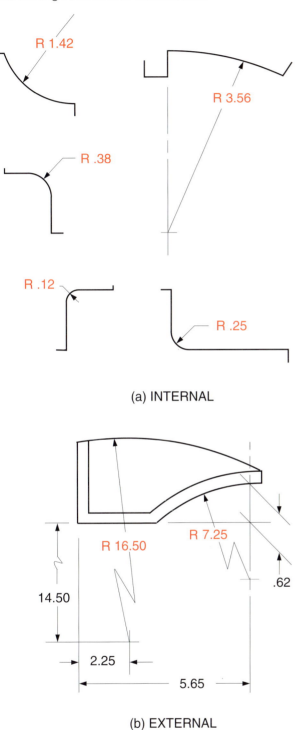

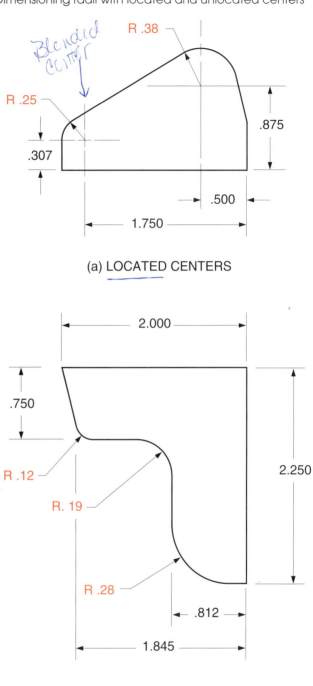

8.6.3 Chords, Arcs, and Angles

Care needs to be taken when viewing dimensions for chords, arcs, and angles. Each is dimensioned in a different manner. For a *chord*, the dimension line is perpendicular to its extension lines and parallel to the chord. For an **arc**, the dimension line follows the contour of the arc's curve. In addition, the arc symbol, ⌒, appears above the dimension. When an *angle* is dimensioned, the extension lines project from the sides forming the angle, and the dimension line forms an arc. Examples of the dimensioning of chords, arcs, and angles are provided in Figure 8.18.

FIGURE 8.17
Spherical radius symbol

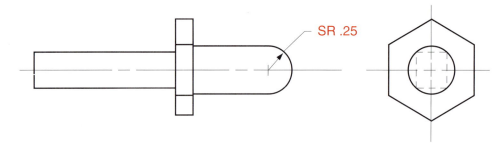

FIGURE 8.18
Dimensions for chords, arcs, and angles

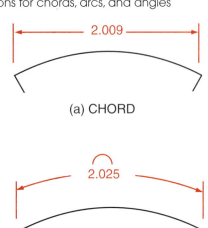

(a) CHORD

(b) ARC

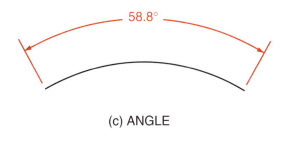

(c) ANGLE

FIGURE 8.19
Dimensioning for fully and partially rounded ends

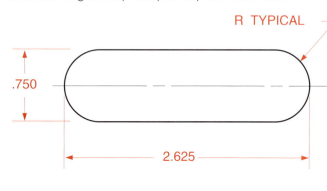

(a) FULLY ROUNDED END

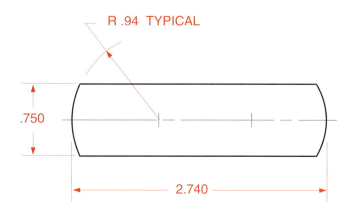

(b) PARTIALLY ROUNDED END

■ 8.6.4 Rounded Ends

Overall dimensions, that is, dimensions over the entire length of a part, are used on parts with rounded ends. On fully rounded ends, radii are indicated but not dimensioned, as illustrated in Figure 8.19a. For parts with partially rounded ends, radii are dimensioned, as depicted in Figure 8.19b.

■ 8.6.5 Holes

When interpreting engineering drawings, a technician will have an opportunity to view many different types of holes and how they are dimensioned. Some holes are round, others are not. Some are slanted, others require different depths and contours. Holes are one of the most commonly dimensioned features on a drawing, and they need to be interpreted correctly.

8.6.5.1 The Round Hole

On a drawing, a round hole is shown dimensioned as illustrated in Figure 8.20. Note that when a hole goes completely through the object, the abbreviation THRU appears after the dimension. When a hole does not extend completely through the object, it is referred to as a **blind hole**. The depth dimension of a blind hole represents the depth of the full diameter as measured from the outer surface of the part, and this depth is indicated by a dimension. When the depth of any hole on a drawing is required, the symbol ⤓ is used, even for a blind hole. On older drawings, the designation DP (meaning *deep* or *depth*) is seen instead.

FIGURE 8.20
Acceptable dimensioning practices for round holes

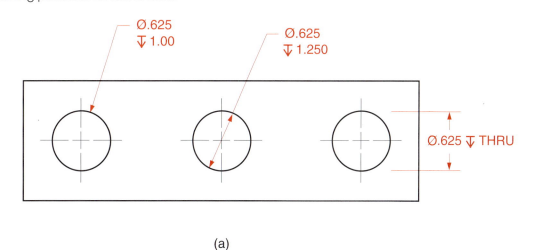

(a)

OR

(b)

8.6.5.2 The Slotted Hole

A **slotted hole,** or **slot**, as it is often called, is an elongated feature with parallel sides and curved ends. It is usually dimensioned by indicating its length and width in one of the three methods shown in Figure 8.21. Slotted holes are normally used in situations where an adjustment in parts or assembly are necessary. In Figure 8.21a, the slotted hole is shown dimensioned using a longitudinal center line and a width dimension. Figure 8.21b identifies the length and width dimension with a local note. Figure 8.21c illustrates the use of dimension lines for the overall width and length. In each case, the end radii are shown but not dimensioned.

8.6.5.3 The Counterbored Hole

A **counterbore** is an enlargement of one or both ends of a hole to provide a below-the-surface mounting for a screw or a nut, or a sunken seat for a fastener. Counterbored holes are seen on drawings of castings as well as on machined surfaces. The appropriate notation for a counterbored hole is the symbol ⌴, as demonstrated in Figure 8.22, although CBORE will be found on older drawings. On a drawing, a counterbored hole is specified with a note giving the diameter, depth, and inside corner radius. If the thickness of the remaining material is of some importance, this thickness, rather than the depth, is dimensioned.

8.6.5.4 Countersunk and Counterdrilled Holes

The purpose for a countersunk hole or a counterdrilled hole is to obtain a flush or a below-the-surface mounting area for a screw or a rivet. Countersinking is similar to counterboring. The difference is that the sides of the **countersink** are at an angle to the cutting surface, while the sides of a counterbore are perpendicular to the cutting surface. The designated symbol used to identify a countersink is ⋁. The designation CSK is frequently seen as well. For the **counterdrill**, the notation is CDRILL. Note that the diameter, depth, and the total included angle for the countersink and the counterdrill are indicated in Figure 8.23. As is true with round holes, the depth dimension is always taken to mean the full diameter of the counterdrill from the outer surface of the object, as illustrated in Figure 8.23b.

FIGURE 8.21
Dimensioning options for slotted holes

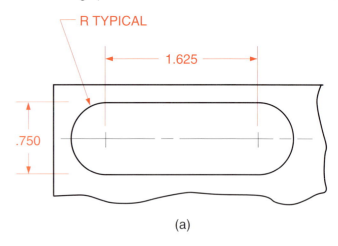

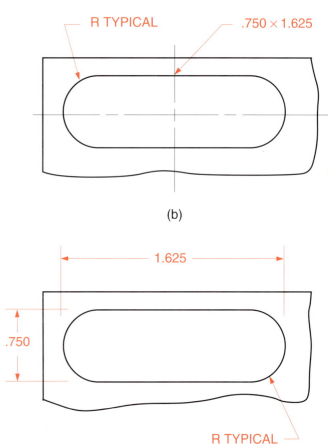

8.6.5.5 The Spotface

Machining for a **spotface** is often performed on a casting to clean up the surface or to provide an accurately machined area for a washer or for the head of a fastener. On a drawing, the diameter of a spotface will

FIGURE 8.22
Dimensioning practices for counterbored holes

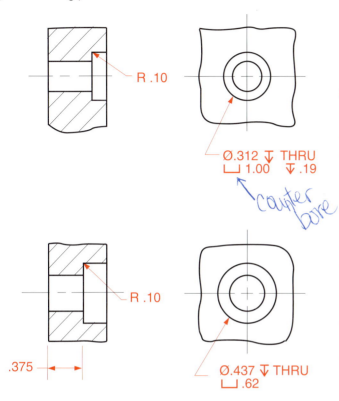

FIGURE 8.23
Examples of countersunk and counterdrilled holes

always be specified. Either the depth or the remaining material thickness may be specified, as shown in Figure 8.24. The spotface, whose symbol is the same as that used for a counterbore or the letters SF, may instead be specified by a note.

■ 8.6.6 The Chamfer

The **chamfer** is an angled, machined relief needed to eliminate sharp edges, corners, or burrs on mechanical parts. There are basically two types of chamfers: external and internal. The *external chamfer* is normally dimensioned showing an angle and a linear dimension, or by two linear dimensions. When an angle and a linear dimension are used, the linear dimension is that distance from the indicated surface of the object to the beginning of the chamfer, as shown in Figure 8.25b. Sometimes, a note is used to specify 45-degree chamfers—but only 45-degree chamfers because both the radial and longitudinal values (the vertical and horizontal dimensions) are the same, as pictured in Figure 8.25a.

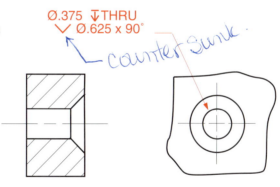

(a) COUNTERSUNK

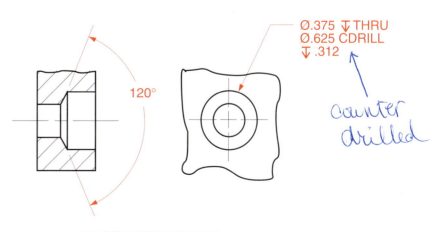

(b) COUNTERDRILLED

FIGURE 8.24
Dimensions for a spotface

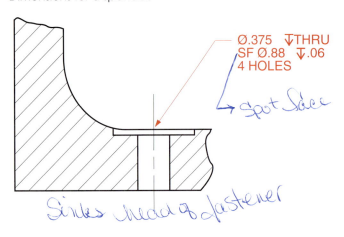

FIGURE 8.25
Dimensioning practices for external chamfers

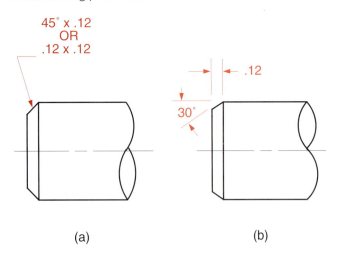

FIGURE 8.26
Dimensioning practices for internal chamfers

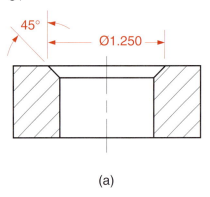

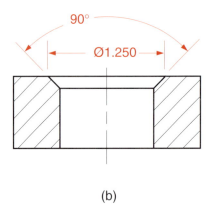

FIGURE 8.27
Dimensioning practices for chamfers for surfaces at other than 90 degrees

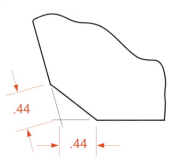

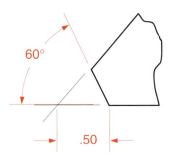

For an *internal chamfer* located on the edge of a round hole, the same condition may be found as for the external chamfer except for when dimensional control of the chamfer is required. This is illustrated in Figure 8.26.

When chamfers are needed for surfaces that intersect at other than 90 degrees, a drafter is likely to use the practices illustrated in Figure 8.27.

■ **8.6.7 The Keyway**

A **keyway** is a seat for a key; it is also called a *keyseat*. It is a slot in a shaft or a hub, or both, to retain a key. It is used, most often, on an assembly of moving parts. The key, which usually takes the form of a

FIGURE 8.28
Dimensions for keyways

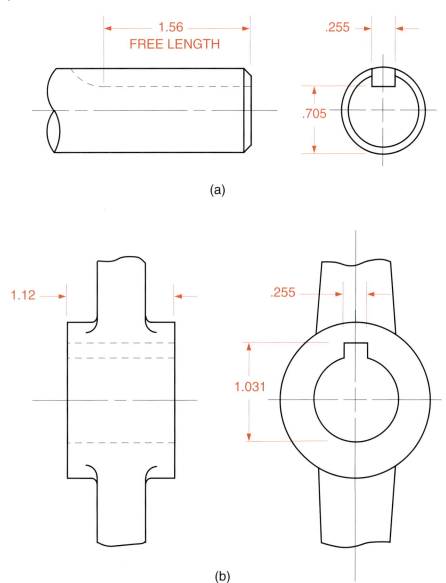

square in cross section, is used to ensure that both the shaft and its mating part will rotate together and not separate during radial movement. The keyway is usually produced with a narrow milling machine cutter or with a saw-blade-type cutter. These methods cause the end of the keyway on a shaft to have a radius on one end, as seen in Figure 8.28a.

A keyway is dimensioned by its width, depth, location, and length, as illustrated in Figure 8.28, which shows dimensioning practices for a keyway (a) in a keyseat and (b) in a hub. Note that the depth of the keyway is normally dimensioned from the opposite side of the keyseat or hole. Various standard types of keys are shown in Appendix J.

8.7 SUMMARY

This section brought under discussion the broad topic of dimensioning. The term *dimension* was defined, and the purpose and fundamental rules for dimensioning were presented. The two most common types of dimensioning systems, unidirectional and aligned, were detailed. Numerous aspects of the application of dimensions were covered, including dimension lines, extension lines, and the various types of leaders.

Examples of dimensioning practices for features such as diameters, radii, chords, arcs, and angles as well as round, and slotted, counterbored, counter-sunk, and counterdrilled holes were covered in depth. In addition, methods for the dimensioning of spot-faces, chamfers, and keyways were outlined, and examples of each were presented.

TECHNICAL TERMS FOR STUDY

actual size   The actual measured size or dimension of a part, as provided by a measuring tool.

aligned system   A form of dimensioning system which requires that numerals be in alignment with dimension lines so that they are placed parallel to the bottom of the drawing for horizontal dimensions and parallel to the right side for vertical dimensions.

arc   A segment of a circle formed by a radius. When a dimension is required for an arc, the arc symbol, ∩, will be found immediately above the arc dimension.

basic size, or basic dimension   A numerical value used to describe the theoretical size of a part or object, presented in decimal form.

blind hole   A hole that does not extend completely through an object or part.

chamfer   An angled relief that is needed to eliminate sharp edges, corners, or burrs on machined parts.

counterbore   An enlargement of one or both ends of a hole, for providing a below-the-surface mounting for a screw, a nut, or both.

counterdrill   Similar to a counterbore except that an angled tool is used to enlarge the hole instead of a flat-ended tool. The symbol for the counterdrill is CDRILL.

countersink   A chamfered or angled end of a hole for the seating of a flat-head screw or countersunk rivet.

datum   A theoretical exact point. A datum is the origin from which the location of the features of a part are established.

diameter   The length of a straight line passing through the center of a circle and ending at the cir-cumference at each end. On a drawing, the diameter symbol, ∅, precedes the dimension. When a spherical diameter is required on a drawing, the spherical diameter symbol, S, appears and precedes the diameter symbol, ∅, as follows: S∅.

dimension   A numerical value expressed in the appropriate units of measure (inches, feet, yards, meters, millimeters, etc.).

dimension line   A line that shows the direction and extension of a dimension. It is terminated at each end, usually, with arrowheads.

extension line   A line that is used to indicate the extension of a surface to a location outside of the object (part) outline.

feature   A general term that refers to a physical attribute of an object or part. It may include an object's shape or form.

foreshortened radius   A jagged line that represents a radius whose center is outside the boundary of the drawing.

keyway   A seat for a key. Also known as a keyseat. It is a slot in a shaft, hub, or both, designed to retain a key. Normally used on assemblies that require moving parts.

leader   A line used to direct a dimension, note, or symbol to the intended location on a drawing. It takes the form of an inclined line except for a short horizontal portion extending to the note or dimension.

limit dimensions   The largest and smallest acceptable dimensions.

nominal size   The approximate size of a part or object; normally described in fractional form.

radius   The length of a straight line originating at the center of a circle and continuing to the perimeter. On a drawing, the radius symbol, R, precedes a radius dimension. When a spherical radius dimension is required on a drawing, the spherical radius symbol, S, appears and precedes the radius symbol, R, as follows: SR.

reference dimension   A dimension, usually without a tolerance, used for informational purposes only.

Section 8: *Dimensioning* **147**

slotted hole A slot or elongated hole whose sides are parallel and whose ends are curved or rounded.

spotface An area of a casting that is machined for the purpose of seating the head of a fastener or flat washer. Its machined surface is perpendicular to the hole.

true position The exact location of a feature, established by basic dimensions.

unidirectional system A dimensioning system in which all dimensions are placed in one way or direction, meaning that all numerals, figures, and text are lettered horizontally and placed parallel to the bottom edge of the drawing. It is the easiest-to-read system for interpreting engineering drawings.

Section 8: *Dimensioning*   **149**

Student _____   Date _____

Section 8: Competency Quiz

PART A COMPREHENSION

1. What is the purpose of dimensions? (5 pts)

2. What is the accepted ANSI specification used for dimensioning in industry? (5 pts)

3. List five fundamental rules for dimensioning on a drawing. (15 pts)

4. In orthographic projection, on which view do most of the dimensions usually appear? (3 pts)

5. Identify the two predominate dimensioning systems. Define each system. (10 pts)

6. What is the purpose of a leader? (5 pts)

7. How is a not-to-scale dimension identified on a drawing? (3 pts)

8. State the purpose of an extension line. (5 pts)

9. What is a chamfer? (5 pts)

©1995 West Publishing Company

150   Section 8: *Competency Quiz*

10. How do a counterbore, countersink, and counterdrill differ? (12 pts)

   Counterbore - straight down to flat edge
   sink - down at an angle
   drill - straight down ... indicated

11. What symbol identifies the diameter dimension? (3 pts)

   ⌀

12. Draw the symbol that identifies an arc. (3 pts)

   ⌒

13. What is the general purpose of a slotted hole? (5 pts)

   when aligning is necessary

14. On a drawing, what is the accepted designation or symbol for the following? (15 pts)

   counterbore: ⊔

   spherical radius: SR

   counterdrill:

   spherical diameter: S⌀

   countersink: ⌵

15. On what parts or objects would a keyway be found? (6 pts)

   shaft or a ...

Section 8: *Dimensioning* **151**

Student_____ Date _____

PART B TECHNICAL TERMS

For each definition, select the correct technical term from the list on the bottom of the page. (10 pts each)

1. _____*arc*_____ A segment of a circle formed by a radius; when dimensioned, has a symbol immediately above it.

2. _____*chamfer*_____ An angled relief needed to eliminate sharp edges, corners, or burrs on machined parts.

3. _____*countersink*_____ A chamfered or angled end of a hole, for the seating of a flat-head screw or a countersunk rivet.

4. _____*dimension*_____ A numerical value expressed in one of the appropriate units of measure.

5. _____*extension line*_____ A line used to indicate the extension of a surface to a location outside of the object outline.

6. _____*leader*_____ A line used to direct a dimension, note, or symbol to the intended location on a drawing.

7. _____*spotface*_____ An area of a casting that is machined for the purpose of seating the head of a fastener or a flat washer.

8. _____*unidirectional*_____ A dimensioning system in which all dimensions are placed in one direction and parallel to the bottom edge of the drawing.

9. _____*blind hole*_____ A hole that does not extend completely through an object or part.

10. _____*counterbore*_____ An enlargement of one or both ends of a hole, providing a below-the-surface mounting for a screw, a nut, or both.

A. arc
B. blind hole
C. chamfer
D. countersink
E. counterbore
F. diameter
G. dimension
H. extension line
I. keyway
J. leader
K. radius
L. slotted hole
M. spotface
N. unidirectional system

©1995 West Publishing Company

152 Section 8: *Competency Quiz*

PART C DIMENSION IDENTIFICATION

1. Each of the blanks indicates an important item when interpreting dimensions on an engineering drawing. Identify each where indicated.

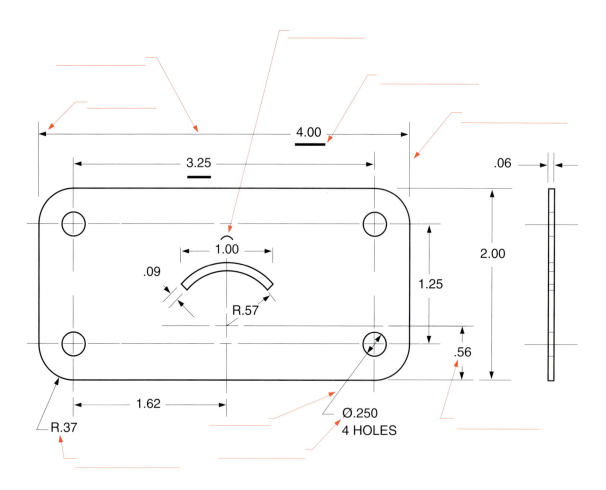

2. In the space provided, identify the various types of holes shown in section.

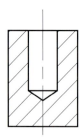 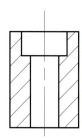 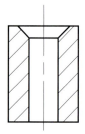 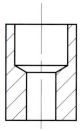

Section 8: *Dimensioning* **153**

Student _____ Date _____

3. Given the pictorial of the lower mold below, provide the requested information in the squares in the two-view drawing.

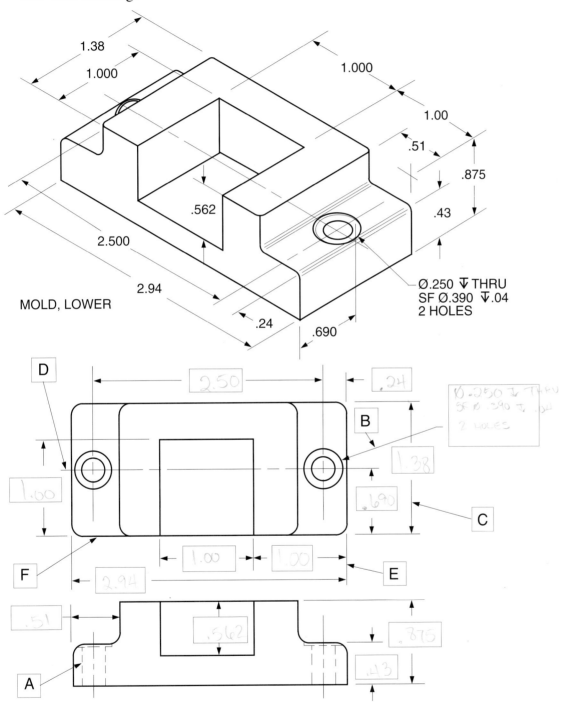

MOLD, LOWER

4. For the drawing in exercise 3, identify the types of lines indicated by the following letters:

A. _hidden_ B. _Leader_
C. _dimension_ D. _center_
E. _extension_ F. _object_

©1995 West Publishing Company

Section 8: *Competency Quiz*

5. Given the two-view drawing of the conveyor guide, convert the metric dimensions to three-place decimal-inch equivalents. Place your answers on the chart.

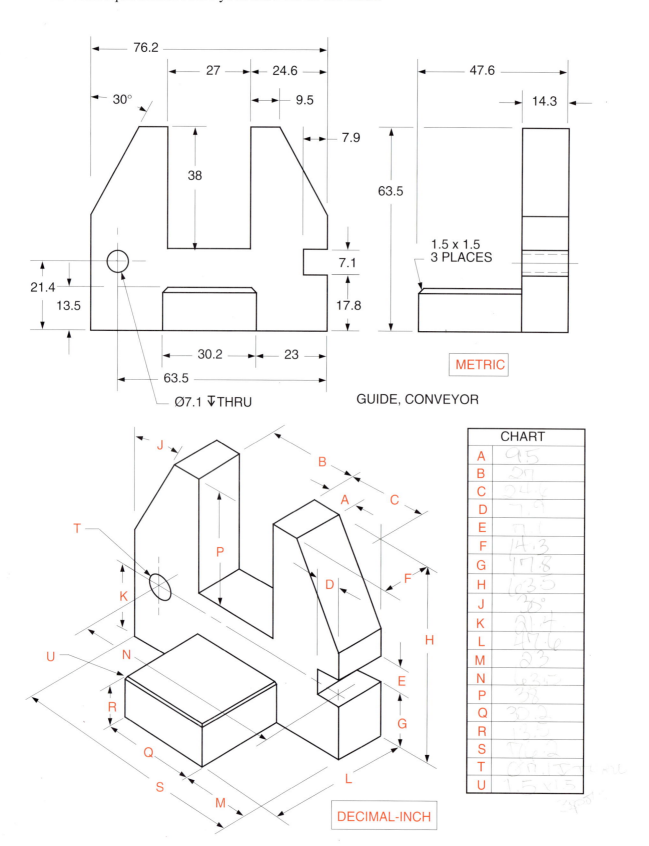

Section 8: *Dimensioning*   **155**

Student _____   Date _____

Competency Quiz continues on Page 156.

©1995 West Publishing Company

Section 8: *Competency Quiz*

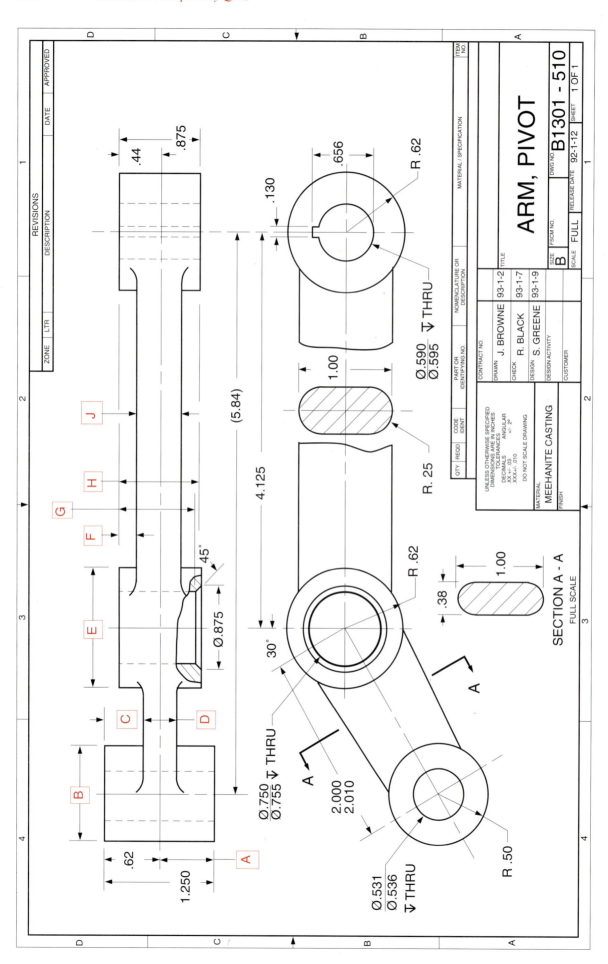

Section 8: *Dimensioning*   **157**

Student _____   Date _____

5. Refer to the PIVOT ARM, part number B1301-51, on page 156, when answering the following questions. Dimensions are decimal-inch.

   a. What is the title of the part? _____*Arm Pivot*_____

   b. From what material is the object made? _____*Meehanite Casting*_____

   c. What is the diameter of the hole in zone B-1 _____*.580 x .590 thru*_____

   d. What kind of section is section A-A? _____*full section*_____

   e. What type of dimensioning system was used on the pivot arm? _____*unidirectional*_____

   f. How may holes are specified on the part? _____*3*_____

   g. What type of section is shown in zone B-2? _____*removed*_____

   h. What type of chamfer is shown in zone C-3? _____*45° internal*_____

   i. What type of section clearly shows the chamfer? _____*broken out*_____

   j. What is the drawing size? _____*B*_____

   k. What is the width of the keyway? _____*.120*_____

   l. What kind of dimension is (5.84)? _____*reference dimension*_____

   m. What is the meaning of the Ø symbol? _____*Diameter*_____

   n. What does the ⊽ symbol indicate? _____*Depth*_____

   o. What does the letter R mean in zone A-4? _____*Radius*_____

Determine the dimensions for the following letters, which are shown in squares on the drawing:

A. _____*.63*_____   **B.** _____*1.00*_____

C. _____*.43*_____   **D.** _____*.38*_____

E. _____*1.24*_____   **F.** _____*.19*_____

G. _____*.813*_____   **H.** _____*.875*_____

J. _____*.50*_____

©1995 West Publishing Company

PART D TECHNICAL MATHEMATICS

1. For the template, determine the missing dimensions and place them on the chart. The answers should be in two-place decimals. Do not scale the drawing.

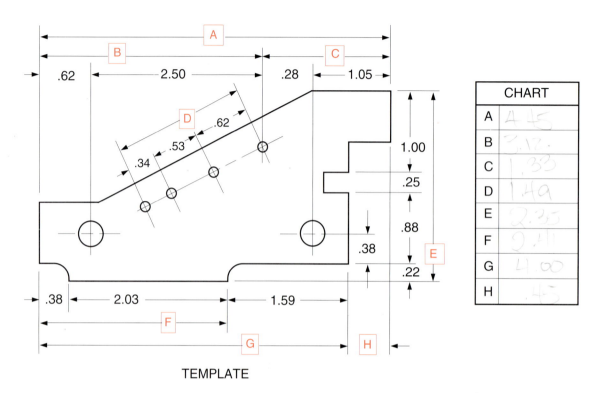

TEMPLATE

2. Find the missing dimensions for the lock plate. Place them on the chart. Answers should be in millimeters. Do not scale the drawing.

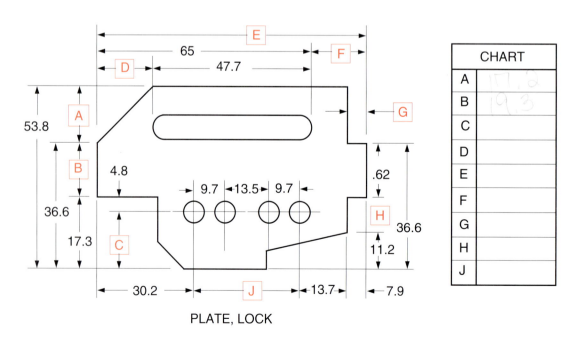

PLATE, LOCK

Section 8: *Dimensioning* **159**

Student _____ Date _____

3. The front and side view of a spacer are shown below. Convert all missing dimensions from decimal-inch to millimeters on the chart. Do not scale the drawing.

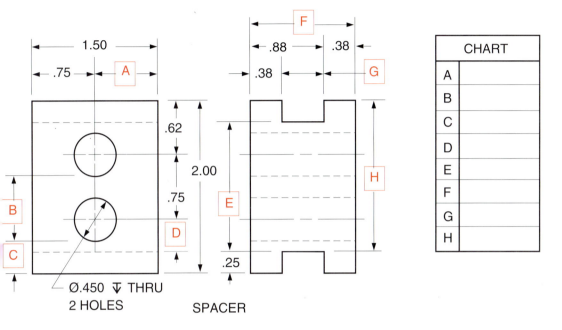

SPACER

4. Answer the following questions relative to the drill plate below. Responses should be in millimeters. Do not scale the drawing.

   a. At the 2.000-inch dimension, there are 9 holes equally spaced. What is the distance between each hole? _____ mm

   b. At the 1.812-inch dimension, there are 4 holes equally spaced. What is the distance between each hole? _____ mm

   c. Three holes are equally spaced at the .625-inch dimension. What is the distance between each hole? _____ mm

   d. If the center distance between each hole is .625 inches, at the location of the 5 equally spaced holes, what is the overall hole dimension? _____ mm

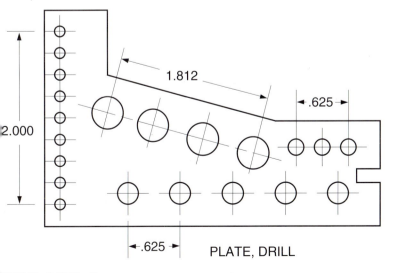

PLATE, DRILL

©1995 West Publishing Company

SECTION 9

Tolerancing

LEARNER OUTCOMES

You will be able to:
- State the purpose of tolerancing.
- Explain different tolerancing methods.
- Define ten tolerancing terms.
- Identify and determine the meaning of geometric tolerancing symbols.
- Explain the accumulation of tolerances.
- Discuss the interpretation of limits.
- Demonstrate your learning through the successful completion of competency exercises.

9.1 THE NEED FOR TOLERANCES

The advent of mass production and the demand for high-quality parts and assemblies have led to a need for interchangeable parts and assemblies. All parts and machines must be produced to some measurable, acceptable standard or size so that when they are assembled, they will fit and operate properly. Even the most proficient machinist operating the most accurate and sophisticated equipment needs to be allowed some variance in dimensions when producing a part. From the effort to achieve high quality and ensure that a part made at one manufacturing facility will be identical to one produced at another facility, the concept of *tolerancing* emerged.

It is virtually impossible to produce parts as precisely and accurately as specified on a drawing. So that parts will fit and function as required while still being economical to manufacture, part producers are given an appropriate variation, or margin of error, on each dimension. This allowable variance is called a **tolerance**.

9.2 THE PURPOSE OF TOLERANCING

The purpose of tolerancing is to control the dimensions of any two or more mating parts so that they will be interchangeable. *Interchangeability* implies that a part can change places with another part of the same size, shape, form, or dimensions. An example of a part whose dimensions need to be controlled can be found in Figure 8.4, Section 8.

162 Section 9: *Tolerancing*

9.3 TOLERANCING TERMS

Although the dimensioning terms referred to in Section 8 are important to a discussion of tolerancing, the following terms are related more specifically to this subject. They are frequently used when the topic of tolerancing arises, and they are known by engineers and drafters. They should be learned by shop personnel working with and interpreting engineering drawings. The most important of these terms are as follows:

- **Mating parts**—Parts that will be fastened or assembled together, such as a shaft in a hole, a screw in a tapped hole, or a key in a slot.
- **Fit**—The looseness or tightness of two mating parts when assembled.
- **Allowance**—The fit between two mating parts. There are two types of allowances: interference and clearance.
- **Interference allowance**—Interference allowance occurs when a shaft is larger than the hole with which it is supposed to mate, causing the shaft to be pressed or forced into the hole.
- **Clearance allowance**—A type of fit that provides clearance, or space, between mating parts. It is the opposite of interference allowance.
- **Tolerance**—The total amount by which a specific dimension is permitted to vary. The tolerance is the difference between the upper (maximum) and lower (minimum) limits of a dimension. Examples of tolerance values are shown in Figure 9.1.
- **Unilateral tolerance**—A tolerance, in inches, in which variation is permitted in one (uni) direction from the specified dimension. Examples of this type of tolerance are shown here:

$$.875 \, {}^{+.000}_{-.005} \quad \text{and} \quad .750 \, {}^{+.003}_{-.000}$$

- **Bilateral tolerance**—A tolerance, in inches, in which variance is permitted in two (bi) directions from the specified dimension, as illustrated here:

$$.375 \, {}^{+.005}_{-.005} \quad \text{and} \quad .250 \, {}^{+.003}_{-.003}$$

- **Geometric tolerance**—A general term applied to the category of tolerances used to control form, profile, orientation, location, and runout dimensions.
- **Limits of size**—These are limits or limit dimensions that describe the extent within which variations of geometric form and size are allowed. They are the specified maximum and minimum sizes of a dimension, as pictured in Figure 9.1.

9.4 TOLERANCING METHODS

When viewing a tolerance applied to a dimension, a technician should take note of the following practices, which are illustrated in Figure 9.2.

FIGURE 9.1
Limits of size with tolerance illustrations

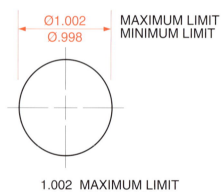

(a)

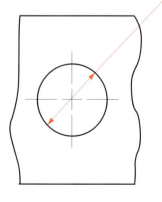

(b)

FIGURE 9.2
Tolerancing limits

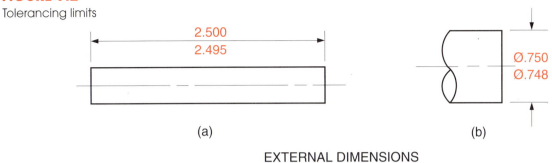

EXTERNAL DIMENSIONS

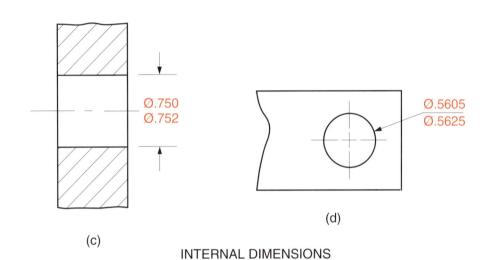

INTERNAL DIMENSIONS

1. For external dimensions, the high limit (maximum value) of a dimension is placed above the low limit (minimum value). This style is used for items such as shafts, keys, outside diameters, overall width and length dimensions, and so forth. Examples are shown in Figures 9.2a and 9.2b. For internal features, such as holes, keyways, and slots, the low limit (minimum value) is located above the high limit (maximum value), as illustrated in Figures 9.2c and 9.2d.
2. Figure 9.3 illustrates how a dimension appears on a drawing when it is followed by a *plus-or-minus tolerance*. Figure 9.3a is a unilateral tolerance, and 9.3b is a bilateral tolerance. Note that in both cases, the positive tolerance is shown above the negative tolerance.

9.5 INTERPRETATION OF LIMITS

On a drawing, all limits are considered absolute—that is, accepted without question. Therefore, dimensional limits, regardless of the number of decimal places, are shown as if they were continued with an infinite number of zeros (0000). For example:

   1.22   equals 1.220… 00
   1.20   equals 1.200… 00
   1.202 equals 1.2020…00
   1.200 equals 1.2000…00

■ 9.5.1 Single Limits

As we have seen, minimum and maximum dimensions control the limits. When only one dimensional limit is needed, the abbreviation MIN or MAX appears immediately following the dimension. This situation often occurs on such part features as the depth of holes, the length of threads, corner radii, and chamfers. Single limits are used where (1) the intent of the limit is clear, (2) the unspecified limit is relatively unimportant, or (3) both. An illustration of how and when a single limit is used is presented in Figure 9.4.

FIGURE 9.3
Plus-or-minus tolerancing

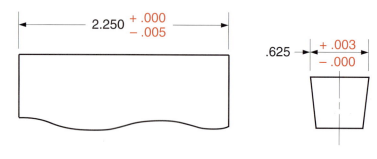

(a) UNILATERAL TOLERANCES

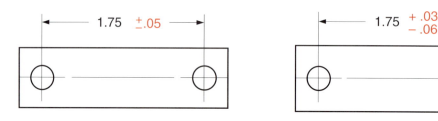

(b) BILATERAL TOLERANCES

FIGURE 9.4
Applying MIN and MAX limits

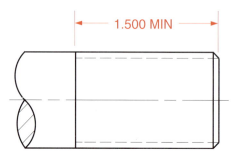

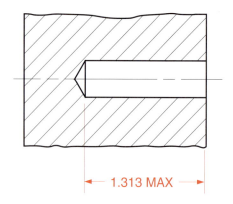

9.6 THE ACCUMULATION OF TOLERANCES

When the location of a surface on an object is affected by more than one tolerance, the tolerances are considered to be accumulative. If this **accumulation of tolerances** has a negative effect on the part—that is, if it makes the part unacceptable because of poor quality—then whomever designed or drafted the part should probably have selected another type of dimensioning system. In the interpretation of dimensioning and tolerancing practices, a tradesperson needs to be aware of the accumulation of tolerances when inspecting a drawing. Hence, three different methods of dimensioning and their effects on tolerances are presented here:

- *Chain dimensioning*, shown in Figure 9.5a, results in the greatest tolerance accumulation as illustrated by the ±.003 variation between holes X and Y.
- *Datum dimensioning*, depicted in Figure 9.5b, reduces the total tolerance accumulation between holes X and Y to ±.002.
- *Direct dimensioning*, shown in Figure 9.5c, results in the least tolerance accumulation, as il-

FIGURE 9.5
The accumulation of tolerances

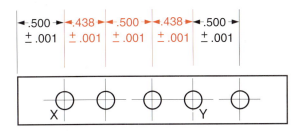

(a) CHAIN DIMENSIONING—GREATEST TOLERANCE ACCUMULATION BETWEEN X AND Y

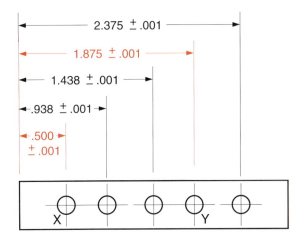

(b) DATUM DIMENSIONING—LESSER TOLERANCE ACCUMULATION BETWEEN X AND Y

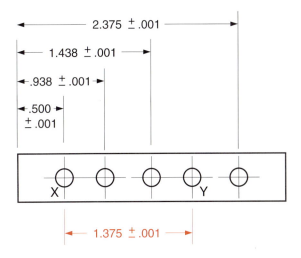

(c) DIRECT DIMENSIONING—LEAST TOLERANCE ACCUMULATION BETWEEN X AND Y

lustrated in the ±.001 variation between holes X and Y.

Obviously, for this particular set of circumstances, direct dimensioning results in the least accumulation of tolerances and is therefore the most desirable practice to use. This, however, is not always the case. Dimensioning and tolerancing practices are tailored to each specific set of circumstances.

9.7 SYMBOLOGY

An important aspect of learning about *geometric tolerancing* is the symbols that make up the system and how to interpret them effectively. Symbols are used to specify geometric characteristics on engineering drawings for shop use. Symbols are preferred to notes and text because they convey the intended meaning in a precise and direct form. In addition, their use tends to reduce errors due to improperly worded notes.

■ 9.7.1 The Geometric Characteristic Symbol

The symbols used when noting geometric characteristics on a drawing are illustrated in Figure 9.6. They are referred to as geometric characteristic symbols. Note how the design of each symbol, for the most part, closely resembles the word it represents. It is critical that personnel working in engineering and the manufacturing sector of industry know and understand these symbols, their interpretation, and their application.

■ 9.7.2 The Feature Control Frame

A tolerance displayed on a drawing is one type of **feature control symbol**. Such symbols are enclosed in a rectangular box referred to as a **feature control frame** and always appear horizontally on the drawing, as in aligned drawing practices. The frame and the information that it contains are a simplified form of what, in the past, took the form of notes. The information in the frame is read from left to right and may contain some or all of the data shown in Figure 9.7. Note how vertical lines separate the various symbols within the frame. The example illustrated identifies a geometric characteristic symbol, a tolerance, a

FIGURE 9.6
Geometric characteristic symbols

GEOMETRIC CHARACTERISTIC		TYPE OF TOLERANCE	PERTAINS TO:
STRAIGHTNESS	—	FORM TOLERANCE	INDIVIDUAL FEATURE
FLATNESS	▱		
ROUNDNESS	○		
CYLINDRICITY	⌭		
PROFILE OF A LINE	⌒		INDIVIDUAL OR RELATED FEATURES
PROFILE OF A SURFACE	⌓		
ANGULARITY	∠		RELATED FEATURES
PERPENDICULARITY	⊥		
PARALLELISM	∥		
POSITION	⌖	LOCATION TOLERANCE	
CONCENTRICITY	◎		
CIRCULAR RUNOUT	↗	RUNOUT TOLERANCE	
TOTAL RUNOUT	↗↗		

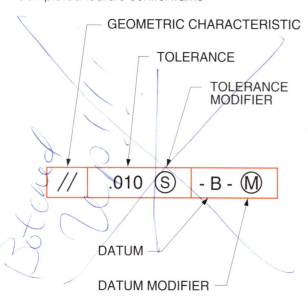

FIGURE 9.7
Completed feature control frame

- GEOMETRIC CHARACTERISTIC
- TOLERANCE
- TOLERANCE MODIFIER
- DATUM
- DATUM MODIFIER

tolerance modifier, a datum reference, and a datum modifier.

A feature control frame is applied to a specific feature of an object and is not repeated. It is seen in only one location, providing a particular geometric tolerance to a specific feature.

■ 9.7.3 The Datum Identification Symbol

Previously, a datum was defined as a theoretically exact point, axis, or plane from which the locations or geometric characteristics of the features of an object are established. A **datum identification symbol** consists of a frame containing a datum reference letter preceded and followed by a dash for emphasis. This symbol is illustrated in Figure 9.8. Any letter of the alphabet, with the exception of *I*, *O*, and *Q*, may be used as a datum reference letter. The symbol indi-

FIGURE 9.8
Datum identification symbol

FIGURE 9.10
Datum target symbol

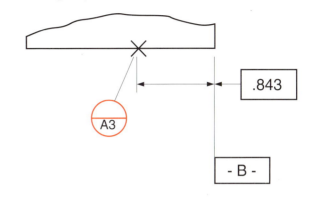

cates the datum on the drawing, and the letter identifies the datum location. When more than one datum is required, each additional datum is identified by the next consecutive letter of the alphabet: the letter *A*, then *B*, then *C*, and so forth.

■ 9.7.4 The Basic Dimension and the Datum Target Symbol

A **basic dimension symbol** is identified by the abbreviation **BSC** and by a frame enclosing the dimension, as shown in Figure 9.9. Theoretically, the basic dimension is considered to be without tolerance, but because this is not possible, a separate tolerance is provided. Basic dimensions may be in decimal-inch, metric, or angular increments.

FIGURE 9.9
Basic dimension symbol

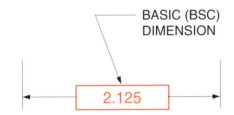

A **datum target symbol** is a circle divided into half circles by a horizontal line. Its purpose is to specify a point, line, or area related to a datum on an object. A leader, without a termination, is used to point to a specific surface, point, or line on the drawing. The lower half of the circle consists of a datum identifying number followed by a target number. Targets are numbered consecutively. Included in the upper half is information referring to the target location, when required. Figure 9.10 illustrates a datum target symbol.

■ 9.7.5 The Tolerance Zone

In geometric tolerancing, the **tolerance zone** is that area bounded by the upper and lower limits, or the total amount of permissible error in a dimension. The geometric characteristics of the object determine the shape of the tolerance zone, which may be in any of the following shapes for two-dimensional and three-dimensional situations:

Two-dimensional
- a circle
- a rectangle
- a space between two concentric circles

Three-dimensional
- a cylinder
- a rectangular solid
- a space between two concentric cylinders

■ 9.7.6 Tolerancing Modifiers

The use of a **modifier** is a basic aspect of geometric tolerancing. It is used to change or alter the interpretation of a symbol. The following items are the four modifiers used when applying geometric tolerances.

- **Maximum material condition (MMC)**—This is the condition in which a feature of size contains the maximum amount of material within the stated limits. For example, the MMC of the *external feature* in Figure 9.11b is ∅.520 inch because it is where the most material exists for the dimension. The MMC for the *internal feature* is ∅.480 inch in Figure 9.11a because the material is the largest size when the hole is the smallest allowable size. The identifier symbol is Ⓜ.

FIGURE 9.11
Maximum material condition for an internal and an external feature

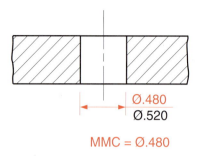

(a) INTERNAL FEATURE

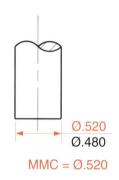

(b) EXTERNAL FEATURE

FIGURE 9.12
Least material condition for an internal and an external feature

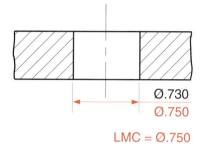

(a) INTERNAL FEATURE

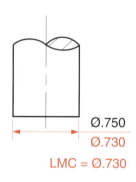

(b) EXTERNAL FEATURE

- **Least material condition (LMC)**—This is the opposite of MMC; it is the condition in which a feature of size contains the least amount of material within the stated limits. Examples of internal and external conditions of LMC are given in Figure 9.12. Its identifier symbol is Ⓛ.
- **Regardless of feature size (RFS)**—This modifier is also referred to as Ⓛ, and it instructs shop personnel that a tolerance of form or position must be held or maintained regardless of the actual finished, machined size of the object.
- **Projected tolerance zone**—This modifier allows a tolerance to be extended beyond the original surface. It is known as Ⓟ.

9.8 THE APPLICATION AND INTERPRETATION OF GEOMETRIC CHARACTERISTIC SYMBOLS

The application and interpretation of geometric characteristic symbols is a fairly new area of creating and reading blueprints. The applications discussed in this worktext conform to the most recent revised standard, designated as ANSI Y 14.5M-1982. What follows is a detailed description of the various geometric characteristic symbols, how they are applied to the drawing by the drafter, and how they should be interpreted by shop personnel.

9.8.1 Straightness

The term **straightness** implies that a line, an edge, or a surface of an object does not curve or bend. It is assumed that the object being viewed is straight for its entire length. But how straight is straight? Certainly we know by now that straight can be straight only within some acceptable tolerance. It is impossible for something to achieve perfect straightness. For a plane surface to be controlled for straightness, a feature control frame along with its leader is applied to an edge of the object, as illustrated in Figure 9.13a. The figure shows the appropriate symbol that is applied to a drawing. The interpretation in Figure 9.13b shows that the longitudinal element of the surface must be straight within the allowed .005-inch tolerance.

FIGURE 9.13
Application and interpretation of the straightness characteristic symbol

FIGURE 9.14
An object that is straight but not flat

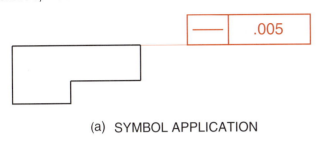
(a) SYMBOL APPLICATION

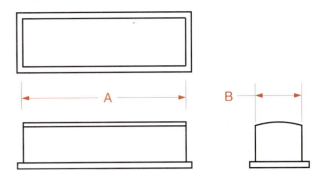

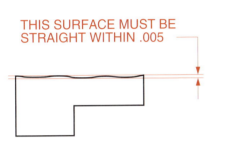
(b) SYMBOL INTERPRETATION

For example, Figure 9.14 shows multiple views of an object and illustrates that the direction of plane surface A of the object is straight and may be flat, but plane surface B is not straight or flat.

The flatness symbol is used when size tolerances alone are not sufficient to control the quality and form of a surface. A flatness tolerance zone is the space that lies between two parallel planes. Figure 9.15a shows that in the symbol application, the feature control symbol is directed to the edge view of the surface of the object. The symbol interpretation, Figure 9.15b, indicates that flatness must be held to within the .005-inch tolerance to be acceptable.

■ 9.8.2 Flatness
While straightness affects a plane surface in only one direction, **flatness** affects an object in all directions. A surface may be straight in one direction but not flat.

FIGURE 9.15
Application and interpretation of the flatness characteristic symbol

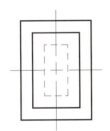
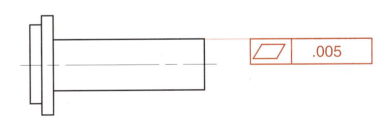

(a) SYMBOL APPLICATION

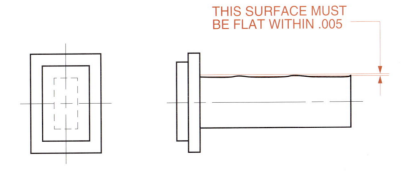

(b) SYMBOL INTERPRETATION

■ 9.8.3 Roundness

We have learned that a circle is a closed plane curve of which all points are equidistant from the center. **Roundness**, which is also referred to as **circularity**, is measured as the radial space between two perfectly concentric circles. When a round or circular object is in an out-of-round condition or has an eccentric shape, it will appear as in Figure 9.16a. The tolerance zone, which is sometimes referred to as the roundness error, is also shown. Roundness is not a radius or a circle because it is not measured from the center. It is a feature control of revolution, as may be found in a cone, cylinder, or spherical object. The tolerance zone for objects containing roundness is taken at any cross section of the feature. Figure 9.16b, identifies how a roundness characteristic symbol appears on a drawing, and Figure 9.16c shows how the symbol is to be interpreted.

■ 9.8.4 Cylindricity

While roundness refers to any cross-sectional area on a given circular surface, **cylindricity** affects the entire length of an object. It is a combination of straightness and roundness. The tolerance zone for this characteristic is that area within two hypothetically perfect cylinders. The zone consists of an inner (lower) limit and outer (upper) limit. Any deviation from this

FIGURE 9.16
Application and interpretation of the roundness characteristic symbol

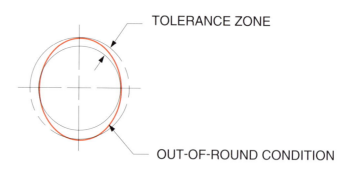
(a) ROUNDNESS ERROR

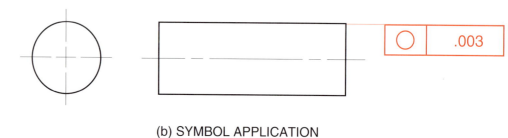
(b) SYMBOL APPLICATION

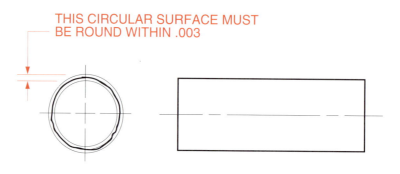
(c) SYMBOL INTERPRETATION

Section 9: *Tolerancing* 171

FIGURE 9.17
Application and interpretation of the cylindricity characteristic symbol

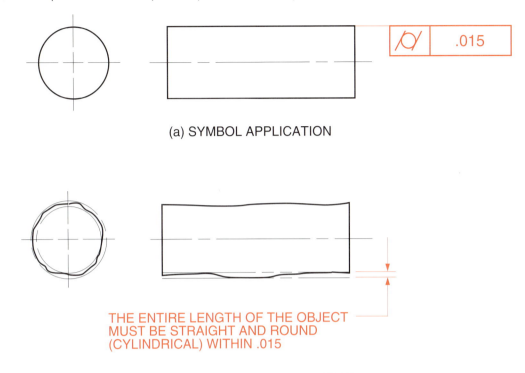

(a) SYMBOL APPLICATION

THE ENTIRE LENGTH OF THE OBJECT MUST BE STRAIGHT AND ROUND (CYLINDRICAL) WITHIN .015

(b) SYMBOL INTERPRETATION

zone is called a cylindricity error. Figure 9.17a illustrates the drawing callout for cylindricity, and Figure 9.17b offers the interpretation of the characteristic symbol.

9.8.5 Line Profiles and Surface Profiles

The outline, shape, or contour of a line or surface is its profile. The **profile of a line** controls only the elements of a surface or a plane in a specific direction. The tolerance zone for the line profile is defined as the space between two imaginary lines perfectly parallel to the element that requires profiling.

The **profile of a surface**, on the other hand, appears when the entire surface of an object must be controlled in both directions. The tolerance zone for the surface profile is the space between two imaginary curved surfaces perfectly parallel to the surface that requires profiling. Three different types of profile tolerances may be found on an engineering drawing. They include bilateral, unilateral (down or inside), and unilateral (up or outside) tolerances. A feature control symbol appears close to the profile for a bilateral tolerance, as shown in Figure 9.18a. An inside unilateral tolerance is recognized by seeing a portion of the boundary of the tolerance inside the profile, as illustrated in Figure 9.18b. Figure 9.18c identifies an outside unilateral tolerance, where a portion of the tolerance boundary is shown outside the profile. If a tolerance type is not specified on a drawing, it is assumed that the tolerance is bilateral, and it is divided equally on the inside and outside of the basic profile. For surface profiles, a datum normally appears. Figure 9.19 illustrates how the characteristic symbols appear on a drawing for the profile of lines, and how such a drawing is interpreted.

Figure 9.20a shows the symbol application for the profile of surfaces, while Figure 9.20b shows the interpretation of the profile-of-surfaces characteristic symbol.

9.8.6 Angularity

Angularity is the condition of a surface or axis at a specified angle (other than 90 degrees) from an axis or a datum plane. Therefore, an angularity tolerance

FIGURE 9.18
Types of line and surface profile tolerances

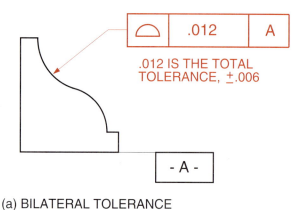

(a) BILATERAL TOLERANCE

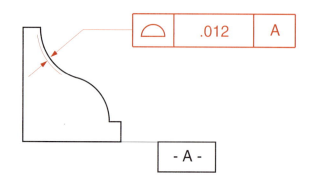

(b) UNILATERAL TOLERANCE (INSIDE)

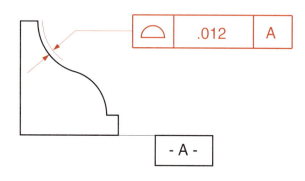

(c) UNILATERAL TOLERANCE (OUTSIDE)

Note: The ⌒ represents a profile of surface

FIGURE 9.19
Application and interpretation of the profile-of-a-line characteristic symbol

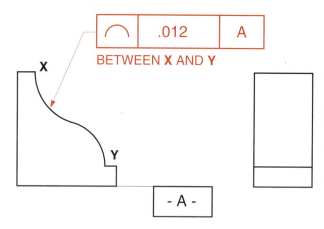

(a) SYMBOL APPLICATION

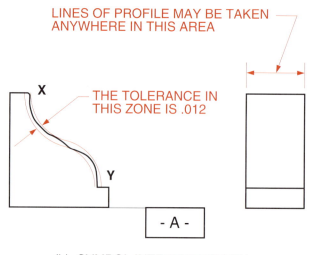

(b) SYMBOL INTERPRETATION

requires a datum reference. The angularity tolerance consists of two imaginary parallel planes that form the specified angle with the datum. All the points along the angular axis or on the angular surface must lie within these parallel planes. The angular dimension providing the required angle is basic (**BSC**), without tolerance, and the feature control frame is normally used in conjunction with a leader that points to the area representing the controlled feature. Figure 9.21 illustrates the application and interpretation of the angularity characteristic symbol. Note that the tolerance is identified in decimals or millimeters, not in angular increments. This is because it is more desirable to have a tolerance zone of uniform width for the entire surface than one that would have a tendency to "fan out" from a zero point of origin.

FIGURE 9.20
Application and Interpretation of the profile-of-a-surface characteristic symbol

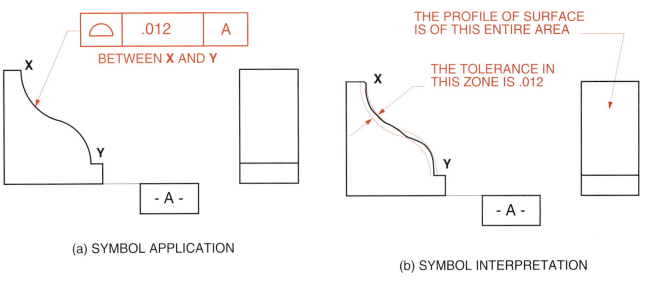

FIGURE 9.21
Application and interpretation of the angularity characteristic symbol

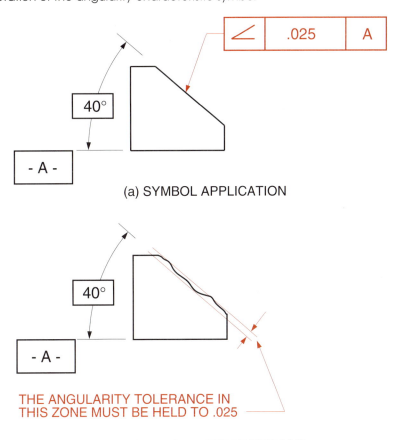

9.8.7 Perpendicularity

Perpendicularity occurs when the meeting of a given line or surface is at a right angle (90 degrees) with another line or surface. It is a feature control specifying that all elements of a surface, axis, line, or plane be at (90 degree) with a datum. Therefore, the perpendicularity tolerance requires a datum reference. The tolerance zone for this characteristic is formed by two imaginary parallel lines, as shown in Figure 9.22. The elements of the surface requiring the tolerance must fall within the size limits bordered by the two imaginary parallel lines. Figure 9.22 also shows the application and the interpretation of the perpendicularity characteristic symbol.

FIGURE 9.22
Application and interpretation of the perpendicularity characteristic symbol

FIGURE 9.23
Application and interpretation of the parallelism characteristic symbol for a plane surface

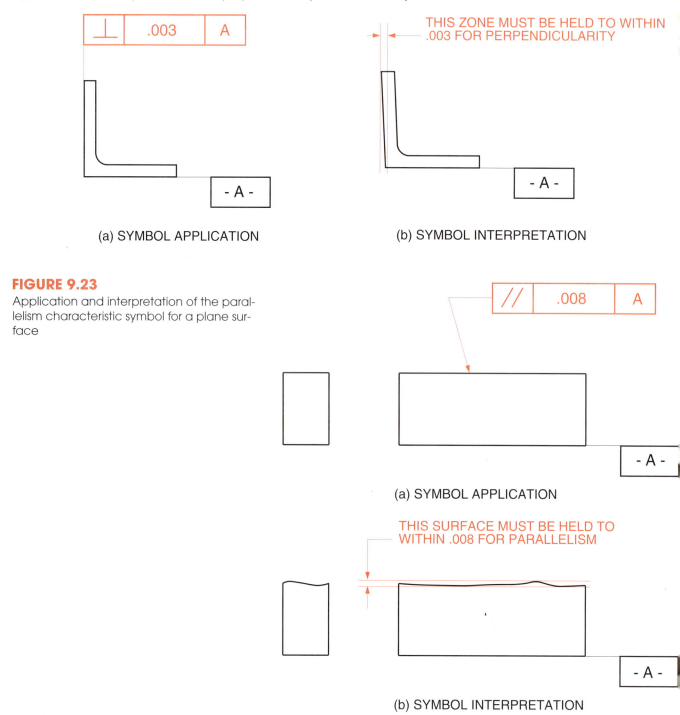

Section 9: *Tolerancing* **175**

9.8.8 Parallelism

Parallelism exists when a surface, line, or axis is equidistant from a datum plane or axis. A parallelism tolerance zone for a surface feature is formed by two imaginary planes that are parallel to a specific datum at a distance equal to the parallelism tolerance. The parallelism tolerance has a datum reference. The parallelism tolerance for a cylindrical feature, such as a cone or cylinder, is a zone that is parallel to an axis. In all cases, all elements of the area to be toleranced must fall within the stated size limits. Figure 9.23 illustrates both the application and the interpretation of the parallelism characteristic symbol for a plane surface, while Figure 9.24 portrays the same characteristics for circular features.

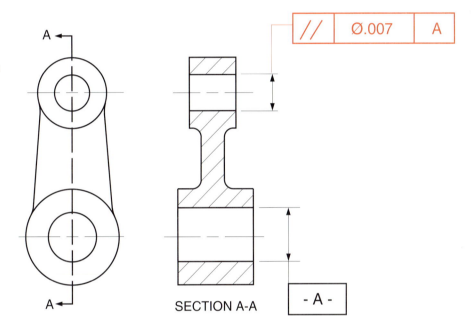

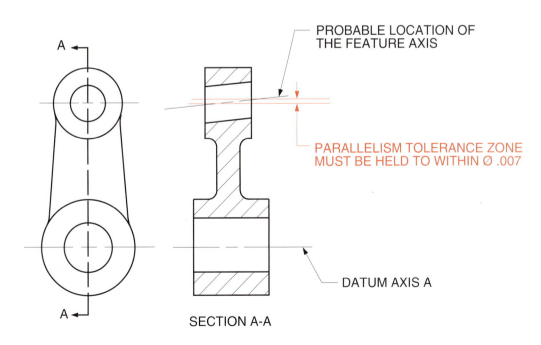

FIGURE 9.24
Application and interpretation of the parallelism characteristic symbol for a circular feature

■ 9.8.9 Positional Tolerance

A **positional tolerance** takes the form of a zone within which the center plane or center axis of a feature of size is allowed to vary from its true, basic (theoretically exact) position. The position is established by the basic dimensions from specific datum features as well as between other interrelated features. The symbol for a positional tolerance consists of a circle with crosshairs through the center. Following this symbol is the tolerance, a modifier when required, and the appropriate datum reference(s). (Tolerancing modifiers were covered earlier, in Section 9.7.6.) Figure 9.25 provides an example of typical positional tolerance symbology.

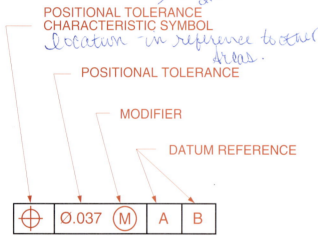

FIGURE 9.25
Positional tolerance symbology

The advantage to using positional tolerancing versus the previously used true positioning methods is that positional tolerancing allows for a larger tolerance zone and, therefore, allows the machine operator more room for error. A simple example of the application and interpretation of a positional tolerance is shown in Figure 9.26.

■ 9.8.10 Concentricity

In most cases, it is important that parts and assemblies occupy the same center line if they are coaxially dependent. Cylinders and other regular objects do possess axes and should be termed as being coaxial. In engineering practice, the terms coaxial and concentric are used interchangeably. Two cylindrical objects, for example, have **coaxiality** if they possess the same axis or are on one line. When this condition occurs, **concentricity** exists, and a concentricity tolerance needs to be applied. The concentricity tolerance, which is an imaginary cylinder about the exact axis of the datum, is only concerned with a center line relationship. It creates a cylindrical tolerance zone in which must lie all center lines for each successive part or subpart of an overall object or assembly. When an unacceptable concentricity condition exists, it is referred to as being eccentric, or off center. The concentricity characteristic symbol consists of two concentric circles, as is illustrated in Figure 9.27, which provides an example of the application and interpretation of the symbol. Note, too, that the tolerance shown in the feature control frame is preceded by the diameter symbol (Ø).

■ 9.8.11 Runout

Runout can be defined as any deviation in the position or location of a surface of rotation as an object is revolved about its datum axis. A runout tolerance is a feature control that limits this deviation on surfaces or circular objects. Runouts are most frequently called out on objects that are concentric cylinders or other shapes of revolution that possess circular cross sections. It is a composite tolerance that may involve errors in roundness, straightness, concentricity, and perpendicularity. There are two types of runout conditions. One is called **circular runout**, and the other is referred to as **total runout**.

Circular runout involves tolerance deviation as it relates to circular elements of an object. The circular runout tolerance applies to any area of a part where a single-line element section is cut. An illustration of the application and interpretation of the circular runout characteristic symbol is shown in Figure 9.28.

Total runout involves the deviation of the elements of an entire surface, not just circular elements. Figure 9.29 shows how the total runout characteristic symbol is applied to a drawing, and how it is to be interpreted. Note the difference in interpretation between the circular and total runouts.

Section 9: *Tolerancing* **177**

FIGURE 9.26
Application and interpretation of a positional tolerance

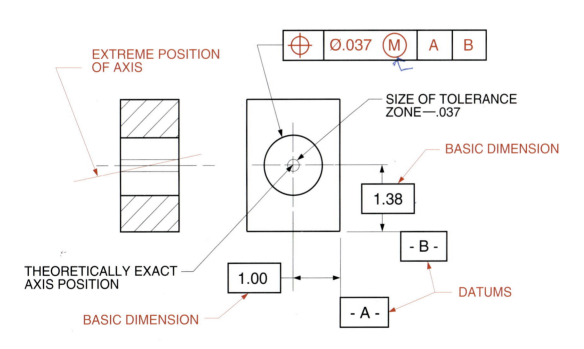

(a) SYMBOL APPLICATION

(b) SYMBOL INTERPRETATION

FIGURE 9.27
Application and interpretation of the concentricity characteristic symbol

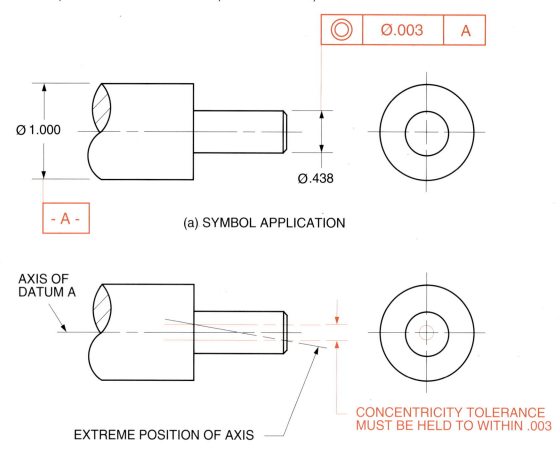

9.9 SUMMARY

In today's marketplace, with high production, microminiaturization, multinational corporations, foreign competition, and new and emerging technologies, the need for high-quality parts and assemblies has never been greater. To produce these parts accurately and with a degree of interchangeability, there needs to be some sort of acceptable variance on engineering drawings. This variance is called a tolerance.

This section explained the need for and purpose of tolerances, and it defined several standard tolerancing terms. Examples of dimensions that need to be controlled and methods for controlling them were identified. Tolerancing methods and practices were discussed, including plus-or-minus, unilateral, and bilateral tolerancing. The meaning and intent of tolerancing limits, tolerance expressions, and the accumulation of tolerances were detailed. An in-depth coverage on the usage of the American National Standards Institute (ANSI) specification Y 14.5 M-1982 as it relates to geometric tolerancing was presented. Detailed information on geometric characteristic symbols was covered by identifying each symbol, how and where it is placed on an engineering drawing, and how it is to be interpreted by the drawing reader. Characteristic symbols and practices discussed were straightness, flatness, roundness, cylindricity, profile of lines and surfaces, angularity, parallelism, position, concentricity, and circular and total runout.

FIGURE 9.28
Application and interpretation of the circular runout characteristic symbol

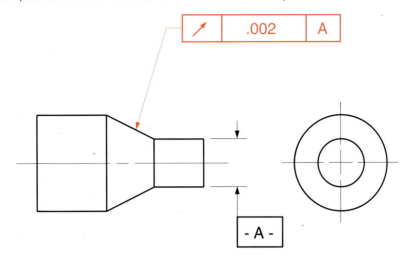

(a) SYMBOL APPLICATION

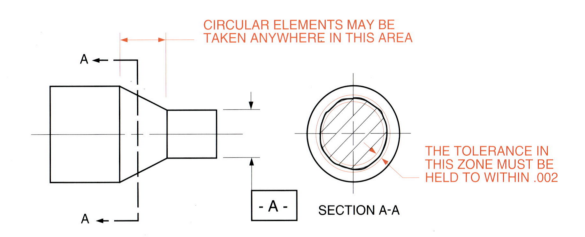

(b) SYMBOL INTERPRETATION

TECHNICAL TERMS FOR STUDY

accumulation of tolerances The adding together of tolerances for given dimensions on an object to determine the effect on the part.

allowance The fit between two mating parts.

angularity The condition of a surface or axis at a specified angle (other than 90 degrees) from an axis or a datum plane.

basic dimension symbol A symbol that consists of the letters BSC and a dimension enclosed in a frame. It indicates a dimension that is considered to be without tolerance.

bilateral tolerance A tolerance in which variance is permitted in two (bi) directions from the specified dimension.

circularity See **roundness**.

circular runout The tolerance deviation as it relates to circular elements of an object.

clearance allowance A type of fit that provides clearance or space between mating parts. It is the opposite of an interference allowance.

FIGURE 9.29
Application and interpretation of the total runout characteristic symbol

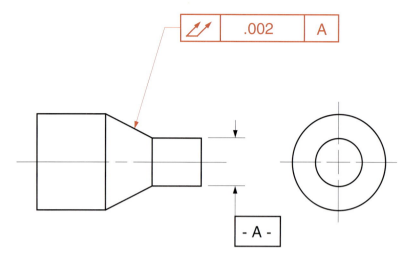

(a) SYMBOL APPLICATION

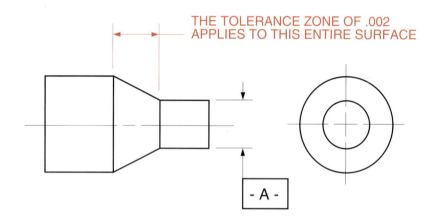

(b) SYMBOL INTERPRETATION

coaxiality A general term used to describe objects whose axes are in line or on the same line.

concentricity A term used to describe parts and assemblies that occupy the same center line. Same as **coaxiality**.

cylindricity A combination of **straightness** and **roundness**. Cylindricity affects the entire length of an object.

datum identification symbol A symbol consisting of a frame that includes a datum reference letter preceded and followed by a dash for emphasis. The symbol indicates the datum on the drawing, and the letter identifies the datum location. Any letter of the alphabet except for I, O, and Q may be used as a datum reference letter.

datum target symbol A symbol that specifies a point, line, or area related to a datum on an object. The symbol consists of a circle divided into half circles by a horizontal line.

feature control frame A horizontally located rectangular box that houses information relative to feature control symbols and other geometric tolerancing information.

feature control symbol   A symbol that shows the tolerance required for a particular dimension; housed in a feature control frame.

fit   The looseness or tightness between two mating parts when assembled.

flatness   A geometric condition that affects an object in all directions.

geometric characteristic symbol   Symbols used to show various geometric characteristics on an engineering drawing.

geometric tolerance   A general term applied to the category of tolerances used to control form, profile, orientation, location, and runout dimensions.

interference allowance   A type of fit in which a shaft is larger than a hole in which it is supposed to mate, requiring the shaft to be pressed or forced into the hole.

least material condition (LMC)   The condition in which a feature of size contains the least amount of material within the stated limits.

limits of size   Limits or limit dimensions that describe the extent within which variations of form and size are allowed. They are the maximum and minimum sizes of a dimension.

mating parts   Parts that are fastened together, such as a shaft in a hole, a screw in a tapped hole, or a key in a slot.

maximum material condition (MMC)   The condition in which a feature of size contains the maximum amount of material within the stated limits.

modifier   A condition that changes or alters the interpretation of a geometric characteristic symbol. There are four modifiers used in geometric tolerancing. They are maximum material condition, least material condition, regardless of feature size, and projected tolerance zone.

parallelism   A condition in which a surface, line, or axis is equidistant from a datum plane or axis;

for example, when two lines are equidistant for their entire length.

perpendicularity   A situation in which the meeting of a given line or surface is at a right angle (90-degree angle) with another line or surface.

positional tolerance   A zone within which the center plane or center axis of a feature of size is allowed to vary from its true, basic position.

profile of a line   The geometric feature that controls the elements of a surface or a plane in a specific direction.

profile of a surface   The geometric feature that controls the entire surface of an object in both directions.

projected tolerance zone (P)   A condition that allows a tolerance to be extended beyond the original surface.

regardless of feature size (RFS)   A condition in which a tolerance of form or position must be held or maintained regardless of the actual finished, machined size of the object.

roundness   The radial space between two perfectly concentric circles. Also referred to as **circularity**.

straightness   A geometric condition in which a line, edge, or surface of an object does not curve or bend.

tolerance   The total amount by which a specific dimension is permitted to vary; the difference between the upper and lower limits of a dimension.

tolerance zone   In geometric tolerancing, the area bounded by the upper and lower limits, or the total permissible error in a dimension.

total runout   The deviation in position or location of the elements of an entire surface, not just circular elements.

unilateral tolerance   A tolerance in which variation is permitted in one (uni) direction from a specified dimension.

Student _____   Date _____

Section 9: Competency Quiz

PART A COMPREHENSION

1. Define *tolerance*. (5 pts)

2. What American National Standards Institute (ANSI) specification is used throughout industry for dimensioning and tolerancing practices? (5 pts)

3. What is the purpose of tolerances on a drawing? (5 pts)

4. Explain the difference between a bilateral tolerance and a unilateral tolerance. (8 pts)

5. Explain the difference between an interference allowance and a clearance allowance. (8 pts)

6. Explain the difference between a least material condition (LMC) and a maximum material condition. (8 pts)

184 Section 9: *Competency Quiz*

7. Draw freehand the following geometric characteristic symbols. (12 pts)

   a. **perpendicularity** ⊥

   b. **concentricity** ◎

   c. **parallelism** //

   d. **flatness** ▱

   e. **profile of a surface** ⌒

   f. **roundness** ○

8. What is a tolerance zone? (5 pts) An amount a object is
   allowed to vary
   amount or error allowed (limits)

9. Sketch a feature control frame. (5 pts)

10. List four modifiers used in geometric tolerancing. (8 pts)
    (M) max material cond.
    (L) least "   "
    (S) regardless of feature size
    (P) projected

11. What is the advantage of using geometric tolerancing practices on engineering drawings? (10 pts)
    allow for a greater allover tolerance
    actually allow amount of error to be more

12. What alphabet letters cannot be used in a datum identification symbol? (6 pts)
    I O Q

13. Identify the information contained in a completed feature control frame. (5 pts)

14. Identify the kind of information normally seen in a positional tolerance frame. (5 pts)

15. Define the term *runout*. (5 pts)

Section 9: *Tolerancing*   **185**

Student _____   Date _____

PART B TECHNICAL TERMS

For each definition, select the correct technical term from the list on the bottom of the page. (10 pts each)

1. _*Angularity*_ The condition of a surface or axis at a specified angle (other than 90 degrees) from an axis or a datum plane.

2. _*Unilateral*_ A tolerance in which variance is permitted in only one direction from a specified dimension.

3. _*clearance Allowance*_ A type of fit that provides clearance or space between mating parts.

4. _*feature Control frame*_ A horizontally located rectangular box that houses information related to feature control symbols and other geometric tolerancing information.

5. _*limits of Size*_ Limits or limit dimensions that describe the extent to which variations of form and size are allowed.

6. _*Modifier*_ A condition that changes or alters the interpretation of a geometric characteristic symbol.

7. _*perpendicularity*_ A situation in which the meeting of a given line or surface is at a right angle (90-degree angle) with another line or surface.

8. _*RFS*_ A symbol that instructs shop personnel that a tolerance of form or position must be held or maintained regardless of the actual finished, machined size of the object.

9. _*tolerance*_ The total amount by which a specific dimension is permitted to vary; the difference between the upper and lower limits of a dimension.

10. _*Straightness*_ A geometric condition in which a line, edge, or surface of an object does not curve or bend.

A. angularity
B. bilateral tolerance
C. clearance allowance
D. concentricity
E. feature control frame
F. feature control symbol
G. fit
H. limits of size
I. modifier
J. perpendicularity
K. RFS
L. straightness
M. tolerance
N. unilateral tolerance

©1995 West Publishing Company

186   Section 9: *Competency Quiz*

PART C TOLERANCING APPLICATION

1. What is the upper limit for the following dimensions? Place your answer in the spaces provided.

a. .251 _____   b. .375 _____
    .250   .380

c. 1.505 _____   d. 17.6875 _____
    1.498   17.6876

2. What is the lower limit for the following dimensions?

a. .406 _____   b. 3.750 _____
    .409   4.515

c. 7.654 _____   d. 12.012 _____
    7.650   11.998

3. In the blank spaces, indicate whether the dimension has a unilateral or bilateral tolerance.

a. .750 $^{+.002}_{-.000}$ _____   b. 1.281 $^{+.010}_{-.010}$ _____

c. 2.6125 $^{+.0000}_{-.0005}$ _____   d. .031 $^{+.001}_{-.001}$ _____

e. .375 $^{+.005}_{-.003}$ _____

Student _____ Date _____

4. For the following situations, change each dimension shown to maximum (upper) and minimum (lower) limit dimensions.

a.

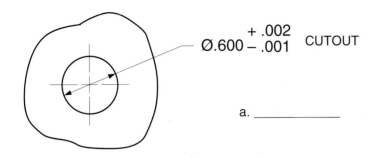

a. _____

b.

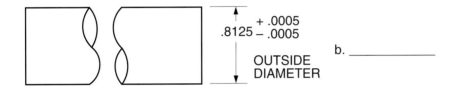

b. _____

c.

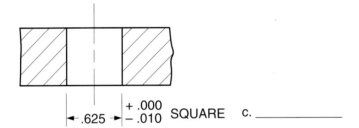

c. _____

d. _____

d.

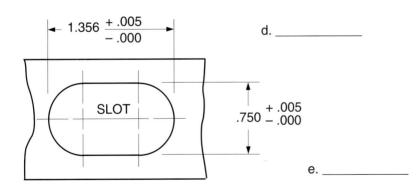

e. _____

e.

©1995 West Publishing Company

5. Identify the meaning of the symbols, letters, and characters inside the following feature control frames.

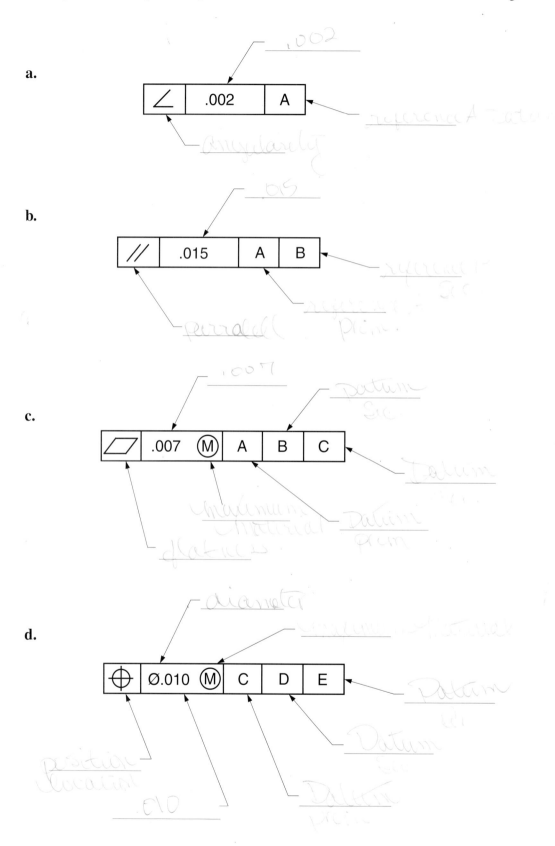

Section 9: *Tolerancing* **189**

Student _____ Date _____

6. Apply geometric tolerances to the knurled sleeve so that the outside diameter is round to within .003 inch and the ends are parallel to within .001 inch.

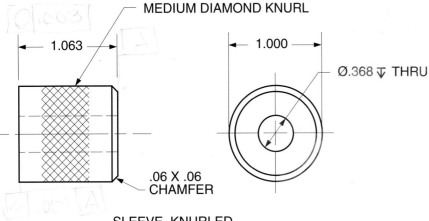

SLEEVE, KNURLED

7. Apply flatness and perpendicularity tolerances to the spacer. Flatness tolerance is .004, and perpendicularity is .010. Select and add appropriate datums.

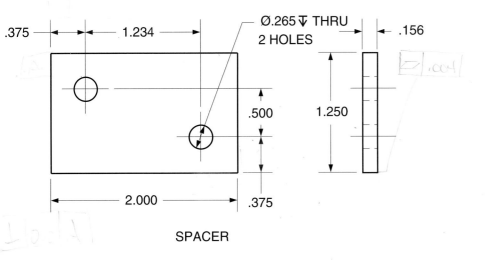

SPACER

8. Apply tolerances to the stepped shaft so that the two different diameters have a total runout of .015 inch with the datum.

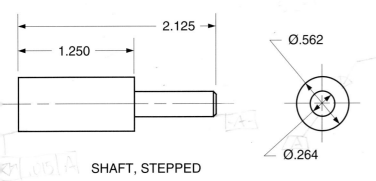

SHAFT, STEPPED

©1995 West Publishing Company

Section 9: *Competency Quiz*

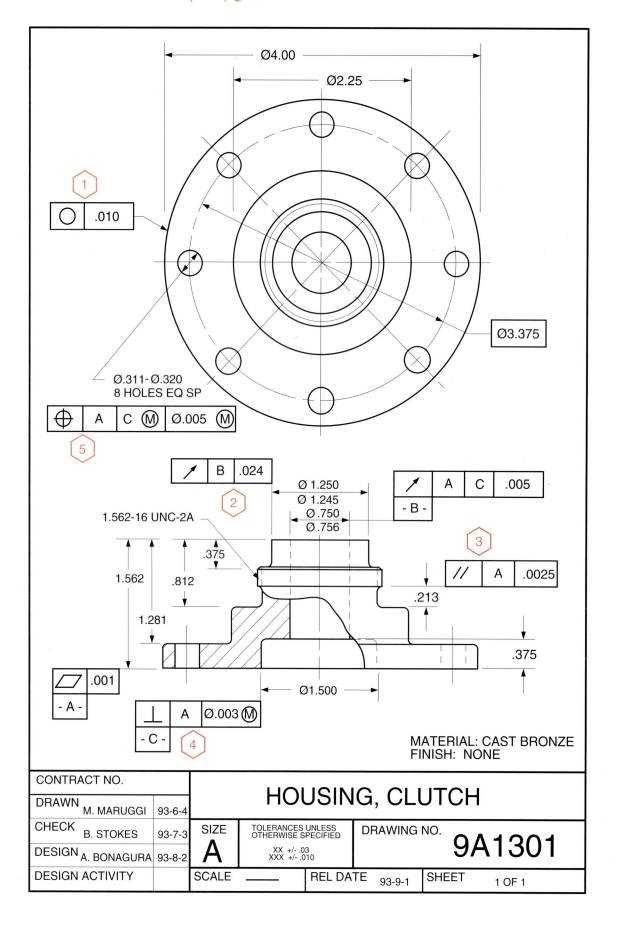

Section 9: *Tolerancing*   **191**

Student _____   Date _____

PART D TOLERANCING INTERPRETATION

The following questions are related to drawing 9A1301, a clutch housing (see page 190).

1. What is the tolerance for the Ø4.00-inch dimension?

   ±.03

2. How many basic dimensions are there?

   1

3. What is the tolerance for the Ø.812 dimension?

   +/-.010

4. How many datums are on the drawing? ___3___
   What are they? __A, B, C__

5. What is the diameter of the eight equally spaced holes?

   Ø.311-.320

6. Interpret all the symbols at ① . roundness .010

7. How many different geometric characteristic symbols are shown? 6

8. Interpret all the symbols at ② . circular runout .004 at point B

9. How many features use datums A, B, and C as a reference? 0

10. Interpret all the symbols at ③ . parallel .0005 at point A

11. What is the material for the part? Cast Bronze

©1995 West Publishing Company

192   **Section 9:** *Competency Quiz*

12. Interpret all the symbols at ④.

13. What is the size of the drawing?

14. What is the depth of the Ø.750/Ø.756 hole?

15. What is the overall tolerance for the Ø1.250/Ø1.245 dimension?

16. Interpret all the symbols at ⑤.

17. How many runout symbols are on the drawing?

18. What does the symbol with the ⊥-A-⊥ datum reference and ⊥.001⊥ tolerance indicate? _____

SECTION 10

Fasteners and Joining Methods

LEARNER OUTCOMES

You will be able to:

- Explain the difference between fastening and joining methods.

- Identify six types of screw threads and their general applications.

- Name ten threaded fasteners and their general characteristics.

- Identify three general types of unthreaded fasteners.

- Define twelve fastening and joining technical terms.

- List the component parts of a weld symbol.

- Identify the meaning and intent of ten weld symbols.

- Apply weld symbols to various types of joints.

- Explain the difference between soldering and brazing processes.

- List six forms of industrial adhesives.

- Demonstrate your learning through the successful completion of competency exercises.

10.1 INTRODUCTION

To design and develop new products, an engine department needs to determine how parts, subasse blies, and assemblies are fastened or joined togethe. Two categories of fastenings to accomplish this task are the threaded and unthreaded types. On a drawing, the threaded type is recognizable by the appearance of a screw thread.

Some parts need to be fastened together on a permanent basis; certain parts are put in place for the purpose of remaining permanently installed and are not removed or disassembled for routine or periodic maintenance. Other designs call for a semipermanent fastening; some parts need to be easily removable but can cause damage to the fastener. Still other fastenings are removable, which means that they can be disassembled from their mating parts easily, on a regular basis, and without damage to the fastener or its surrounding parts.

10.2 THE SCREW THREAD

The screw thread is the predominant method for fastening two or more parts together. Simply defined, the screw thread is a spiral-shaped groove cut into the surface of a bar, rod, or cylinder. For proper operation, there must be two mating, movable components: one with an internal thread, and one with an external thread. The most basic form of mating parts having internal and external threads is a screw and a nut. Figure 10.1 illustrates hardware in the form of a hexagonal nut (internal thread) and a hexagonal head cap

Section 10: *Fasteners and Joining Methods*

internal and external threads

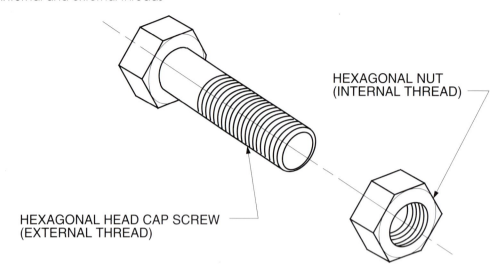

screw (external thread), which are commonly used in industrial fastening applications.

The screw thread is used in the design, fabrication, production, and operation of most products. It is, therefore, necessary for a shop technician to be able to recognize various screw threads on a drawing and to interpret their representation. There are several different types of thread forms, including the acme, American National, square, unified, buttress, knuckle, worm, and sharp V threads. These thread forms are depicted in Figure 10.2. Each serves a specific purpose:

- *Acme screw thread*—The acme screw thread is a thread form designed for areas of high stress in traversing motion and for power transmission.
- *American National thread*—This type of thread is suitable for use on bolts, screws, and nuts as well as for general use on some plastics and soft metals. American National threads are designated as N, NC, NF, NEF, or NS.
- *Square thread*—The square thread is used primarily for the transmission of power because its teeth are square and at right angles to the axis. This is an older thread form that has been replaced by the acme form.
- *Unified thread*—This is the most widely used of all the thread forms and is usually referred to as the *Unified National* thread. It has been approved by standards organizations in Canada,

FIGURE 10.2
Screw thread forms

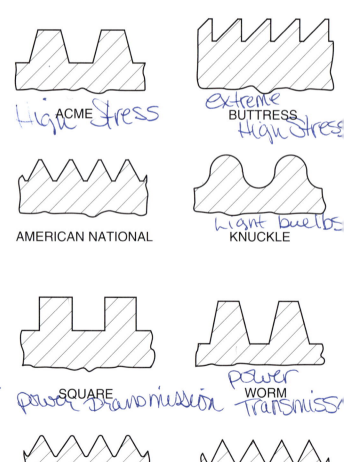

Great Britain, and the United States. Its applications are the same as for the American National threads, and it is mechanically interchangeable with the American National threads of the same diameter and pitch. Unified threads are designated by UN, UNC, UNF, UNEF, UNS, or UNM.

- *Buttress thread*—This form of thread is used in applications involving exceptionally high stress, in one direction only, along the thread axis.
- *Knuckle thread*—The knuckle thread is usually found on glass and plastic jars and on electric light bulbs and sockets.
- *Worm thread*—As with the acme and square threads, this type of thread is used for power transmission purposes.
- *Sharp V thread*—The sharp V thread has a sharp or pointed crest that makes it useful for applications involving assembly, adjustment, or both.

Because multinational interchangeability of parts and assemblies is required for large companies, the newest thread standard being used throughout the world is the ISO metric thread form. It has the same profile as that of the Unified National thread except that its thread depth is not as great, and it is used on the same applications.

10.3 SCREW THREAD TERMINOLOGY

To speak intelligently about screw threads and fastening hardware, a shop technician needs to be familiar with some basic terms used when referring to threaded parts. Most are shown in Figure 10.3 and are defined here:

- **Crest**—The surface that joins the sides of a thread and is the farthest from the cylinder or cone from which it projects.
- **Depth of thread**—The distance between the crest and the root of the thread.
- **External thread**—A thread located on the external periphery of a rod, cylinder, or cone.
- **Internal thread**—A thread located on the internal periphery of a part, such as those found on a nut.

FIGURE 10.3
Thread terminology

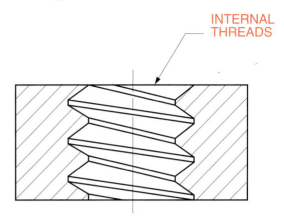
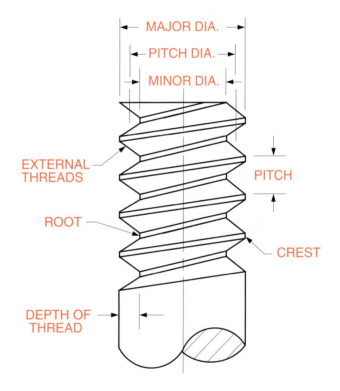

- **Left-hand thread**—A thread that, when viewed axially, winds in a counter-clockwise and receding direction.
- **Major diameter**—The largest, or external, diameter of a screw thread.
- **Minor diameter**—The smallest, or root, diameter of a screw thread.
- **Nominal thread size**—The general designation given for the identification of threads.

196   **Section 10:** *Fasteners and Joining Methods*

- **Pitch**—The distance between corresponding points on adjacent thread forms, measured parallel to the axis.
- **Pitch diameter**—The equivalent of an imaginary cylinder whose surface cuts the thread forms where the width of the thread and the groove are equal.
- **Right-hand thread**—A thread that, when viewed axially, winds in a clockwise and receding direction. On drawings, threads are always considered to be right-hand unless otherwise specified.
- **Root**—The edge or surface that joins the sides of adjacent thread forms and coincides with the cylinder or cone from which the side projects.
- **Thread class**—The amount of tolerance or tightness of fit between internal and external threads (also referred to as *class of fit*). A designation of fit consists of a *1*, a *2*, or a *3*, followed by an *A* (external thread) or a *B* (internal thread). Refer to Appendix E for information relative to thread class.
- **Threads per inch (TPI)**—The number of threads in one inch of length of screw threads. This measurement is accomplished with either a steel rule or a thread gage.
- **Thread series**—For the Unified National thread, this is the classification of a thread that indicates the number of threads per inch for a specified diameter. For example, the words *coarse*, *fine*, and *extra fine* and the numbers *8*, *12*, and *16* are used to describe the thread series. Appendix E provides information on standard thread series.

10.4 DRAWING APPLICATION OF SCREW THREADS

For shop personnel, there are two major areas that must be understood in the application of threads on an engineering drawing: (1) how threads are represented or shown graphically on a drawing, called **thread representation**; and (2) **thread designation**, which is how threads are identified through the use of text information, according to some accepted standard.

10.4.1 Thread Representation

There are three types of representations used by industry to portray all the screw thread forms on engineering drawings: the *detailed representation*, the *schematic representation*, and the *simplified representation*. Of the three, only the simplified representation will be covered here because it is the most commonly used method and is preferred by the United States Department of Defense specification MIL-STD-9 as well as the American National Standards Institute specification ANSI Y 14.6. Figure 10.4 illustrates how the simplified representation appears on a drawing for both internal and external threads.

10.4.2 Decimal-Inch Thread Designation

Decimal-inch thread designation is text information on a drawing. It provides for the identification of standard series threads and consists of three groups of alphanumeric characters (letters and numerals), separated by dashes, which indicate nominal size, the number of threads per inch, the thread series symbol, and the thread series. Figure 10.5 shows a typical external thread designation in the decimal-inch system of measurement.

The *nominal size* of threads is designated in decimal form. No tolerance is noted, and no tolerance is applied in the following example, in which .250 is the nominal size:

.250-20 UNC-3A (EXTERNAL THREAD)
.250-20 UNC-3B (INTERNAL THREAD)

The number of threads is, in fact, the number of threads per inch (TPI). In the following example, the 18 denotes the number of threads per inch:

.312-**18** UNC-2A
.312-**18** UNC-2B

Of the several Unified National thread series mentioned earlier, four have been standardized. They are the UNC, UNF, UNEF, and the UN series. Table 10.1 identifies each and their specific use for industrial applications.

Thread classes differ from each other by the amount of tolerance or by the amount of tolerance and allowance. For those threads that are categorized under the unified form, classes 1A, 2A, and 3A apply only to external threads, and classes 1B, 2B, and 3B

Section 10: *Fasteners and Joining Methods* **197**

FIGURE 10.4
Simplified representation of internal and external screw threads

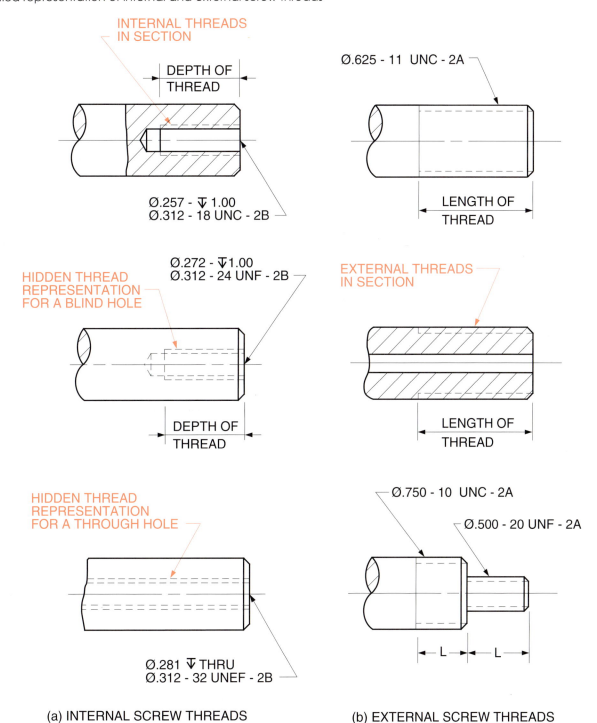

(a) INTERNAL SCREW THREADS (b) EXTERNAL SCREW THREADS

FIGURE 10.5
Typical external decimal-inch thread designation

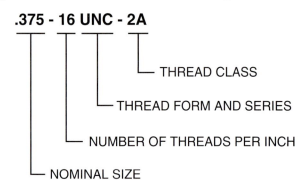

TABLE 10.1
Unified thread series usage

SERIES	USAGE
UNC (Unified National Coarse)	Screws, nuts, bolts, threads in soft metal, and polymers for general use
UNF (Unified National Fine)	Screws, nuts and bolts that require a better fit; on automotive parts; also where adjustment is required
UNEF (Unified National Extra Fine)	Tubing, thin nuts; applications where the length of thread is quite short
UN (Constant Pitch)	Special applications for which large diameters and other threads do not apply

apply only to internal threads. Note that some of these classes are identified in Figure 10.4.

10.4.3 Metric Thread Designation

Basic *metric thread designation* differs from the decimal-inch method. ISO metric threads are designated by the letter *M* followed by the nominal size of the thread (in millimeters), the letter *X*, the pitch (in millimeters), and the tolerance class. The examples shown here identify typical internal and external thread designations:

M 8 X 1.50-5g 6g (EXTERNAL THREADS)
M 6 X 0.8 7H (INTERNAL THREADS)

The tolerance system for ISO metric screw threads provides for allowances and tolerances indicated by a *tolerance class*. There are three tolerance grades that are used—grade 4, grade 6, and grade 8—all of which reflect the size of the tolerance, categorized as follows:

- Grade 8 tolerance is closest to the unified class 1A and 1B fits.
- Grade 6 tolerance is closest to the unified class 2A and 2B fits.
- Grade 4 tolerance is closest to the unified class 3A and 3B fits.

ISO has also established the amount of tolerance by providing a series of tolerance position symbols for external and internal threads:

- For external threads (screws, bolts), the letter *e* indicates a large allowance *g* a small allowance, and *h* is no allowance.
- For internal threads (nuts, threaded holes), the letter *G* indicates a small allowance, while *H* indicates no allowance.

The letter symbols just shown will appear after the *tolerance grade*. For example, *6g* means that what is needed is a medium tolerance grade with a small allowance for an external thread. The ISO *class of fit* is determined by selecting one of three qualities (fine, medium, or coarse) combined with one of three lengths of engagement (short, or *S*; normal, or *N*; or long, or *L*). However, the use of tolerance positions *g* for external threads and *H* for internal threads are more commonly shown. Complete internal and external designations for ISO metric screw threads are illustrated in Figure 10.6.

10.5 FASTENERS

Fasteners, both threaded and unthreaded, are a fundamental method for joining parts together, which makes them an essential component of almost every design. There are three basic categories of fasteners: the *removable type*, the *semipermanent type*, and the *permanent type*. Removable fasteners are classified as those that can be removed easily using hand tools, without damaging mating parts. Typical examples in-

FIGURE 10.6
ISO metric screw thread designations for internal and external threads

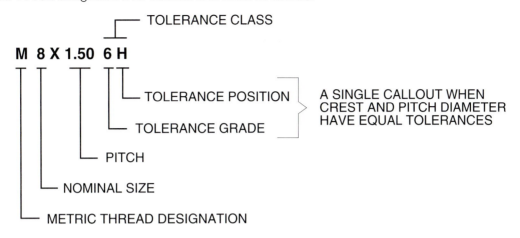

(a) INTERNAL METRIC THREADS

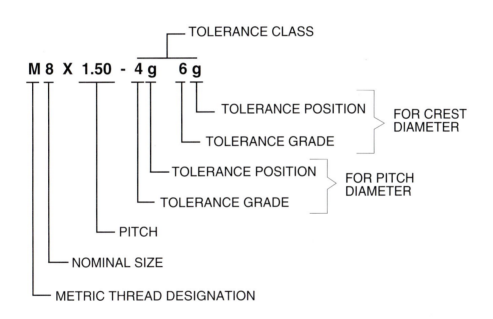

(b) EXTERNAL METRIC THREADS

clude the many types of bolts and nuts used around the house and in industry. Semipermanent fasteners can also be removed easily, but this may result in some damage to the fastener. Examples of this type of fastener include cotter pins, grooved pins, dowel pins, clevis pins, and special self-locking nuts with nylon inserts. Permanent fasteners are those that are put in place to remain permanently installed; they are not removed for routine or periodic maintenance work. A rivet is an example of a permanent fastener.

10.5.1 The Threaded Fastener

Threaded or screw-type fasteners are the most commonly used fastening devices and are generally considered to be removable-type fasteners. Figure 10.7

provides examples of threaded fasteners used in industry as well as in the building trades.

A *through bolt*, shown in Figure 10.7a, usually has a square or hexagonal head at one end and threads at the other. It is used to join two parts together by being passed through each of them and secured with a nut. A washer is sometimes required at the head end or the nut end. A wrench is normally used to tighten or loosen the nut.

A *stud* is required when space is needed between two mating parts or when a through bolt is not suitable for parts that must be removed often. Note that in Figure 10.7b, the stud is threaded on both ends.

A *cap screw* is used mostly in applications that call for it to be passed through a clearance hole in the nearest part and screwed into a threaded hole in the other part. Popular head styles include the hexagonal head, fillister head, and the socket head (the last two are shown in Figure 10.7c). Various head styles are produced to accommodate specific applications.

A *machine screw* is a small threaded fastener used in light to medium applications. It functions with or without a nut, as does the cap screw. Several head styles are available, as shown in Figure 10.7d.

A *stove bolt* is an inexpensive fastener manufactured to loose tolerances. This allows it to assemble easier to a nut. An example of a stove bolt is shown in Figure 10.7e.

The *carriage bolt* identified in Figure 10.7f is characterized by a square section immediately under its head. The head is plain—that is, it doesn't have a groove or a slot to accommodate a tool for turning. A

FIGURE 10.7
Threaded fasteners

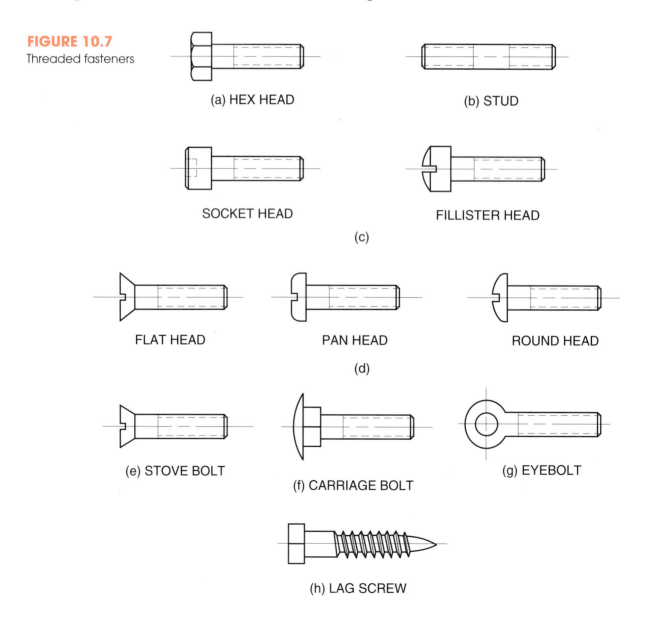

square nut is usually used at the threaded end. As the nut is tightened, the carriage bolt is held stationary by the square section. This section seats itself in a square cutout when used in metal applications and wedges itself when used for fastening wood to metal or wood to wood.

An *eyebolt* is used for lifting purposes—specifically, for hoisting pieces of equipment. It is often used for lifting transportable, mobile, or air-dropped military equipment. The eyebolt is attached with a through hole and a nut or screwed into a threaded hole. It is identified by a hole in the end called an eye, as seen in Figure 10.7g.

A *lag screw*, shown in Figure 10.7h, has several uses, but a major industrial application is the fastening of machinery to pallets or bases prior to shipment. A lag screw resembles a bolt at the head end and a wood screw at the threaded end. Heads are either square or hexagonal in shape and, therefore, require a wrench for tightening.

10.5.1.1 The Set Screw

A *set screw* is another form of threaded fastener. It is used to prevent relative motion between two parts, such as a shaft and a pulley. Set screws can be easily removed and are classified according to the type of head, point, or a combination of the two. Various common types of set screws used in industry are shown in Figure 10.8.

When selecting a set screw, the requirements for a specific application usually must be met, and the machine trades person should be able to identify each of the set screws described here. The *cone point* set screw is used in applications that require two parts to be joined in a permanent position relative to each other. The *flat point* type is used when frequent adjustment is necessary between the two parts being secured. The *cup point* is the most widely used type. It allows for rapid assembly of parts without prior preparation. The *full dog* and *half dog points* are used where members are to be joined permanently at a given, fixed location. For either type, a hole must be drilled to accommodate the point. The *oval point* is utilized in applications similar to those that use the cup point.

10.5.1.2 The Nut

As stated earlier, a nut is a fastener with internal threads that is used in conjunction with some sort of threaded screw or bolt for attaching parts together. Nuts are available in two wrench-type styles: hexagonal and square. Various nut styles are used in industry for the many applications required. Several different types are illustrated in Figure 10.9.

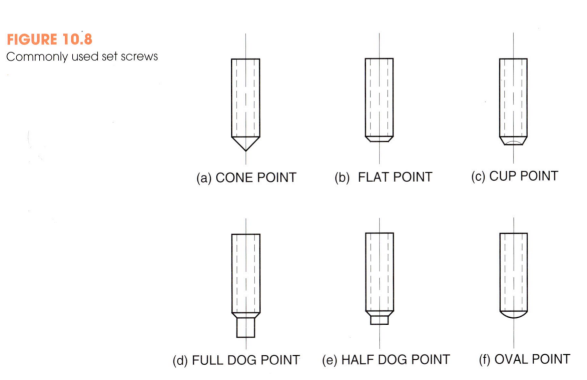

FIGURE 10.8
Commonly used set screws

(a) CONE POINT (b) FLAT POINT (c) CUP POINT

(d) FULL DOG POINT (e) HALF DOG POINT (f) OVAL POINT

FIGURE 10.9
Nuts

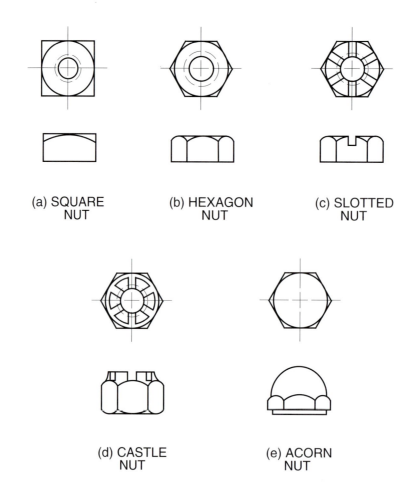

10.5.2 The Unthreaded Fastener

An **unthreaded fastener** is considered to be any one of the three types of fasteners: removable, semipermanent, or permanent, depending on its use and function within an assembly or subassembly. The following sections identify the more common uses for unthreaded fasteners in the workplace.

10.5.2.1 Pins and Rings

Pin and retaining ring-type fasteners have two different functions or uses within an assembly. *Pins* are normally used as retainers where the load is primarily in shear (a cutting action), and they can be used as removable or semipermanent hardware. **Retaining rings** are used to prevent axial movement (in and out along a shaft), and they are usually removable and replaceable in an assembly.

A *clevis pin* is one whose function is to fasten parts that are free to move, such as a joint requiring a fork or a yoke to be attached to a pneumatic cylinder that has an eye, as illustrated in Figure 10.10a. The fit between mating parts for this type of pin is usually a running or a free fit. Because a hole exists toward the end of the barrel of the pin, a cotter pin is used to fasten it to the assembly. Clevis pins are limited in size and shape to whatever the designer or engineer feels is required for a given situation. In Appendix F, standard-size clevis pins are shown.

A *cotter pin* is often used with a clevis pin, but it is also used to lock nuts in place, such as slotted or castle nuts. The attachment of the cotter pin consists of inserting it into a hole and then bending the end over to prevent it from loosening during movement. The cotter pin, which is a removable-type fastener, appears in Figure 10.10b. Cotter pins are commercially

FIGURE 10.10
Clevis pin and cotter pin usage

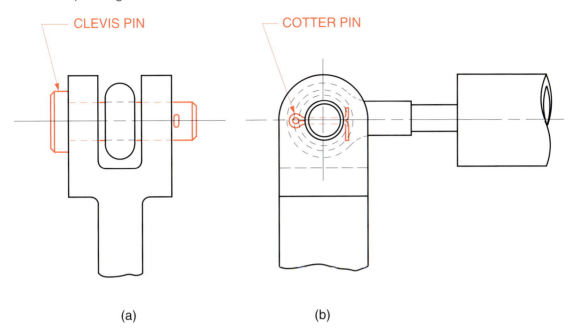

(a)   (b)

available in many sizes to accommodate various needs. Appendix F identifies the more popular sizes of cotter pins.

A *dowel pin*, which is available in a straight form, is used to hold parts in a fixed position or to ensure the alignment of parts in an assembly. The size of the dowel pin is determined by its application. A *taper pin* is used to attach components to a shaft, such as attaching a pulley or a sheave to a shaft. Both types of pins require a press fit at insertion to maintain rigidness in assembly. These pins are shown in Figure 10.11. Standard-size dowel and taper pins can be seen in Appendix F.

A *grooved pin* is used in shear applications. It possesses excellent holding power and is resistant to shock and vibration. A grooved pin has three parallel radial grooves; the grooves, because of their design, allow for a press fit in assembly. This fastener can be used in a permanent or semipermanent situation; however, care must be used when removing this type fastener because it could damage surrounding parts. Type A has tapered full-length grooves, and types C and F have three straight grooves, each with a slightly different starting and ending point. Three of the seven types of standard grooved pins are shown in Figure 10.12.

FIGURE 10.11
Dowel and taper pins

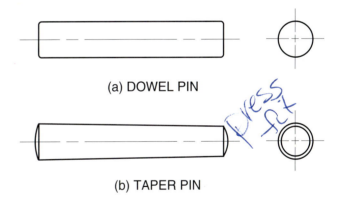

(a) DOWEL PIN

(b) TAPER PIN

Another type of pin used in the assembly of parts is a *spring pin*, sometimes referred to as a *roll pin*. These pins are usually made from high carbon steel, corrosion resistant steel, or a beryllium copper alloy. They are called spring pins because they have a springlike resiliency, are compressible, and are larger than the hole in which they are supposed to fit. They are available in two forms: with a slot throughout the entire length, or shaped in the form of a coil. Examples are provided in Figure 10.13. Further examples of standard grooved pins are offered in Appendix F.

FIGURE 10.12
Grooved pins

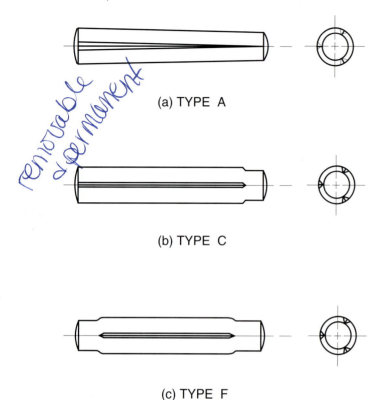

(a) TYPE A

(b) TYPE C

(c) TYPE F

FIGURE 10.13
Spring pins

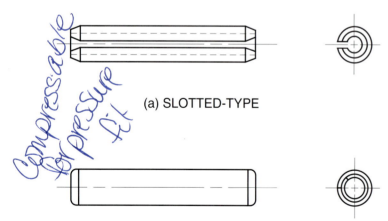

(a) SLOTTED-TYPE

(b) COILED-TYPE

A **retaining ring** is a precision-type fastener that is used to secure components to shafts or to limit the movement of parts in an assembly. Applications vary from miniature assemblies to large construction equipment. Retaining rings are used in military ap-plications and in space exploration equipment. In addition, one could find retaining rings in business machines, automobiles, appliances, and other consumer products. This type of ring is produced in two basic styles: external and internal. Figure 10.14 illustrates both styles.

FIGURE 10.14
External and internal retaining rings

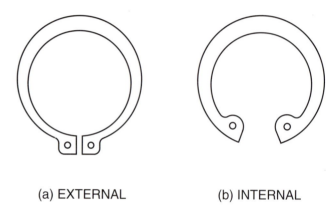

(a) EXTERNAL   (b) INTERNAL

An *external retaining ring* for axial assembly is expanded over a shaft, stud, or similar part with a plier-type tool and is seated in a groove in a shaft. An *internal retaining ring* for axial assembly is compressed for insertion into a groove in a housing or similar installation. Retaining rings are made of materials having excellent springlike properties, such as carbon steel, stainless steel, or beryllium copper. They are often used as an alternative to screws, cotter pins, rivets, and machined shoulders on parts. Annular grooves of the appropriate width and depth must be provided to accommodate these rings for assembly. Figure 10.15 displays several different sizes of commercially available retaining rings.

10.5.2.2 Rivets

A **rivet** is a permanent-type fastener. A rivet cannot be removed without destroying the rivet and possibly damaging mating parts. Riveting is one of the oldest and most reliable methods of attaching parts together. The automotive, appliance, electronic, furniture, structural steel, hardware, boating, and aircraft industries are among the many fields in which the rivet is a popular fastening method. One of the advantages

FIGURE 10.15
Examples of retaining rings

Photo courtesy of Rotor Clip Company, Inc.

FIGURE 10.16
Fastening by riveting

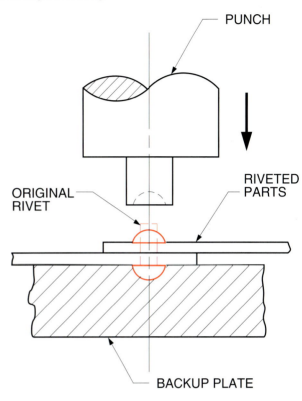

of using rivets is that they are capable of fastening many parts at the same point and can join dissimilar metals of various thicknesses.

A rivet consists of a short length of rod with some style of head at one end. Typical rivet materials include wrought iron, mild steel, aluminum alloys, and copper alloys. A common method of fastening by riveting is illustrated in Figure 10.16. Note that a head is formed at the shank end (bottom) of the rivet, which provides the method of fastening sheets or plates of metal together. Types of rivets used in industry include the solid rivet, semitubular rivet, tubular rivet for through holes, and pop rivet for blind holes. Several head styles for solid rivets are shown in Figure 10.17.

FIGURE 10.17
Solid rivet head styles

(a) COUNTERSUNK HEAD   (b) BUTTON HEAD

(c) ACORN HEAD   (d) CONE HEAD   (e) PAN HEAD

10.6 JOININGS

For the purpose of this worktext, **joining** may be defined as the combining or uniting of similar materials through the use of heat, or as the bonding or adhesion of similar or dissimilar materials through the use of chemical materials. Several methods are used in

206 **Section 10:** *Fasteners and Joining Methods*

industry for connecting parts other than by threaded and unthreaded fasteners. They include older methods, such as welding, brazing, and soldering, as well as bonding with industrial adhesives, which is a more recent method of joining. In fact, bonding frequently replaces riveting, brazing, soldering, and welding as the primary fastening method because it is less expensive. In addition, because there is no heat involved in the bonding process, there are no heat distortion problems.

■ 10.6.1 Welding

Welding is an industrial process whereby two metals are permanently joined together to form a whole. A good weld is a combination of the appropriate materials, temperature, pressure, and metallurgical conditions. All welding is done using metal electrodes. The **electrodes** provide the necessary filler material to fill the voids (spaces) between the two pieces of metal to be joined. As electrical heat is applied to the electrodes, droplets from the welding rod form a **bead** between the two parts, thus joining the two (which really become one continuous piece when the welding is complete). Welding processes may be categorized into six major areas that include arc, gas, resistance, thermit, induction, and forge types. Welding is, very often, the most economical, efficient, and effective method of fabrication in the manufacturing process.

There are very specific symbols and letter codes provided by the **American Welding Society (AWS)** Standard A2.4-86 for the purpose of identifying different welding processes. The processes and letter codes are shown in Table 10.2.

10.6.1.1 The Basic Welding Symbol

Just as geometric tolerancing has symbols to simplify and standardize processes, the field of welding uses certain symbols. These symbols ensure that all welding is done to some approved, acceptable, unified standard. The basic welding symbol approved by both AWS and ANSI consists of several component parts. These parts are illustrated in Figure 10.18.

The *reference line*, the *arrow* (termination), and when needed, the *tail*, are the three major characteristics of the basic symbol. Each one has a particular function. For example, the reference line is where the weld symbol makes contact. The arrow, or termina-

tion, is much like a leader and points to the area to be welded. The tail is required when a reference or a specification is noted.

There are two important aspects to the reference line for a welding symbol. The lower side of the line is referred to as the arrow, or near side, and the upper side of the line is called the other, or far side when referring to the location of a weld with regard to a joint.

TABLE 10.2
Common Welding Processes

PROCESS	LETTER CODE
Arc Welding	**(W)**
Flux cored arc welding	(FCAW)
Gas metal arc welding	(GMAW)
Pulsed arc	(GMAW-P)
Short circuit arc	(GMAW-S)
Electrogas	(GMAW-EG)
Spray transfer	(GMAW-ST)
Gas tungsten arc welding	(GTAW)
Plasma arc welding	(PAW)
Submerged arc welding	(SAW)
Shielded arc welding	(SMAW)
Stud welding	(SW)
Oxyfuel Gas Welding	**(OFW)**
Oxyacetylene welding	(OAW)
Pressure gas welding	(PGW)
Resistance Welding	**(RW)**
Resistance spot welding	(RSW)
Resistance seam welding	(RSW)
Projection welding	(RPW)
Solid State Welding	**(SSW)**
Cold welding	(CW)
Explosion welding	(EXW)
Forge welding	(FOW)
Friction welding	(FRW)
Roll welding	(ROW)
Ultrasonic welding	(USW)
Other Welding Processes	
Electron beam welding	(EBW)
Electroslag welding	(ESW)
Flash welding	(FW)
Induction welding	(IW)
Laser beam welding	(LBW)
Thermit welding	(TW)

Section 10: *Fasteners and Joining Methods*   207

FIGURE 10.18
Component parts of the basic welding symbol

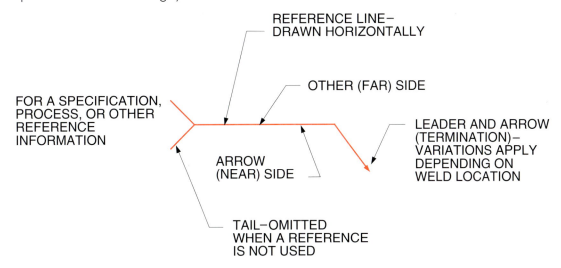

The arrow has no relationship as to a weld being on the near or far side. It merely indicates that there is a welded joint. To clarify this feature, refer to Figure 10.19, which illustrates an *arrow-side* and an *other-side* welded joint. Note the symbol △, for a fillet weld.

FIGURE 10.19
Fillet weld: arrow side and other-side

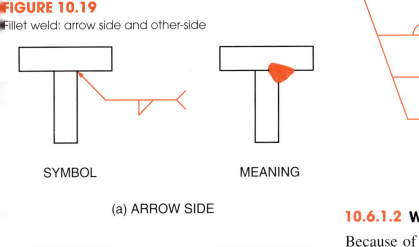

At times, multiple reference lines may be used to identify a sequence of welding operations. This is shown in Figure 10.20.

FIGURE 10.20
Multiple reference lines

10.6.1.2 Welding Symbol Elements

Because of the amount of information that may be required in the joining of parts through the welding process, it is important to recognize and understand the elements added to the welding symbol and to know where the symbol is placed on a drawing. Each of the elements applied to the basic welding symbol has a specific location and a relationship to the other components of the symbol. Figure 10.21 identifies the welding symbol, its elements, and their standard location.

FIGURE 10.21

Standard location of the elements of the basic welding symbol

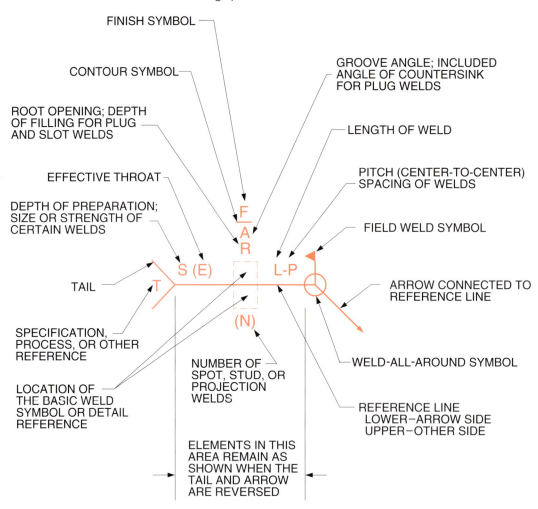

Examples of the various weld symbols that may be applied to the basic welding symbol are illustrated in Figure 10.22. A standard group of supplementary symbols may also be added to the basic welding symbol. They are shown in Figure 10.23.

Figure 10.24 provides several examples of the more common types of welds, their desired application, joint preparation, and appropriate symbology. On an engineering drawing, the person interpreting the drawing will only expect to see the symbol, not the weld itself. The darkened area of the weld is not shown.

10.6.1.3 Weld Symbol Locations

Normally, the person interpreting an engineering drawing will be dealing with a drawing that consists of three or more views in orthographic projection.

The welding symbol or symbols usually appear in the view in which the welded joint is most clearly shown. When a welding symbol appears in one view, it is not necessary to duplicate the symbol in another view. Usually, the welding symbol will be found in the front, or principal view of the drawing. Figure 10.25 illustrates various locations of the welding symbol.

■ 10.6.2 Resistance Welding

Resistance welding is commonly referred to as **spot welding**. In resistance welding, an electric current passes through the precise location where the required parts are joined together. It is usually used on thin sheet metal parts and accessories only. The predominant use is when overlapping members need to

Section 10: *Fasteners and Joining Methods*   **209**

FIGURE 10.22
Examples of weld symbols

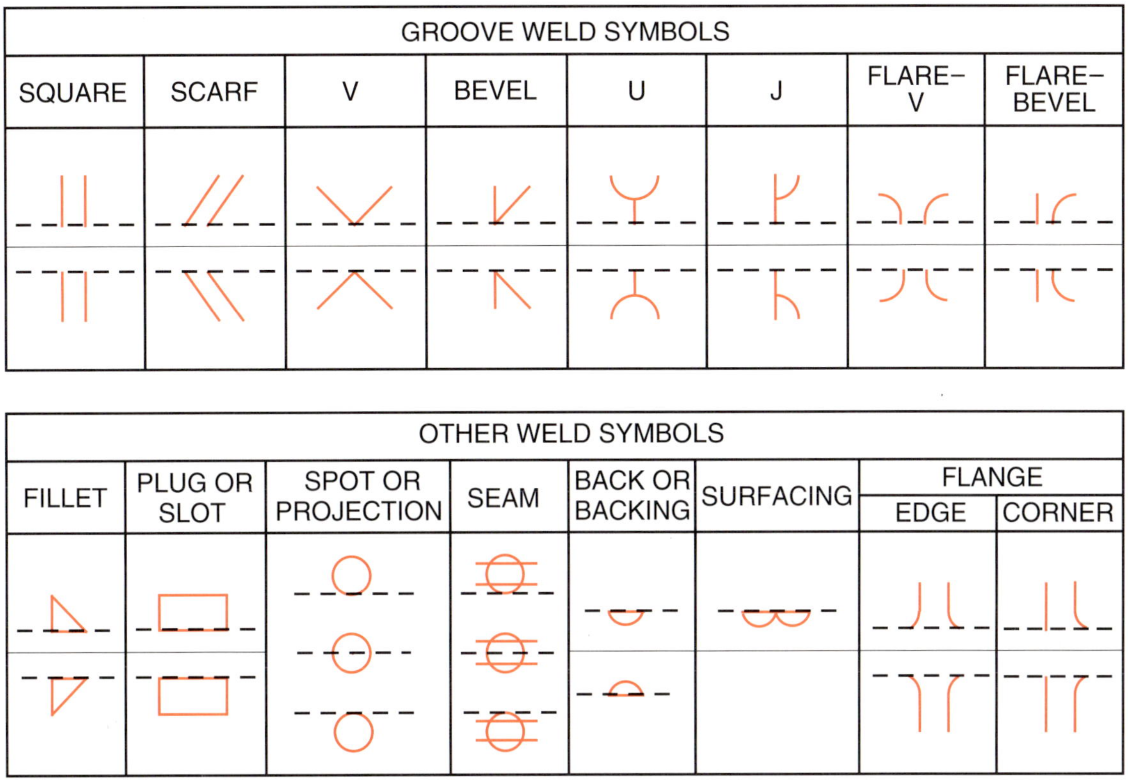

GROOVE WELD SYMBOLS							
SQUARE	SCARF	V	BEVEL	U	J	FLARE–V	FLARE–BEVEL

OTHER WELD SYMBOLS						FLANGE	
FILLET	PLUG OR SLOT	SPOT OR PROJECTION	SEAM	BACK OR BACKING	SURFACING	EDGE	CORNER

FIGURE 10.23
Supplementary weld symbols

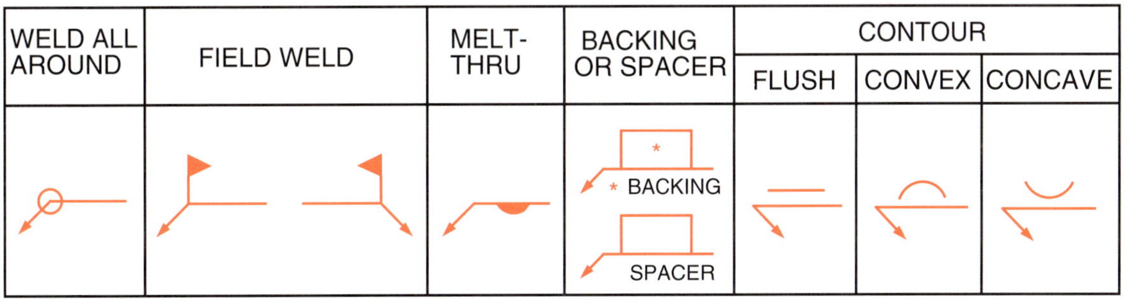

WELD ALL AROUND	FIELD WELD	MELT-THRU	BACKING OR SPACER	CONTOUR		
				FLUSH	CONVEX	CONCAVE

* BACKING

SPACER

Section 10: *Fasteners and Joining Methods*

FIGURE 10.24
Common types of welds

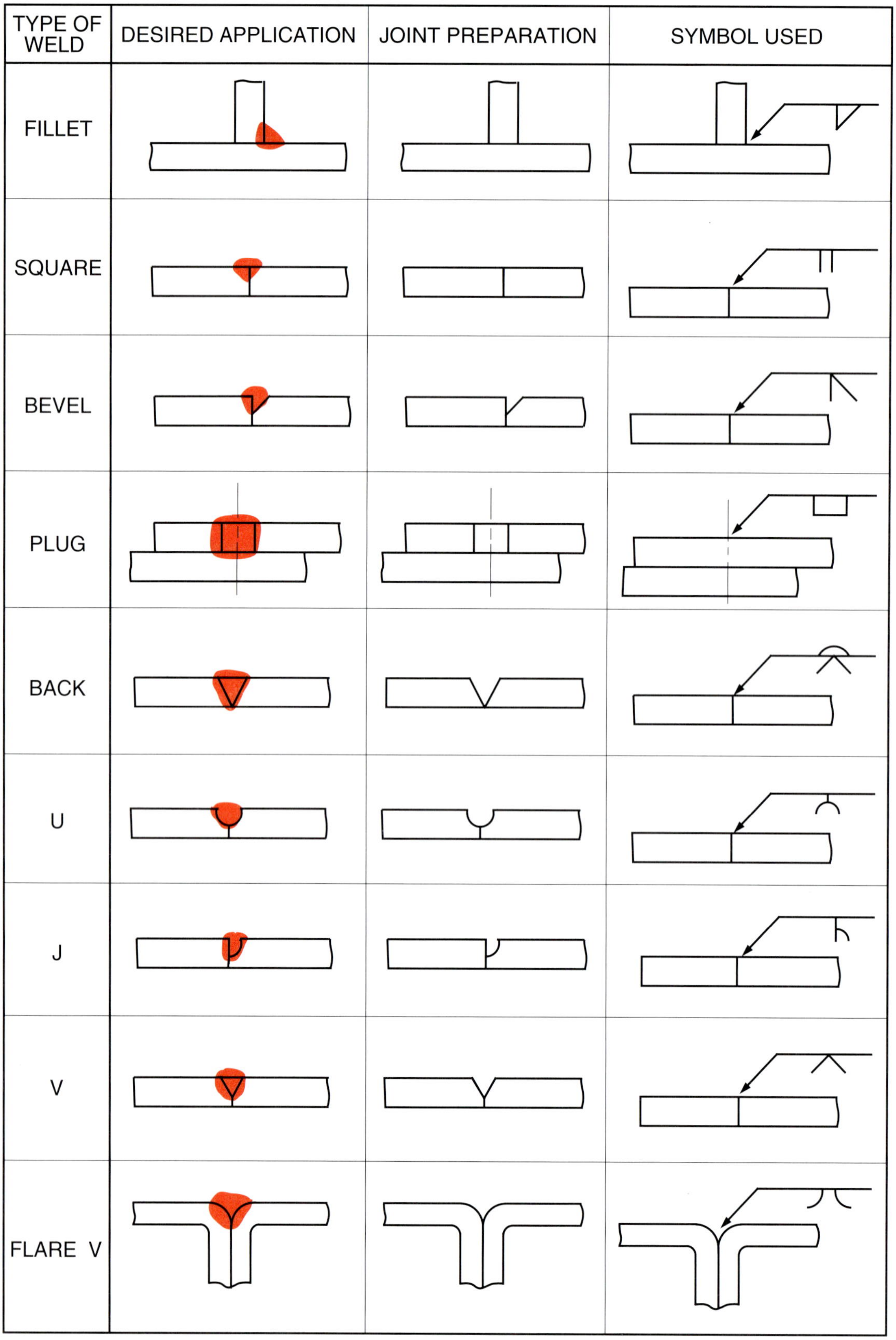

TYPE OF WELD	DESIRED APPLICATION	JOINT PREPARATION	SYMBOL USED
FILLET			
SQUARE			
BEVEL			
PLUG			
BACK			
U			
J			
V			
FLARE V			

FIGURE 10.25
Weld symbol locations

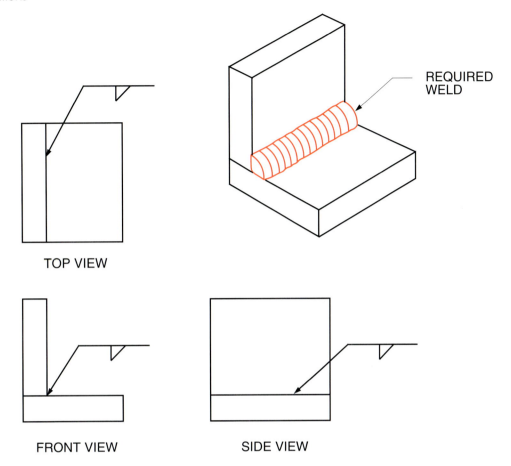

be joined. One would likely find spot welds on products such as television, radio, and VCR chassis, portable and home heating units, and electronics racks for commercial and military use.

In the resistance welding process, the appropriate amount of heat, pressure, and time in which the electrical current is allowed to pass are required for a good joint, one that will not break loose. Heat is a consequence of the electrical resistance of the workpiece and the interface between the electrodes as the electrical current passes through. Pressure varies throughout the weld cycle. An example of resistance welding is provided in Figure 10.26.

The preferred symbol for a spot weld is ○. This symbol is similar to the weld-all-around symbol except that it appears on the reference line, not at the junction of the reference line and the arrow line. The circle symbol is located tangent to and below the reference line when an arrow-side weld is required, and tangent to but above the reference line when an otherside condition exists. However, it is centered about the reference line when no side requirement is needed. Figure 10.27 identifies the location of the spot weld symbol in relation to the reference line. Note, too, that the tail is always included.

Common resistance weld symbols are used for spot, projection, and seam welding. They were previously illustrated in Figure 10.22. Examples of the complete symbols are shown in Figure 10.28.

■ 10.6.3 Brazing

Brazing is similar to welding in the sense that it forms a strong metallurgical bond where the joining takes place. Brazing is the joining of metals through the use of heat and a filler material in which the melting point is above 800 degrees Fahrenheit (427 degrees Celsius), but below the melting temperature of

FIGURE 10.26
Example of how resistance welding occurs

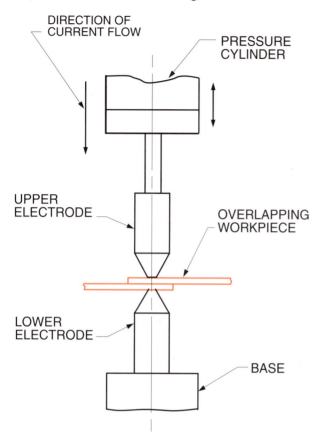

FIGURE 10.27
Location of spot weld symbols

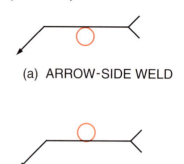

(a) ARROW-SIDE WELD

(b) OTHER-SIDE WELD

(c) NO ARROW- OR OTHER-SIDE SIGNIFICANCE

FIGURE 10.28
Complete spot, projection, and seam weld symbols

TYPE OF WELD	SPOT OR PROJECTION WELD	SEAM WELD
WELD ARROW SIDE	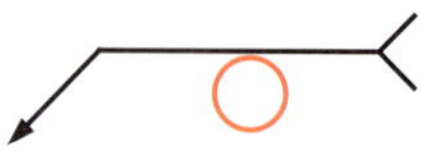 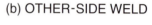	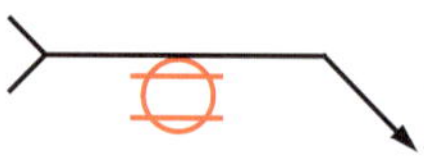
WELD OTHER SIDE	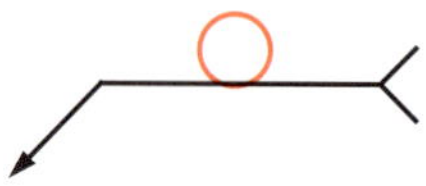	
NO ARROW- OR OTHER-SIDE SIGNIFICANCE	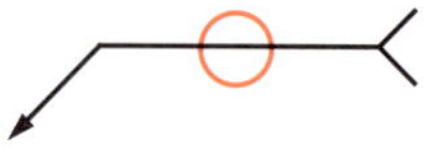 	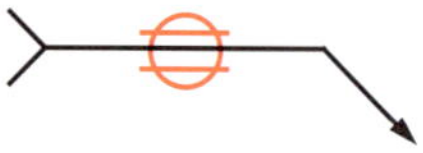

the metals being joined. It is a process whereby filler material is distributed to the joint through capillary attraction. When silver is used as the filler material, it is referred to as *silver soldering*.

Brazing has several advantages over welding. They include the following:

- Almost all metals are capable of being brazed.
- It is suitable for the joining of dissimilar metals.
- Since less heat is required to braze than to weld, the process can be performed more quickly and economically.
- It is adaptable to high production and is also capable of automation.

The most common brazing materials are silver, silver alloys, aluminum, aluminum alloys, copper, and copper alloys. Typical brazing applications for consumer products include heat exchangers for aircraft, air-conditioning units, refrigeration systems, evaporators, and condensing systems. All the symbols used for brazing are also used for welding with one exception, the scarf joint.

■ 10.6.4 Soldering

Soldering involves the use of nonferrous metals whose melting points are below 800 degrees Fahrenheit (427 degrees Celsius). The process is often referred to as *soft soldering* because the joining method for temperatures above 840 degrees is considered hard soldering or brazing. Soft solders are available in wire, bar, slab, ribbon, ingot, cake, and powder forms. The form with which most people are familiar is the wire form. Most of us have done some soldering around the home for a specific purpose, such as making a joint for an electrical connection or preparing a copper water pipe. Probably the most common use for solder bars in industry is for the dip soldering of electronics equipment, particularly integrated circuits and printed circuit board assemblies. In other instances, solder is applied using a soldering iron or gun, a torch, or a stream of hot neutral gas; by electric induction; or by wiping.

In the soldering process, it is important that the surfaces to be soldered be free from dirt, oil, scale, and oxides so that a good bond is ensured. For the cleaning of joints, a *flux* is used. Soldering fluxes are available in two basic forms, either corrosive or noncorrosive.

■ 10.6.5 Adhesive Bonding

Great strides have been made over the past fifteen years in the chemical industry in the development, use, and reliability of **adhesive bonding**. By definition, an adhesive bond occurs when a substance joins or adheres materials by means of surface attachment. The term *structural bonding* is used to describe adhesive bonding when it is used to support heavy loads, as is often the case when it is used in the automotive and aircraft industries. Current bonding methods are often used to replace older joining methods, such as welding, brazing, soldering, riveting, and fastening with screws.

The adhesives used today include a wide range of types and forms. A few of the more frequently used are thermoplastic and thermosetting resins, elastomers, and in some instances, ceramics. For the bonding of metal products, adhesives with an epoxy resin base seem to provide the best results. The different forms that the adhesives take include tapes, beads, drops, coatings, gels, liquids, paste, and solids. Adhesive formulations are applied by brush, roller, trowel, or spatula. In addition, they may be sprayed on using power-fed flow guns.

Two important factors must be considered when bonding with adhesives. The first one is the shelf life of the product; that is, for the product to give the best results, the adhesive must be used within the time frame recommended by the manufacturer. Second, the area to be bonded must be free of dirt, grime, and foreign particles. Directions need to be followed for best results.

10.7 SUMMARY

Industries use various methods to attach parts together. One of the methods discussed in this section is the use of threaded and unthreaded fasteners. The other method is the process of joinings.

Threaded fasteners covered include the through bolt, stud, cap screw, machine screw, stove bolt, carriage bolt, eyebolt, lag screw, set screw, and nut. Unthreaded fasteners that fall into the categories of removable, semipermanent, and permanent types were also identified; these include the clevis pin, cotter pin, dowel pin, taper pin, grooved pin, spring pin, retaining ring, and rivet.

214   **Section 10:** *Fasteners and Joining Methods*

The term *joining* was defined as the combining or uniting of similar materials through the use of heat, or the bonding or adhesion of similar or dissimilar materials through the use of chemical materials. The traditional joining methods of welding, brazing, and soldering were covered, and bonding with industrial adhesives was addressed in detail. Several figures provided insight into welding symbols, their intent, and how they are used and interpreted on an engineering drawing. Examples of where the various types of joinings are used were also provided.

TECHNICAL TERMS FOR STUDY

adhesive bonding   Joining materials by means of a surface treatment with a bonding substance.

American Welding Society (AWS)   An organization whose Standard A2.4-86 is used throughout industry for the symbology and processes used in welding.

bead   The result of a heated welding rod (electrode) forming a welded joint.

brazing   Similar to welding, a joining method that uses heat and a filler material to provide a metallurgical bond in which the melting point is above 840 degrees Fahrenheit but below the melting temperature of the metals being joined.

crest   The surface that joins the sides of a thread and is the farthest from the cylinder or cone from which it projects.

depth of thread   The distance between the crest and the root of a thread.

electrode   A rod form of filler material to perform welding.

external thread   A thread located on the periphery of a rod, cylinder, or cone.

fastener   A piece of hardware whose purpose is to join parts together. Fasteners are categorized as removable, semipermanent, or permanent, and they are either threaded or unthreaded.

internal thread   A thread located on the internal periphery of a part, such as those found on a nut.

ISO metric thread   A thread form used throughout most of the world, but not formally in the United States.

joining   The combining or uniting of similar materials through the use of heat, or the bonding or adhesion of similar or dissimilar materials through the use of chemically based materials.

lead   A term used to describe the distance a threaded part moves axially with respect to a fixed mating part, in one complete revolution.

left-hand thread   A thread that, when viewed axially, winds in a counter-clockwise and receding direction.

major diameter   The largest, or external, diameter of a screw thread.

minor diameter   The internal, smallest, or root diameter of a screw thread.

nominal thread size   The general designation given for the identification of threads.

nut   A fastener with internal threads, used in conjunction with some sort of threaded screw or bolt for attaching parts together.

pitch   The distance between corresponding points on adjacent thread forms, measured parallel to the axis.

pitch diameter   The equivalent of an imaginary cylinder whose surface cuts the thread form where the width of the thread and the groove are equal. It is a radial dimension.

resistance welding   Commonly referred to as spot welding; a process in which an electric current passes through the precise location where the required parts are joined together.

retaining ring   A precision-type fastener used to secure components to shafts or to limit the movement of parts in an assembly. They are available in both internal and external types.

right-hand thread   A thread that, when viewed axially, winds in a clockwise and receding direction.

rivet   An unthreaded, permanent-type fastener that is capable of fastening many parts at the same point, including the joining of dissimilar metals.

Section 10: *Fasteners and Joining Methods* **215**

root The edge or surface that joins the sides of adjacent thread forms and coincides with the cylinder or cone from which the side projects.

screw thread The predominant method of fastening two or more parts together. It is a spiral-shaped groove cut into the surface of a bar, rod, or cylinder.

soldering The joining of nonferrous metals whose melting point is below 800 degrees Fahrenheit. Metals are joined through the use of a filler material. Sometimes referred to as *soft soldering*.

spot welding *See* **resistance welding**.

thread class The designation of the amount of tightness of fit between internal and external threads; also referred to as the class of fit. It is signified by the numbers 1, 2, or 3, followed by the letter *A* for external threads, or *B* for internal threads.

thread designation The identification of threads on an engineering drawing through the use of text information.

threaded fastener The most commonly used fastening device. Examples include the through bolt, stud, cap screw, machine screw, stove bolt, carriage bolt, eyebolt, lag screw, set screw, and nut.

thread representation How threads are represented or shown graphically on an engineering drawing.

thread series The thread classification that indicates the number of threads per inch for a specified external diameter.

threads per inch (TPI) The number of threads in one inch of length of screw threads.

unthreaded fastener Fasteners that do not have threads, such as retaining rings, clevis pins, cotter pins, dowel pins, taper pins, grooved pins, spring pins, and rivets.

welding An industrial process whereby two metals are permanently joined together to form a whole. It involves the combination of the appropriate materials, temperature, pressure, and metallurgical conditions.

welding symbol A series of graphic symbols used to simplify and standardize the welding process. These symbols ensure that all welding is accomplished according to some approved, acceptable, unified standard authorized by the American Welding Society (AWS) and the American National Standards Institute (ANSI).

Student _____   Date _____

Section 10: Competency Quiz

PART A COMPREHENSION

1. What is a screw thread? How is it used? (5 pts)

2. List six types of screw threads and their general application. (6 pts)

3. Insert the proper screw thread terms in the following illustration. (8 pts)

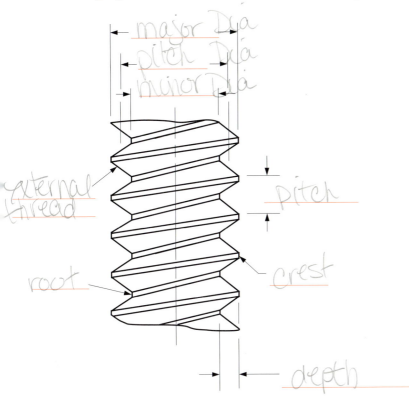

- major Dia
- pitch Dia
- minor Dia
- external thread
- pitch
- root
- crest
- depth

Section 10: *Competency Quiz*

4. What is meant by *class of fit* for screw threads? (5 pts)

   tolerance or tightness of fit "thread Class"

5. Explain the difference between thread representation and thread designation. (6 pts)

   representation- shown on a drawing -graphically

   designation- text information

6. What do the following thread designations signify? Be specific. (6 pts)

   a. .250-28 UNF-2B _____   b. .500-13 UNC-3A _____

   c. M 8 X 1.25-4g6g _____

7. What are the three basic categories of fasteners? (6 pts)

   permanent
   removeable
   semi-permanent

8. List ten kinds of fasteners. (5 pts)

9. What is the purpose of a set screw? Identify six types of points used on set screws. (5 pts)

Section 10: *Fasteners and Joining Methods* **219**

Student _____ Date _____

10. When are rivets an appropriate fastener? (5 pts)

to join permanently / different types of material

11. What is the purpose for a retaining ring? Identify two types. (5 pts)

to retain movement   Shafts etc .

external / internal

12. List five methods of fastening and joining mechanical components and parts together. (5 pts)

13. Define the term *joining*. (5 pts)

14. Explain the welding process. (5 pts)

15. Name the three parts of the welding symbol. (3 pts)

16. What is another term for *resistance welding*? (5 pts)

17. What are the basic differences between the brazing process and the soldering process? (5 pts)

©1995 West Publishing Company

Section 10: *Competency Quiz*

18. List six different forms in which adhesives are available. (3 pts)

19. What term is used when referring to the adhesive bonding used to support heavy loads? (3 pts)

Structural Bonding

20. What two important factors must be considered when bonding with adhesives ? (4 pts)

Shelf life of adhesive
must be Clean - free of dirt, oil, etc.

Section 10: *Fasteners and Joining Methods*   **221**

Student _____   Date _____

PART B TECHNICAL TERMS

For each definition, select the correct technical term from the list on the bottom of the page. (10 pts each)

1. _Adhesive bonding_ Joining materials by means of a surface treatment with a bonding substance.

2. _Brazing_ A joining process that uses heat and a filler material to produce a metallurgical bond in which the melting point is above 840 degrees Fahrenheit. but below the melting temperature of the metals being joined.

3. _pitch_ The distance between the crest and the root of a thread.

4. _fastener_ A piece of hardware whose purpose is to join parts together.

5. _ISO metric_ A thread form used throughout the world, but not formally in the United States.

6. _welding_ The combining or uniting of similar materials through the use of heat, or the bonding or adhesion of similar or dissimilar materials through the use of chemically based materials.

7. _resistance or spot welding_ A process in which an electric current passes through the precise location where the required parts are joined together.

8. _Soldering_ The joining of nonferrous metals whose melting point is below 800 degrees Fahrenheit.

9. _representation_ How threads are represented or shown graphically on an engineering drawing.

10. _joining_ An industrial process whereby two metals are permanently joined to form a whole. It involves the combination of the appropriate materials, temperature, pressure, and metallurgical conditions.

A. adhesive bonding
B. brazing
C. depth of thread
D. external thread
E. fastener
F. ISO metric thread
G. joining
H. major diameter
I. resistance welding
J. soldering
K. thread designation
L. thread representation
M. unthreaded fastener
N. welding

©1995 West Publishing Company

222 Section 10: *Competency Quiz*

PART C INTERPRETING FASTENERS

1. Identify the following threaded fasteners:

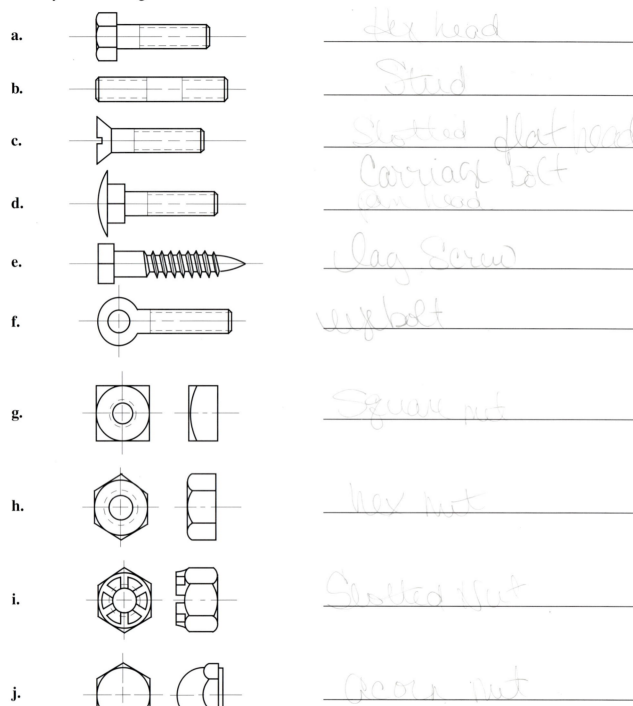

a. Hex head

b. Stud

c. Slotted flat head

d. Carriage bolt / pan head

e. Lag screw

f. Eye bolt

g. Square nut

h. Hex nut

i. Slotted nut

j. Acorn nut

Student _____ Date _____

Section 10: *Fasteners and Joining Methods*   **223**

2. In the space provided, identify the unthreaded fasteners shown here:

a.   dowel pin

b.  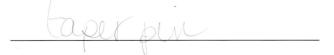 taper pin

c.   grooved pin

d. 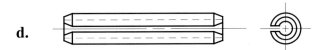  roll pin — spring pin

e. 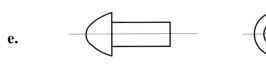  Acorn head rivet

f. 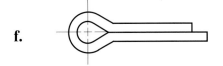  Cotter pin

©1995 West Publishing Company

Section 10: Competency Quiz

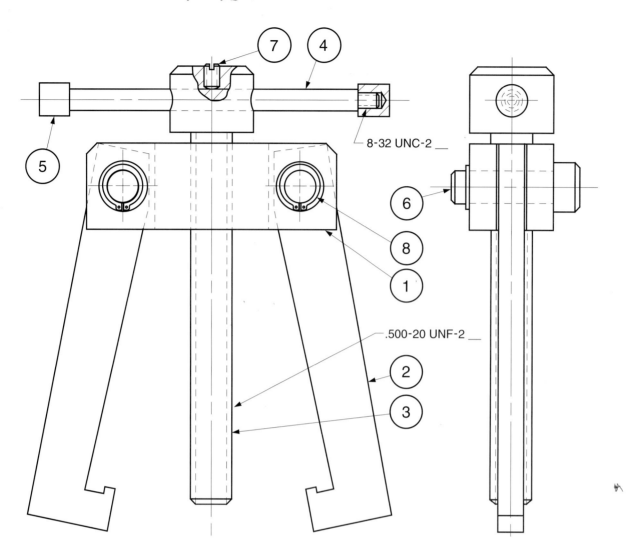

A14534 - 000 ASSEMBLY, GEAR PULLER

QTY REQD	PART NO.	DESCRIPTION	MATERIAL/SPECIFICATION	ITEM
2	A14534 - 008	RETAINING RING	EXTERNAL - TYPE SH - 31 STN STL	8
1	A14534 - 007	SET SCREW	HEADLESS - 10-32 x .31 CONE PT STN STL	7
2	A14534 - 006	CLEVIS PIN	Ø.625 x 1.50 STN STL	6
2	A14534 - 005	END CAP	Ø.375 x .38 STN STL ROD	5
1	A14534 - 004	HANDLE	Ø.250 x 4.00 STN STL ROD	4
1	A14534 - 003	SCREW	Ø1.00 x 4.75 STN STL ROD	3
2	A14534 - 002	FINGER	.312 x .875 x 4.25 STN STL BAR	2
1	A14534 - 001	YOKE	1.00 x 1.00 x 3.12 STN STL BAR	1

Section 10: *Fasteners and Joining Methods*   **225**

Student _____   Date _____

3. Refer to the gear puller assembly, drawing A14534-000 (see page 224), when responding to the following questions.

a. What is the title of the assembly drawing? ___*Gear Puller Assy*___

b. How many different parts comprise the assembly? ___*8*___

c. How many different types of fastening devices are used to hold the parts together? ___*4*___
Name them. *Retaining ring   Screw*
Set Screw

d. Fill in the blank space for the external thread shown on the screw, .500-20 UNF-___*2*___.
What does UNF mean? *Unified National*

e. Is the retaining ring an internal or external type? ___*external*___

f. What holds the retaining ring in place? ___*yoke*___

g. What type of point does the headless set screw have? ___*flat point*___

h. Complete the thread designation for the internal thread for the end cap, 8-32 UNC-2 ___*8-32*___.
What does UNC mean? *Unified National*

i. What is the thread size for the set screw? ___*10-32 x .31*___
Is it a UNC, UNF, UNEF, or some other thread series? *ISO Metric*

j. What does the -20 signify in the screw thread designation? ___*TPI ?*___

k. What is the part name for item ②? ___*Finger*___

©1995 West Publishing Company

Section 10: *Competency Quiz*

l. What is the part number for item ⑤? A 14534-005
    How many are required on the assembly? 2

m. From what kind of material is the clevis pin made? Steel

n. What size material is required to produce the yoke? Steel Bar

o. By what method are the end caps attached to the handle? press fit

p. What is the diameter of the handle? Ø.250

q. Does the set screw have internal or external threads? external

r. What total number of parts comprise the assembly? 12

s. What is the part number for item ②? A14534-002

t. What type of ends are machined on the clevis pin? flat point

Student _____   Date _____

PART D INTERPRETING JOINING METHODS

1. Identify the following common welding process letter designations:

   a. W _____   b. GMAW _____

   c. OAW _____   d. RW _____

   e. RPW _____

2. Express the meaning of the following weld symbols:

   a. 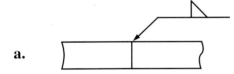 _____

   b. 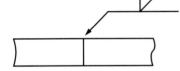 _____

   c. 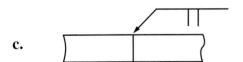 _____

   d. 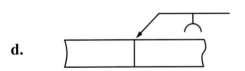 _____

   e. 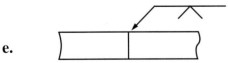 _____

228 Section 10: *Competency Quiz*

f.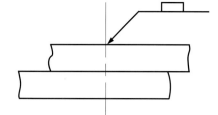

g.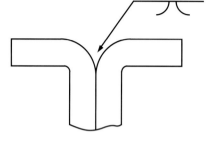

h.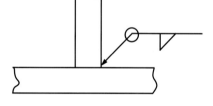

i.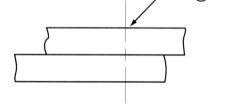

j.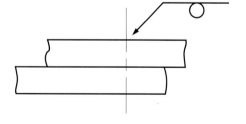

Section 10: *Fasteners and Joining Methods*   **229**

Student _____   Date _____

Competency Quiz continues on next page.

©1995 West Publishing Company

Section 10: Competency Quiz

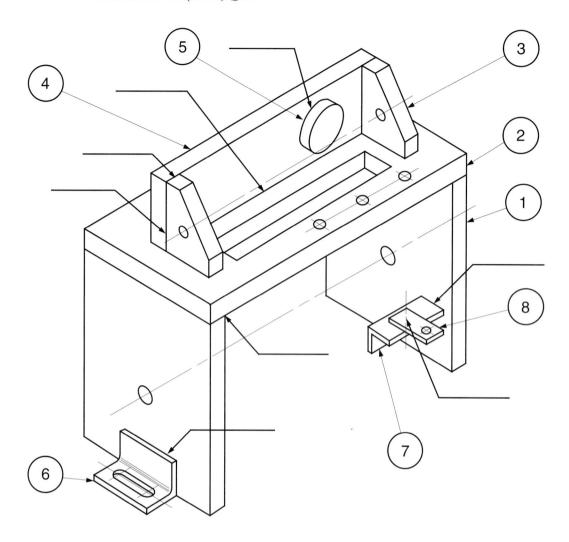

A14618 - 000 ASSEMBLY, FIXTURE

QTY REQD	PART NO.	DESCRIPTION	MATERIAL/SPECIFICATION	ITEM
1	A14618 - 008	STRAP	.12 x .38 x 1.00 STEEL BAR	8
1	A14618 - 007	ANGLE	.18 x .62 x .62 x 1.38 STEEL ANGLE	7
2	A14618 - 006	FOOT	.25 x .75 x .75 x 1.50 STEEL ANGLE	6
1	A14618 - 005	STOP	Ø1.00 x .31 STEEL ROD	5
1	A14618 - 004	SUPPORT	.38 x 1.50 x 5.25 STEEL PLATE	4
2	A14618 - 003	GUSSET	.38 x 1.00 x 1.50 STEEL PLATE	3
1	A14618 - 002	TOP, TABLE	.38 x 3.12 x 6.38 STEEL PLATE	2
2	A14618 - 001	PLATE, SIDE	.38 x 3.12 x 4.00 STEEL PLATE	1

Section 10: *Fasteners and Joining Methods*   **231**

Student _____   Date _____

3. Refer to the assembly fixture, drawing A 14618-000 (see page 230), when responding to the following questions.

   a. What method was used to join the parts together for this assembly?

   b. What is the title of the assembly drawing?

   A14618-000   Fixture Assy.

   c. On the drawing, there are several incomplete weld symbols. Complete them with the appropriate symbol based on the following information:

   (1) Join item ⑤ to item ④ with a fillet weld, arrow side, all around.
   (2) Join item ⑧ to item ⑦ with a plug weld, arrow side.
   (3) Join item ② to item ① with a double bevel weld.
   (4) Join item ④ to item ② with a fillet weld, other side.
   (5) Join item ③ to item ④ at the top with a J-groove weld on the arrow side.
   (6) Join item ③ to item ④ at the vertical joint with a bevel groove, arrow side.
   (7) Join item ⑥ to item ① with a fillet weld, arrow side, all around.
   (8) Join item ⑦ to item ① with a V weld, arrow side, all around.

   d. What is the part number for item ⑤? _____
      How many are required on the assembly?

   e. What is the material and size for the gussets? _____

   f. What is the part name for item ④? _____

   g. What are the total number of parts that make up the assembly? _____

   h. To what part are the side plates joined? _____

   i. What are the part names for the parts made from steel angle? _____

©1995 West Publishing Company

232   **Section 10:** *Competency Quiz*

j. How many holes are on the assembly? _____

k. What is the part name for part number A14618-004? _____

l. How many slots are there? _____

m. Is there any brazing on the drawing? _____

n. Are any of the parts spot welded together? _____

o. What is the material size for the side plates? _____

p. What parts are joined to the table top? _____

q. Name all the parts that have holes or slots in them. _____

SECTION 11

Power Transmission Elements

LEARNER OUTCOMES

You will be able to:

- List four kinds of power transmission elements.
- Identify three types of gears and their use.
- List and define ten technical terms for bevel gears.
- Explain the difference between a worm gear and its pinion.
- Identify and define six technical terms for worm gearing.
- List three types of motion that cams are capable of producing.
- Name two major categories of cams.
- Name three types of belt drives.
- List the two most common types of belts and their application.
- List four applications for roller chain.
- Identify four types of sprockets.
- Demonstrate your learning through the successful completion of competency exercises.

11.1 INTRODUCTION

The mechanical transmission of power is an important aspect of every facet of the industrial community. It touches all parts, components, elements, and details in the movement of objects from one point to another. **Power transmission** refers to the moving of power or energy from the point where it is generated to the point or place where it is utilized. An example of mechanical power transmission—one with which we are all familiar—is the automobile. In an automobile, power is generated in the power plant (called the engine) and transmitted to the drive wheels through mechanical assemblies, such as the transmission, the driveshaft, the differential, and the universal joints. The individual moving parts within the assembled engine, transmission, driveshaft, and differential are considered *power transmission elements*.

Mechanical power is transmitted through elements such as gears, cams, belts, chains, sheaves, sprockets, valves, clutches, hoses, brakes, and shafts. The elements covered in this section are gears, cams, belts, and chain drives.

11.2 GEARS

A **gear** is a mechanical element that transmits motion through the continuous engagement of teeth at a constant angular velocity. A set of gears transmits positive (nonslip) power from one shaft to another or through a rack of teeth. This type of motion is often used where power needs to be transmitted efficiently; examples include the increasing of torque and the increasing or decreasing of speed. When two or more gears are meshed, they form a **gear train**. When gears in mesh are of different diameters, the smaller of the two is called a **pinion**, which is illustrated in Figure 11.1. Note that when two gears are in the proper arrangement, they will rotate in opposite directions.

FIGURE 11.1
Two gears in mesh

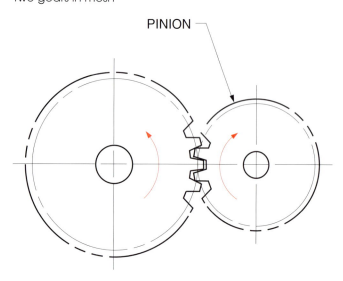

FIGURE 11.2
Various types of gears

Photos courtesy of Boston Gear

There are many types of gears, and they are identified by the method by which each connects to a shaft. A **spur gear** connects shafts that are parallel. A **worm gear** connects shafts that do not intersect, while a **bevel gear** connects shafts that do intersect. Other types of gears include the helical, hypoid, internal, and herringbone gears. Examples of each are given in Figure 11.2.

■ 11.2.1 The Spur Gear

The spur gear is probably the most commonly used of all the various types. It is cylindrical in shape, with straight teeth cut parallel to the axis of the shaft. Its primary use is to transmit power between parallel shafts only. The gear tooth shape most often used is the involute. This shape is formed by an **involute curve**, which is defined as the contour traced by a point on a taut string as it unwinds from a fixed cylinder or circle. Figure 11.3 visually describes the involute curve.

11.2.1.1 Spur Gear Nomenclature

Although this particular section is devoted to spur gears, the terminology that follows can be extended to other types of gears. There are in existence more than fifty definitions of gear-related terms. Only the major ones will be covered here. These terms are illustrated in Figure 11.4.

FIGURE 11.3
The involute curve

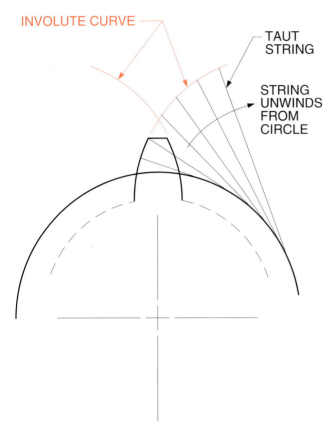

- *Outside diameter*—Also referred to as the addendum circle. It is the circle that coincides with the extreme outer edges of the gear teeth.
- *Root diameter*—The circle that coincides with the bottom of the teeth.
- *Tooth space*—The distance between adjacent teeth at the pitch circle.
- *Tooth thickness*—The thickness of a single tooth as it is measured along the pitch circle.
- *Addendum*—The radial distance between the pitch circle and the outside diameter.
- *Dedendum*—The radial distance between the pitch circle and the root diameter.
- *Circular pitch*—The distance measured along the pitch circle from one point on one tooth to the corresponding point on an adjacent tooth.
- *Clearance*—The amount of space between the top of one tooth and the bottom of the mating tooth when the gears are in mesh. Also, the amount that the dedendum exceeds the addendum.

Section 11: *Power Transmission Elements* **235**

- *Pitch circle*—Also called the pitch diameter. It is an imaginary circle that runs through the midpoint of all teeth. When gears are in mesh, the pitch circles of the gears meet.
- *Center distance*—The distance between the parallel axes of spur gears. Also, the sum of the radii of the two mating gears.
- *N*—The total number of teeth on one gear.
- *Pressure angle*—The angle between the contacting teeth and the line tangent to the pitch circle. Standard pressure angles are 14.5 degrees, 20 degrees, and 25 degrees.

11.2.1.2 Gear Production

Gears can be made through various machining and casting processes. Casting methods include sandcasting, die casting, investment casting, permanent-mold casting, or centrifugal casting. Machining processes include (1) gear **hobbing**, which is performed on a type of milling machine designed for both horizontal and vertical spindles; (2) gear **shaping**, which is a machine capable of making internal, external, spur, and helical gears; and (3) **broaching**. Gears are made from many different metals, including polymers. Smaller gears made of copper alloys, aluminum, zinc, and tin are capable of being produced to high quality and accuracy.

In the production process, the gear arrives at the machine in blank form, complete except that it lacks teeth. During the machining of the teeth, the outside diameter of the gear blank needs to be accurate, and gear tooth dimensions are held to very tight tolerances to ensure that the intended use's requirements are met. After the cutting process is complete, the gear is normally hardened, and the newly formed gear then may be ground, lapped, or burnished for extreme accuracy, or for the desired surface finish.

11.2.1.3 Engineering Drawings for Spur Gears

A few points need to be made relative to the interpretation of engineering drawings for gears. First, the gear teeth are not shown on a drawing. Although they can be drawn very quickly and accurately with a CAD system, it really is not necessary and would be a waste of time to draw them. Instead, the use of phantom lines is intended to portray that the gear

236 Section 11: *Power Transmission Elements*

FIGURE 11.4
Involute spur gear nomenclature

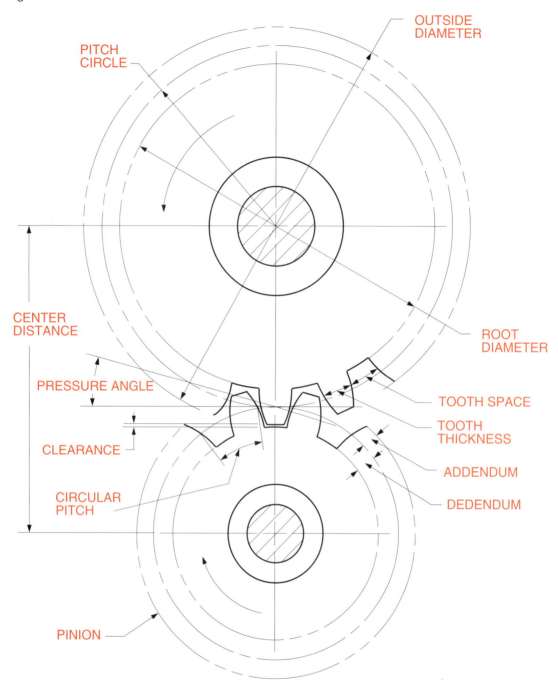

teeth are continuous around the outer periphery. Second, the gear drawing actually shows only the gear blank. Because the machining of the blank and the actual cutting of the teeth are separate operations in the factory, gear blank dimensions are shown in orthographic views, usually including a section view, and the gear cutting data is provided in a separate table, as visually described for a spur gear in Figure 11.5.

■ 11.2.2 The Bevel Gear

A **bevel gear**, when in mesh with another gear, is like a rotating cone that meets the other gear at some defined point called the *apex*. This type of gear connects shafts at an angle to each other, usually at 90 degrees, where they intersect. The point of intersection of the two shafts forms the apex of both pitch cones. Two meshed bevel gears of the same size that

FIGURE 11.5
Engineering drawing of a spur gear

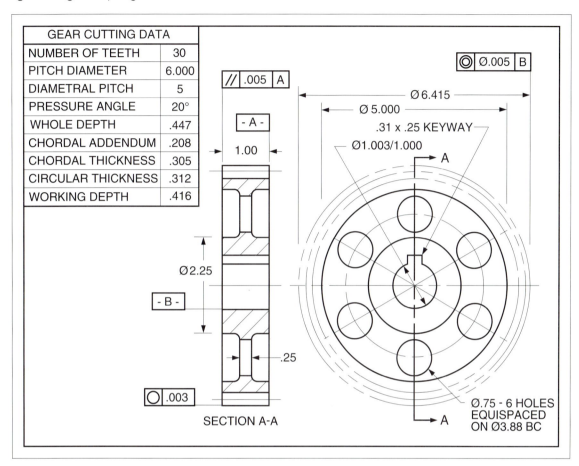

connect shafts at a 90-degree angle are referred to as **miter gears**. In this arrangement, the tooth section becomes smaller as the apex of the pitch cone is approached. As Figure 11.6 illustrates, only a small length of cone, located at the base of the gear, has teeth, usually not more than 30 percent of the full length of the cone. Note, too, that measurements for the pitch diameter, addendum, and dedendum are taken at the large diameter of the bevel gear.

Types of bevel gears include the *straight tooth*, which is shown in Figure 11.6; the **spiral bevel**, which has curved teeth on which contact begins gradually and continues smoothly from end to end; the **zero bevel**, a curved gear but whose teeth lie in the same general direction as teeth of straight bevel gears; and the **hypoid bevel**, one that resembles the spiral bevel gear except that the axis of the pinion is offset relative to the gear axis.

11.2.2.1 Bevel Gear Nomenclature

The design features and nomenclature for bevel gears are generally similar to those shown in Section 11.2.1.1 for spur gears. Some important additional terms include the following, which are visually described in Figure 11.6:

- *Addendum angle*—An angle formed by the sides of the face angle and the pitch angle.
- *Apex*—The point where both pitch cones of two bevel gears meet. The point of two intersecting shafts.
- *Dedendum angle*—An angle formed by the sides of the pitch angle and the root angle.
- *Face angle*—The angle formed by the gear mounting centerline and the outside of the gear teeth.

FIGURE 11.6
Right-angle, straight tooth bevel gears

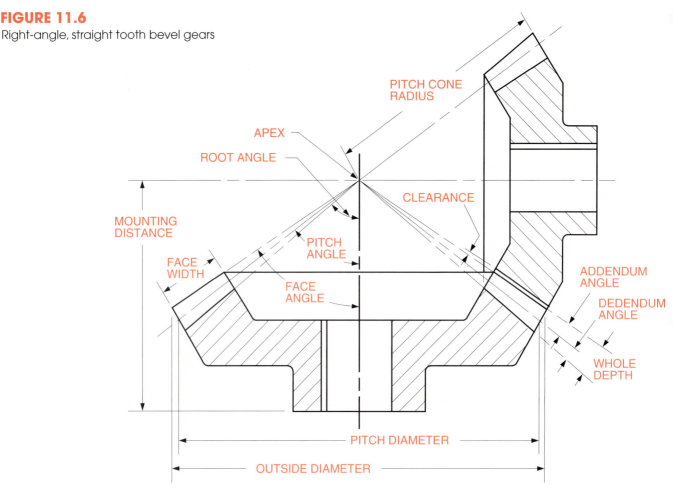

- *Face width*—The width of a tooth, approximately one-third of the cone distance.
- *Mounting distance*—The distance from the apex to the outside of the hub for one or a set of bevel gears.
- *Pitch angle*—The angle between the gear mounting centerline and the centerline of the teeth.
- *Pitch cone radius*—The radial distance from the apex to the outside of the teeth.
- *Root angle*—The angular distance from the center of the hub to the root of the teeth.
- *Whole depth*—The sum of the addendum and dedendum.

11.2.2.2 Engineering Drawings for a Bevel Gear

A bevel gear or a pair of bevel gears usually appears as a single, full-section view on an engineering drawing. The gear teeth are not normally shown on the detail drawing but are often portrayed on the assembly drawing. As is true with the spur gear, the section view includes bevel gear blank dimensions as well as an accompanying chart or table identifying key gear tooth data. A representative gear cutting data chart is shown in Table 11.1.

TABLE 11.1
Bevel gear cutting data chart

BEVEL GEAR CUTTING DATA	
Number of teeth	27
Tooth form	14.5°
Diametral pitch	3
Root angle	52°14'
Whole depth	.719
Chordal addendum	.341
Chordal thickness	.523
Addendum	.333

11.2.3 The Worm and Worm Gear

The **worm** and **worm gear** are the type of gears that lend themselves to gear reduction assemblies and are capable of transmitting power efficiently. They work in tandem and operate at right angles to each other. A large speed reduction is possible with worm gearing because a single-thread worm, in one revolution, advances the worm wheel only one tooth and one space. A worm is considered a special form of a crossed helical gear, and it resembles a screw thread; in fact, it is sometimes called *threads*. In worm gear terminology, the worm is referred to as the pinion, and the worm gear as the *worm wheel*. Figure 11.7 shows the engagement of a worm and a worm wheel whose section view is taken through the center of the worm, perpendicular to the worm wheel. This arrangement illustrates that the worm is identical to a rack of gear teeth and that the worm wheel section is identical to a spur gear. Therefore, the tooth shape, addendum, and dedendum will be the same as for a spur gear.

11.2.3.1 Worm Gearing Nomenclature

In addition to the nomenclature already discussed for the spur gear, the following additions are required when dealing with the worm and worm wheel. They are defined as follows and illustrated in Figure 11.7.

FIGURE 11.7
Common worm and worm wheel nomenclature

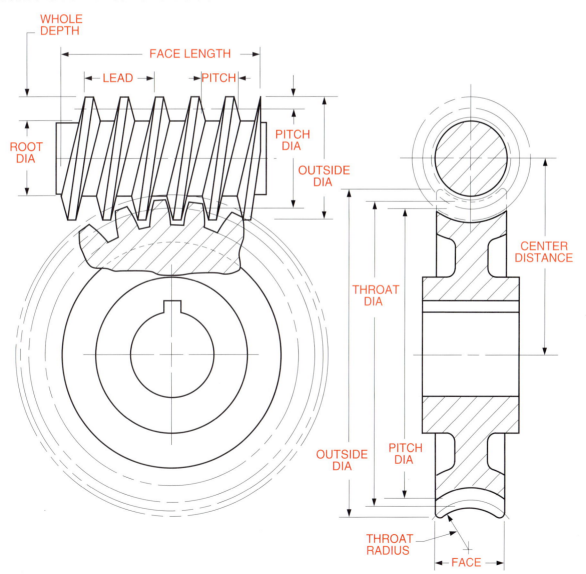

For the worm:

- *Lead of worm threads*—The distance a screw thread advances axially in one turn. For a double-threaded worm, the lead is twice the pitch.
- *Pitch*—The axial distance from a point on a thread to the same point on the next thread.

For the worm wheel:

- *Throat diameter*—A diametral measurement of the gear, measured at the bottom of the tooth arc; the pitch diameter plus twice the addendum.
- *Throat radius*—One-half the pitch diameter of the worm minus the addendum.

11.2.3.2 Engineering Drawings for Worm Gearing

Detail engineering drawings for the pinion and the worm gear appear as separate drawings, as shown in Figures 11.8 and 11.9. Dimensioning practices are dictated by the method of production, and it is common practice to place dimensions on gear blanks and to provide charts or tables on the drawing for gear cutting data.

11.3 CAMS

Another important power transmission element located within an automobile as well as in hundreds of other types of mechanisms is the **cam**. A cam is a mechanical device that provides a rotating, oscillating, or reciprocating motion to a **follower** by means of direct contact. A follower is so named because it literally follows the exterior or interior curved form of the cam. During one revolution of a cam, the follower will rise and fall over a specific time frame, depending on the contour shape of the cam. The shape of a cam is dependent on the motion of the follower.

FIGURE 11.8
Engineering drawing for a pinion

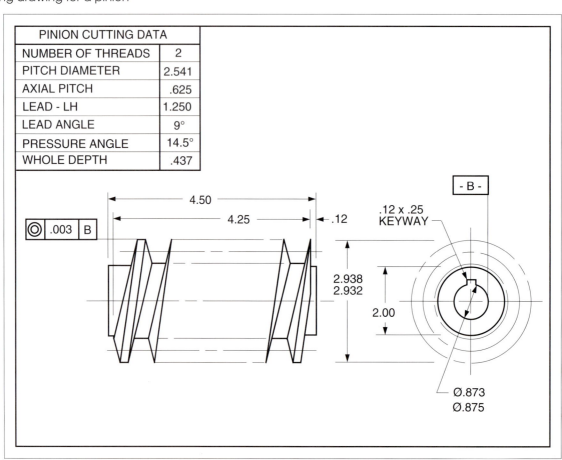

Section 11: *Power Transmission Elements* **241**

FIGURE 11.9
Engineering drawing for a worm gear

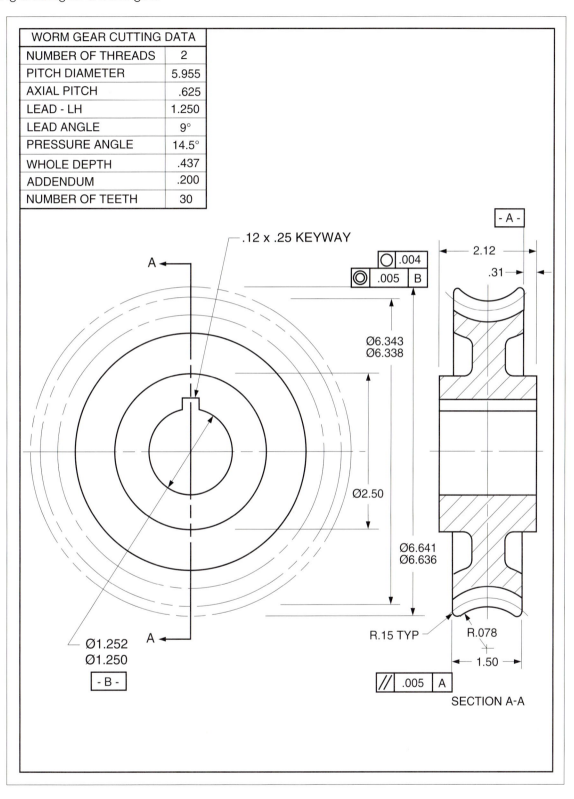

242 Section 11: *Power Transmission Elements*

A cam is actually the result of the movement of its follower. For the most part, cam mechanisms are more desirable than four-bar linkages for engineering design purposes. A simple example of a typical cam and follower arrangement is pictured in Figure 11.10.

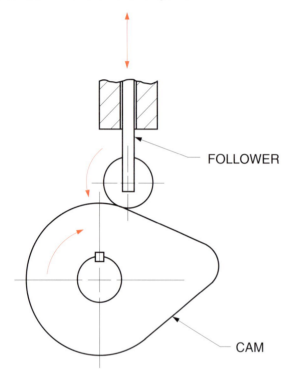

FIGURE 11.10
Typical cam and follower arrangement

■ 11.3.1 Types of Cams

Cams may be classified into two major categories: radial and cylindrical types. The category is determined by the type of follower the cam uses. The **radial-type cam** is also referred to as a **plate cam** or **disk cam** and is characterized by a flat shape along its periphery. The **cylindrical-type cam** is called a **drum cam** because the track for the cam is a machined area around the circumference of the drum. In this type, the line of action is parallel to the axis of the cam. Examples of the radial type, which are the most frequently used type in industry, are illustrated in Figure 11.11. They are shown with various kinds of followers.

In Figure 11.11a, the **roller follower** provides a rolling action between cam and follower and is ideal for reducing friction in motion, while in Figure 11.11b, the **flat** or **spherical-face follower** is used for relatively steep cam applications, such as those required in automobile valve lifters. A **point follower**, shown in Figure 11.11c, is limited to areas that are in need of low speed and low force because of extreme wear. This type is very sensitive to abrupt changes in follower motion. Figure 11.11d shows an **offset follower**, with which the line of action is offset from the camshaft center. A **swinging follower** can operate with a large steepness angle and is illustrated in Figure 11.11e. The **yoke cam** in Figure 11.11f is a positive-acting cam that does not require the use of a follower spring.

When the condition warrants for the follower to move in a plane parallel to the cam shaft, some form of cylindrical cam is used. In Figure 11.12, the follower rod moves vertically, parallel to the axis of the cam, with the attached roller following the contour of the groove in both examples provided.

Figure 11.12a represents a cylindrical cam with a **translating roller follower**, and Figure 11.12b shows a cylindrical drum cam with an exterior groove seat and an **oscillating roller follower**.

■ 11.3.2 Cam Layout and Nomenclature

To understand a cam fully, one needs to examine its construction. Figure 11.13 illustrates the construction of a cam whose *base circle* is held stationary and whose roller is shown around the circle at predetermined radial locations. The various positions of the roller form an **envelope** referred to as the *cam surface*. It is this exterior envelope on which the cam follower rotates. Because the cam is an eccentric element, its center is off center. For this reason, the roller follower will vary in distance from the cam center as it follows the periphery of the cam surface during one *revolution*. This is referred to as the *rise* and *fall* of the cam surface. Note, too, that the cam has a *dwell* (an at-rest) area. This occurs at an area on the cam surface where the roller neither rises nor falls. The terms *rise*, *fall*, *dwell*, and *revolution* are illustrated on the *displacement diagram* in Figure 11.13.

The displacement diagram's purpose is to show, in layout form, the exterior envelope, or contour, of the

Section 11: *Power Transmission Elements* **243**

FIGURE 11.11
Radial-type cams with followers

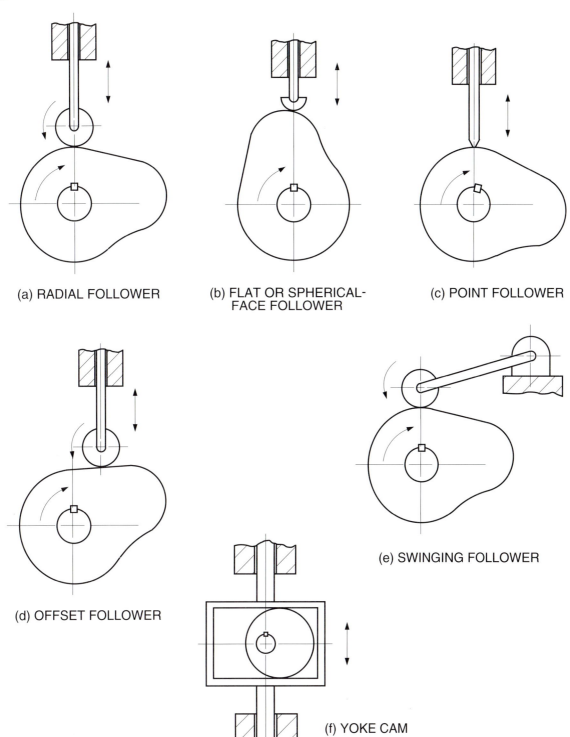

FIGURE 11.12
Cylindrical type cams

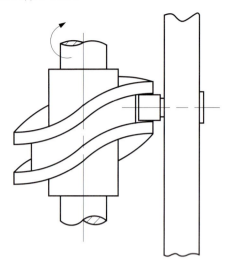

(a) TRANSLATING ROLLER FOLLOWER

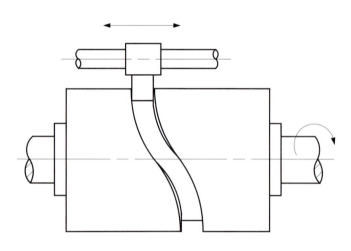

(b) OSCILLATING ROLLER FOLLOWER

cam. It is usually rectangular in shape and shows follower motion related to cam angle or time. The vertical axis refers to follower position or travel, and the horizontal axis indicates the angular position of the cam. The diagram is divided into 30-degree increments (twelve positions) for one entire revolution (360 degrees).

- *Trace point*—The center of the roller follower.
- *Pitch curve*—The path of the trace point. For a radial point follower, it coincides with the cam surface.
- *Rise*—A position on the cam when the follower is moving upward.
- *Fall*—A position on the cam when the follower is moving downward.
- *Dwell*—A position on the cam when the follower is considered to be at rest, when there is no rise or fall to the follower.
- *Displacement diagram*—A diagram that illustrates, in graphical form, the contour of a cam and its follower motion as it relates to cam angle or time.
- *Base circle*—The smallest circle drawn on the cam profile, which emanates from the center of rotation. The base circle determines cam size.
- *Prime circle*—The smallest circle that can be drawn to the pitch curve from the center of rotation.
- *Pitch point*—The position on the pitch curve in which the pressure angle is the greatest.
- *Pressure angle*—The angle between the normal line to the pitch curve and the direction of the follower. For a roller follower, the normal line passes through the roller center and the contact point at the cam surface.
- *Cam surface*—The working curve or contour of the cam profile; the area that the roller follower traces.

11.4 BELTS

Belts (or belting, as they are often called) are an element of a drive system that also includes motors, **sheaves**, and **pulleys**. Belt drives offer flexible, smooth operation and power transmission at a relatively low cost. They are also shock absorbing and resistant to an abrasive atmosphere. Belt drives depend on friction between the belt and pulley surfaces for the transmission of power. Leather, rubberized cord, fabric, or other reinforced materials, such as rayon, nylon, glass, and steel fiber, are materials in which belts are available. Drives of this type, when compared to gear drives and chain drives, offer certain advantages and disadvantages. For example, a belt will slip when an overload is applied, but slippage, while not desirable, is preferable to complete failure, which could occur when gear teeth or a chain break. Belts are subject to creep or slippage, so some

Section 11: *Power Transmission Elements* **245**

FIGURE 11.13
Cam construction and associated nomenclature

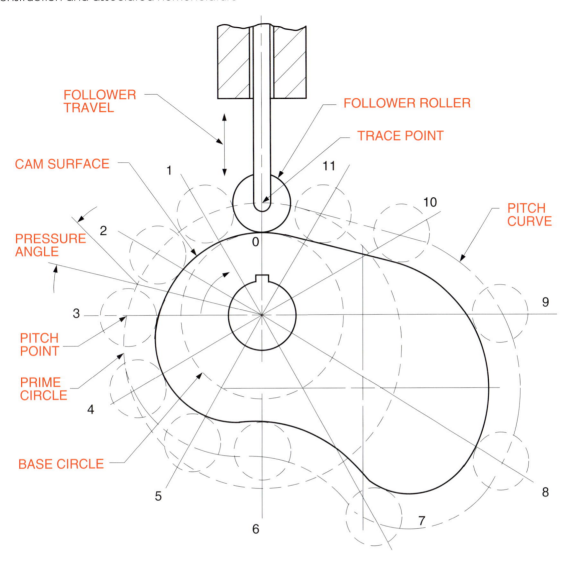

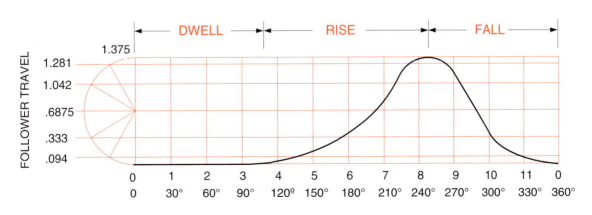

DISPLACEMENT DIAGRAM (ONE REVOLUTION)

246 Section 11: *Power Transmission Elements*

FIGURE 11.14
Types of belt drives

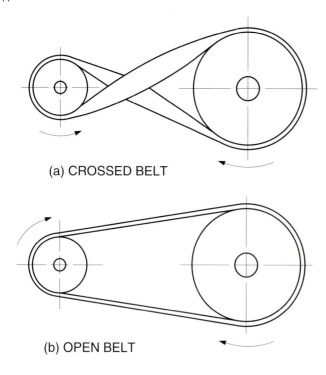

(a) CROSSED BELT

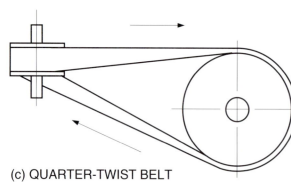

(b) OPEN BELT

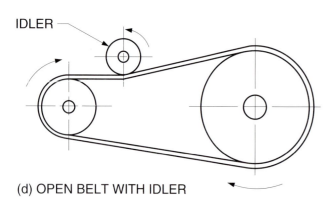

(c) QUARTER-TWIST BELT

(d) OPEN BELT WITH IDLER

form of adjustment needs to occur to ensure that they remain tight during operation. The advantages of belts include low cost, ease of maintenance, and shock-absorbing qualities.

■ 11.4.1 Types of Belt Drives

Basic types of belt drives that have been used for several years are illustrated in Figure 11.14. Because belt drives are similar in nature, the illustrations shown are of belt drives used with flat belts. However, these drives may be used in all belt applications. Included in Figure 11.14 are (a) the **crossed belt**, (b) the **open belt**, and (c), the **quarter-twist belt** drives. An additional illustration identifies the open-belt type with an adjustable **idler** (d) to ensure tension for the drive system to eliminate or to minimize slippage.

■ 11.4.2 Types of Belts

The conventional **flat belt** has been used for more than one hundred years. It is available in an endless type, which means that it has no seam or joint to hinder performance. In addition, flat belts are available in a spliced joint variety in whatever length is required for the application. The joint can either be a vulcanized joint or one in which a mechanical fastener is used. Flat belts are used where high speed rather than power is the primary requirement. This type of belt does not possess the grip strength that is consistent with a V-belt. It has a fairly low efficiency rating at moderate speeds and is generally noisier than other types of belt drives.

When interpreting a drawing for a flat belt, look for the following information which normally appears on a drawing: (1) the total length of the belt around the pulley ends; (2) the type of material, including width and thickness; (3) whether the belt has a seam or is endless; and (4) the type of end, if spliced, and how the splice was accomplished (laced, mechanical fastener, or vulcanized). Flat belts are rectangular in cross section and available in common thicknesses as thin as .06 inch and as wide as 12 inches. Figure 11.15 shows a section of a flat belt.

■ 11.4.3 Flat Belt Pulleys

A **pulley** is a wheel over which a flat belt rides in the transmission of power. A minimum of two pulleys are

FIGURE 11.15
Flat belt section

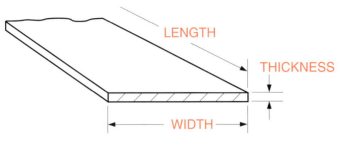

required in a flat belt drive mechanism. When two pulleys are used, one is called the **driver pulley** and the other the **driven pulley**. Obviously, the driver pulley is the one that is attached to the source of power (motor). At times, especially when tightness or tension is critical, an idler pulley is needed, as shown in Figure 11.14d.

Materials for flat belt pulleys include cast iron, fabricated steel, and other materials to a lesser degree. Such pulleys may possess solid, split, or spoked hubs. Usually, the width of a pulley is the same as or slightly larger than the belt width. Pulleys have a crowned rim section on which the belt rides to allow the belt to seat or to center itself during operation. An example of a standard 6-inch flat belt spoked pulley with associated terminology is depicted in Figure 11.16. The terms are defined as follows:

- *Hub*—The material around a shaft hole, for mounting the pulley.
- *Rim*—The outside diameter of the pulley minus twice the crown taper.
- *Spoke (arm)*—The area of support between the hub and rim.

FIGURE 11.16
Standard flat belt spoked pulley

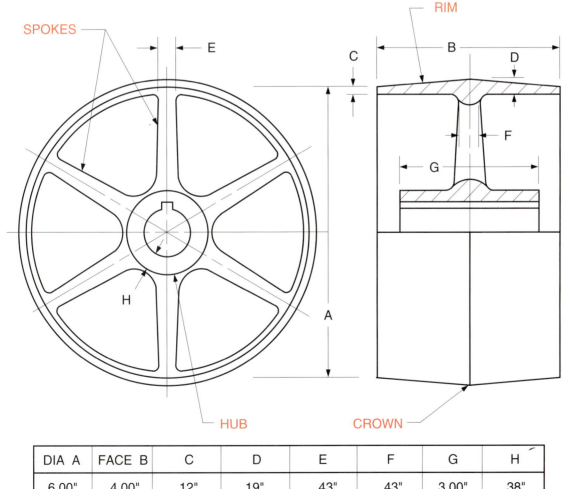

DIA A	FACE B	C	D	E	F	G	H
6.00"	4.00"	.12"	.19"	.43"	.43"	3.00"	.38"

Section 11: *Power Transmission Elements*

■ 11.4.4 V-Belts

Drives that possess a **V-belt** as a power transmission element are used extensively in single and multiple form in automotive, industrial, agricultural, commercial, and consumer product applications. V-belts are used with a wide range of motor horsepowers, from fractional to hundreds of horsepower, and they operate most effectively at speeds ranging from 1,500 to 6,500 feet per minute (fpm). The design of the V-belt, which includes a tapered cross-sectional area, allows it to seat itself firmly into a sheave groove during operation. The driving action is accomplished through the sheave in contact with the sides of the belt, not with the bottom of the belt. Figure 11.17 shows how the V-belt rides in the sheave groove when in motion.

As is true with other types of drive belts, the V-belt has its limitations. The advantage of this type of belt is that it is very versatile, compact, quiet in operation, easily installed and removed, and low in maintenance. The only disadvantage is that one can expect the belt to slip, creep, or both during operation.

11.4.4.1 V-Belt Sizes

V-belt sizes have been standardized throughout the industrial community and are specified by a code that indicates the cross-sectional shape and length. The industrial type designations are classified as *conventional* (A, B, C, D, and E), *narrow* (V), and *light duty* (L) service. Nominal width and thickness dimensions for industrial V-belts are visually described in Figure 11.18.

11.4.4.2 V-Belt Sheaves

A **sheave** is a pulley with grooves, from one groove to several, to accommodate a single V-belt or multiple V-belts simultaneously, depending on the requirement. Most sheaves are cast from iron, which provides stable dimensions and an expected long groove life. Other materials used include formed steel, aluminum, and plastic for lighter applications. Specific sheave materials can be used for speeds up to 10,000 feet per minute.

Sheaves are available with either regular or deep grooves; the deep-groove application is for an angle drive, quarter-twist drive, or vertical shaft drive, or for wherever serious vibration situations occur. The solid hub design is common for sheaves. Figure 11.19a identifies a half-sectional view of a formed steel single-groove sheave with a welded hub, while 11.19b shows a multiple-groove cast iron sheave.

FIGURE 11.17
V-belt and sheave

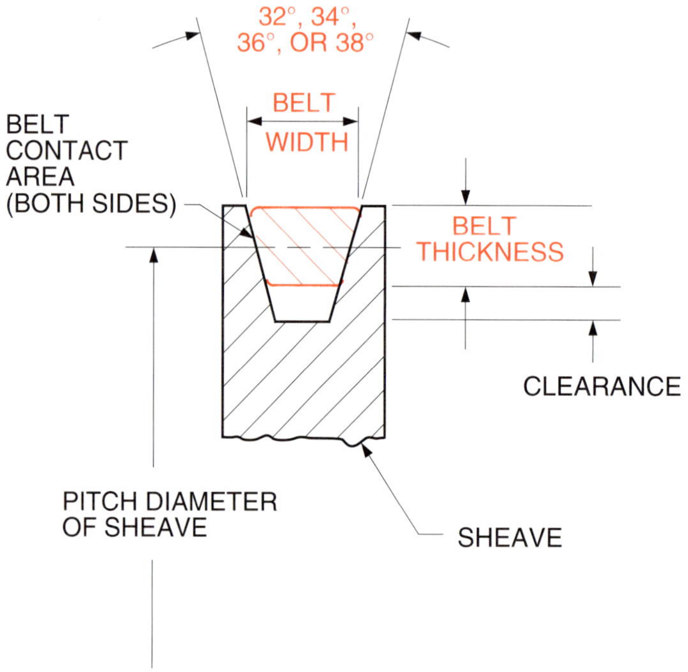

Section 11: *Power Transmission Elements*   249

FIGURE 11.18
Standard industrial V-belt cross sections

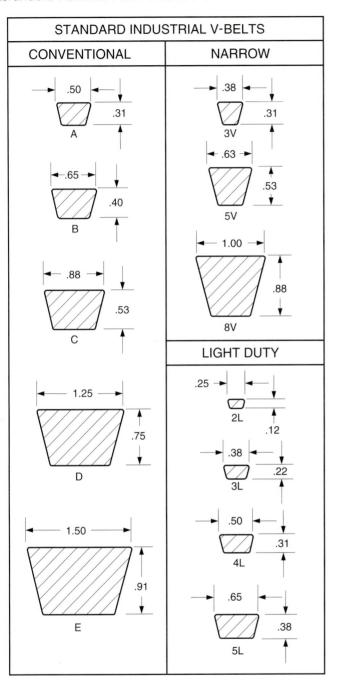

FIGURE 11.19
V-belt sheaves

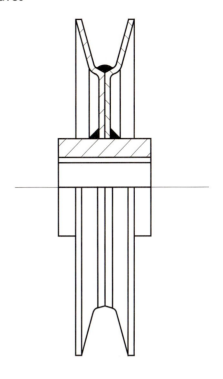

(a) FORMED STEEL GROOVE

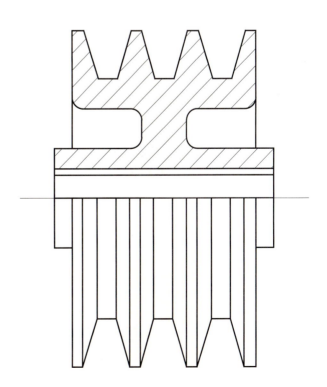

(b) CAST IRON, MULTIPLE GROOVE

11.5 CHAIN

Chain and chain drives combine the versatility of belts with the positive response of gear drives. A chain drive consists of an endless linked chain that mates (meshes) with toothed wheels. The toothed wheel is called a **sprocket**. The teeth of a sprocket are much like those of a gear. There are driver sprockets, driven sprockets, and idler sprockets. The drives are much like belt drives, in that they can operate individually or in multiples. This type of drive provides a positive speed ratio between the driver and the driven sprockets, so that tension on the slack side of the drive is nonexistent. In addition, chain is flexible. It is capable of wrapping around a sprocket from either side, tracking easily around small sprockets because of chordal action, and operating in less-than-clean factory conditions. Chain is compact, and it can be assembled and disassembled easily. Figure 11.20 shows a typical chain arrangement with sprockets and chain links.

the sprocket. Standard roller chains are produced to specifications provided by ANSI B 29.1-1975. This type of chain is commonly used for the transmission of power in industrial machinery, motorcycles, tractors, motortrucks, and in similar applications.

A **roller chain** consists of two kinds of **links**: alternately spaced *roller links* and *pin links* throughout the length of the chain. A roller link is made up of two end plates (which are flat metal stampings), two rollers, and two bushings. Each roller link consists of two end plates attached by two pins. Assembly or disassembly of roller chain is simplified by means of a removable plate secured with a cotter pin or spring clip. Figure 11.21 shows a roller chain assembly.

FIGURE 11.20
Typical chain and sprocket arrangement

FIGURE 11.21
Roller chain assembly

Emerson Power Transmission Corp.

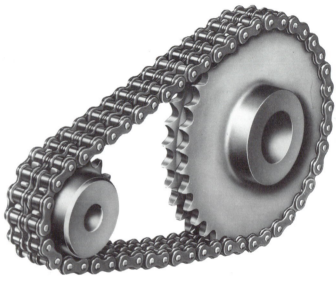

Emerson Power Transmission Corp.

■ 11.5.1 Types of Chain

There are several types of chain in use today, including the *detachable chain*, *pintle chain*, *offset chain*, *bead chain*, *roller chain*, and *silent chain*. The most common roller and silent types will be covered in this section.

Roller chains are produced in several types depending on the kind of service required. They are constructed so that the rollers are spaced equally throughout the chain's length. The advantage of this is that the rollers rotate upon contacting the teeth of

Standard roller chain has four principal dimensions: pitch, chain width, roller diameter, and plate thickness. Chain manufacturers produce catalogs so that the selection process for meeting design requirements is made easy. Roller chain nomenclature is visually illustrated in Figure 11.22.

A **silent chain** is also referred to as an **inverted tooth chain** and includes a series of toothed links alternately assembled with pins. In this type of drive, the chain passes over the face of the sprocket, like a belt. The sprocket teeth do not project through. The chain maintains contact with the sprocket without sliding. As a result, the chain runs quieter than a con-

FIGURE 11.22
Nomenclature for roller chain components

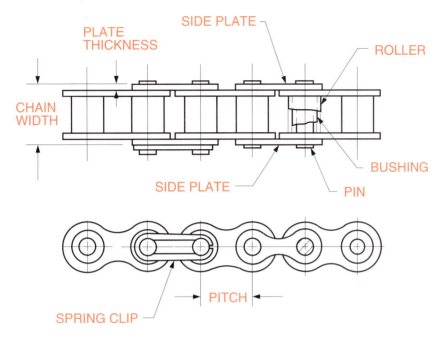

ventional roller chain. Chains of this type may be found in applications that require heavy-duty use, such as in a power take-off. Ratings are provided on the basis of horsepower per inch of chain width for the various pitches available. An example of a silent chain drive arrangement is shown in Figure 11.23.

There are only two components to a silent chain: a chain link and a pin. Figure 11.24 shows how they are assembled and how they mesh with a section of sprocket.

■ 11.5.2 The Sprocket

An important part of any chain drive assembly is the sprocket. Minimally, a chain drive requires a sprocket at the driving shaft and another at the driven shaft. There are four types of sprockets in use today: the plane plate sprocket without hub, with a hub on one side, with a hub on both sides, and with a split hub. Sprocket plates may be solid or with arms or webs. Materials from which sprockets are made include cast iron, steel, and stainless steel. Because sprocket teeth are standardized to accommodate chain, charts and tables are available from manufacturers for sprocket selection by the designer because there are many from which to choose, depending on pitch and power requirements. Examples of sprockets are shown in

FIGURE 11.23
Silent chain drive assembly

Emerson Power Transmission Corp.

252 Section 11: *Power Transmission Elements*

FIGURE 11.24
Silent chain

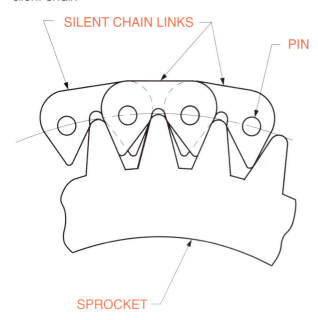

FIGURE 11.25
Types of sprockets

Emerson Power Transmission Corp.

FIGURE 11.26
Sprocket nomenclature

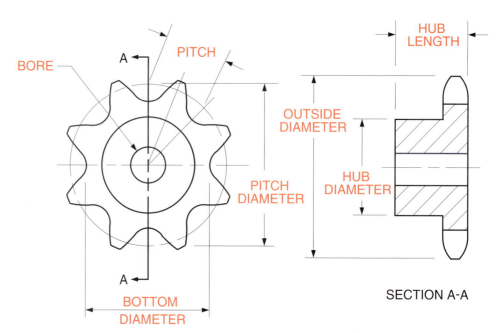

Figure 11.25, and useful terms which are defined below, are visually identified in Figure 11.26.

- *Bore*—The diameter of the sprocket hole that receives the shaft.
- *Bottom diameter*—The root diameter of the sprocket teeth.
- *Hub length*—The mounting length of the hub.
- *Outside diameter*—The diameter that extends over the teeth.
- *Pitch*—The distance from the center of one tooth to the corresponding point on an adjacent tooth.
- *Pitch diameter*—An imaginary circle that runs through the center of all teeth.

Section 11: *Power Transmission Elements* **253**

11.6 SUMMARY

Gears, cams, belts, and chain are considered elements or components of mechanical power transmission. *Power transmission* refers to the moving of power or energy from the point where it is generated to the point where it is utilized. A gear is a mechanical element that transmits motion through the continuous engagement with teeth. The spur gear, which is probably the most commonly used of all gears, and its associated nomenclature were covered in depth. The involute curve and the standard pressure angle for its teeth were described.

It was stated that bevel gears in mesh are very much like two rotating cones that meet at some defined point called the apex. Four types of bevel gears were detailed, as was their nomenclature.

The worm and worm gear are the type of gears that lend themselves to gear reduction assemblies and are capable of transmitting power efficiently. They work in tandem with each other and operate at right angles to each other. Their nomenclature, too, was described.

A cam was described as being a mechanical device that provides a rotating, oscillating, or reciprocating motion to a follower by means of direct contact. During one revolution of a cam, the follower will rise and fall over a specific time frame depending on the contour shape of the cam. The cam shape is dependent on the motion required of the follower. A cam shape is actually the result of the movement of a follower. Two major categories of cams include the radial and cylindrical types. Cam layout, construction, and associated nomenclature were outlined in the section.

Belts provide a useful and relatively inexpensive way of transmitting power. Because of their design, however, exact velocity ratios are difficult to maintain. Belts will slip before failure occurs, while with other methods failure occurs suddenly, resulting in costly repairs. Basic types of belts and belt drives were discussed, including the crossed belt, open belt, and the quarter-twist belt drives. Sheaves and pulleys are also important to belt drive systems. Standard-size pulleys and sheaves are available through manufacturers' handbooks and catalogs as an aid in their selection for a specific application. Standard-size and standard-length V-belts are also available from stock. To simplify the selection process, manufacturers have developed charts and tables identifying ratios, diameters of sheaves and pulleys, and center distances, thus eliminating belt-length calculations.

Chains and chain drives are also members of the power transmission family. Positive velocity ratios are maintained by using the appropriate sprockets. Sprockets are toothed wheels much like gears. Two types of chain were discussed; the roller chain and the silent chain. Roller chain applications include the transmission of power for industrial machinery, motorcycles, tractors, motortrucks, and similar equipment, while the silent chain, because of its characteristics, may be found in a power take-off and in heavy-duty applications. Typical chain and sprocket arrangements were shown, and nomenclature was defined.

TECHNICAL TERMS FOR STUDY

belt   A power transmission element made of leather, rubberized cord, fabric, or reinforced materials, such as rayon, nylon, glass, and steel fiber. It mounts between wheels, pulleys, sheaves, or some combination of these, which are considered drive or driven elements.

bevel gear   A gear in the shape of a cone used to connect shafts that have intersecting axes.

broaching   The process of removing unwanted metal by pushing or pulling a tool that has cutting teeth; a method of cutting gear teeth.

cam   An eccentric mechanical device capable of providing a rotating, oscillating, or a reciprocating motion through direct contact.

chain   A power transmission element consisting of endless linked parts that mesh with toothed wheels called sprockets.

crossed-belt drive   A belt drive in which the belt crisscrosses between the parallel drive and driven pulleys.

cylindrical cam   A cam in the shape of a cylinder, with which the follower moves in a direction parallel to the cam shaft.

disk cam   Also called a plate cam. It is characterized by a flat shape along its periphery.

Section 11: *Power Transmission Elements*

driven pulley   The pulley that receives its turning motion from the drive pulley and its associated belt, gear teeth, chain links, or sprocket teeth.

driver pulley   The pulley that is directly attached to the power source.

drum cam   Also referred to as a cylindrical cam because of its shape.

envelope   The cam surface; the periphery of a cam.

flat belt   A belt that has a rectangular cross section, used where high speed is required. It does not possess a great amount of grip strength.

flat follower   A type of cam follower used for relatively steep cam applications, such as in automobile valve lifters.

follower   A mechanical device that follows the interior or exterior contour form of a cam.

gear   A mechanical element that transmits motion through the continuous engagement with teeth, at a constant velocity.

gear train   A power transmission system that consists of only gears, usually several of them.

hobbing   A method of using a hobbing machine for cutting gear teeth.

hypoid bevel gear   A gear originally developed for automotive use. It is similar to a spiral bevel gear except that its centerline is offset.

idler   A device used in gear trains and in belt drives. It is used in a gear train to change the direction of the last gear or to connect input or output shafts. In belt drives, it is used to take up the slack in the drive system.

inverted tooth chain   *See* **silent chain**.

involute curve   The shape of a gear tooth. It is defined as the curve traced by a point on a taut string as it unwinds from a fixed cylinder or circle.

link   An element of a chain. Two kinds of links are needed for a chain; a roller link and a pin link, which connect to form a chain.

miter gears   Two meshed gears of the same size connected with shafts at a 90-degree angle.

offset follower   A cam follower whose line of center is offset from the camshaft center.

open belt drive   A drive with a drive pulley and a driven pulley that are connected by a belt.

oscillating roller follower   A cam follower whose motion is oscillating (moving back and forth) over one full revolution of a cam.

pin   A component of a chain link assembly. It joins chain links together.

pinion   The smaller of two gears in mesh.

plate cam   *See* **disk cam**.

point follower   A type of cam follower whose use is limited to areas that are in need of slow speed and low force.

power transmission   The moving of power or energy from the point where it is generated to the point or place where it is utilized.

pulley   A wheel over which a flat belt is used in the transmission of power.

quarter-twist belt drive   A drive whose drive and driven pulleys are at a 90-degree angle from each other, which creates a twist to the drive assembly.

radial cam   Referred to as a plate cam or disk cam; a cam characterized by a flat shape along its periphery.

roller chain   A chain assembly that consists of alternately spaced roller links and pin links throughout the length of the chain.

roller follower   A type of cam follower that provides a rolling action between the cam and follower and is ideal for reducing friction in motion.

shaping   A machining process on a shaper for the purpose of cutting gear teeth.

sheave   A pulley with grooves, one or several, to accommodate a single belt or multiple belts simultaneously.

silent chain   Also called an inverted tooth chain. It includes a series of toothed links alternately assembled with pins. In this type of drive, the chain passes over the face of the sprocket like a belt, and the sprocket teeth do not project through.

Section 11: *Power Transmission Elements* **255**

spherical-face follower A cam follower with a semicircular spherical face, used for relatively steep cam applications, such as in automobile valve lifters.

spiral bevel gear A gear whose teeth are curved and oblique. It operates quietly and can be used for high-speed operations.

sprocket A toothed wheel, much like that of a gear, used with chain drives.

spur gear A gear that connects parallel shafts. It is cylindrical in shape, with straight teeth cut parallel to the axis of the shaft.

swinging follower A cam follower that can operate with a steep angle, whose connecting arm swings to accommodate the rise and fall of the cam surface.

translating roller follower A follower used with a cylindrical cam, similar in function to an oscillating follower.

V-belt A belt with a tapered cross section that allows it to seat firmly into a sheave groove during operation.

worm A special form of a crossed helical gear which resembles a screw thread. It is referred to as the pinion in worm-gearing terminology.

worm gear Sometimes referred to as a worm wheel; a gear whose tooth shape, addendum, and dedendum is the same as for a spur gear.

yoke cam A positive-acting cam that does not require the use of a follower spring.

zero bevel gear A curved gear whose teeth lie in the same general direction as teeth of straight bevel gears.

Section 11: *Power Transmission Elements*   **257**

Student _____   Date _____

Section 11: Competency Quiz

PART A COMPREHENSION

1. To what does the term *power transmission* refer? (5 pts)

2. Name seven different kinds of gears. (7 pts)

3. Describe the shape and form of a spur gear. (4 pts)

4. Define *involute curve*. (4 pts)

5. Differentiate between the terms *circular pitch* and *pitch circle* for a spur gear. (5 pts)

6. List three different machining processes for cutting gear teeth. (3 pts)

7. Name four types of bevel gears. (4 pts)

©1995 West Publishing Company

258   **Section 11:** *Competency Quiz*

8. What is the dimensional variation between the addendum and dedendum in gears? (4 pts)

9. What are the two major components of a worm gear assembly? (2 pts)

10. What power transmission element is capable of providing a rotating, oscillating, or reciprocating motion? (4 pts)

11. Describe the purpose of a cam follower. (4 pts)

12. Name two major categories of cams. (2 pts)

13. Identify five types of cam followers. (5 pts)

14. What is the purpose of a displacement diagram? (4 pts)

15. Differentiate between the terms *base circle* and *prime circle* on a cam layout. (5 pts)

16. What type of drive uses friction between it and a pulley or sheave for the transmission of power? (4 pts)

17. List three types of belt drives. (3 pts)

Section 11: *Power Transmission Elements*   **259**

Student _____   Date _____

18. What is the function of an idler in a belt drive system? (4 pts)

19. What are the basic differences between a pulley and a sheave? (2 pts) On which type of drive is each used? (2 pts)

20. Identify four materials from which sheaves are made. (3 pts)

21. Produce a freehand sketch of the cross-sectional area of a V-belt. (3 pts)

22. Name six kinds of chains in use in industry. (6 pts)

23. What is a sprocket? (2 pts) How is it used? (2 pts)

24. What are the advantages of a chain drive over belt and gear drives? (4 pts)

25. What is another name for a silent chain? (3 pts)

©1995 West Publishing Company

260   Section 11: *Competency Quiz*

PART B TECHNICAL TERMS

For each definition, select the correct technical term from the list on the bottom of the page. (10 pts each)

1. _____ Gears in mesh that are like two rotating cones which meet at some defined point called the apex.

2. _____ A special form of crossed helical gear; resembles a screw thread.

3. _____ The moving of power or energy from the point where it is generated to the point or place where it is utilized.

4. _____ A power transmission element that consists of alternately spaced roller links and pin links throughout its length.

5. _____ The amount of space between the top of one tooth and the bottom of the mating tooth when gears are in mesh.

6. _____ The area on a cam surface where the follower neither rises nor falls.

7. _____ The smaller gear in a pair of mating gears.

8. _____ The four types of this device include the plane plate type, the hub on one side type, the hub on both sides type, and the split hub type.

9. _____ A gear with straight teeth parallel to the centerline of the bore.

10. _____ A power transmission element with has a tapered cross-sectional area that allows it to seat itself firmly into a sheave groove.

A. bevel gear
B. cam
C. circular pitch
D. clearance
E. dwell
F. mounting distance
G. pinion
H. power transmission
I. roller chain
J. silent chain
K. sprocket
L. spur gear
M. V-belt
N. worm

Section 11: *Power Transmission Elements*   **261**

Student _____   Date _____

Competency Quiz continues on next page.

©1995 West Publishing Company

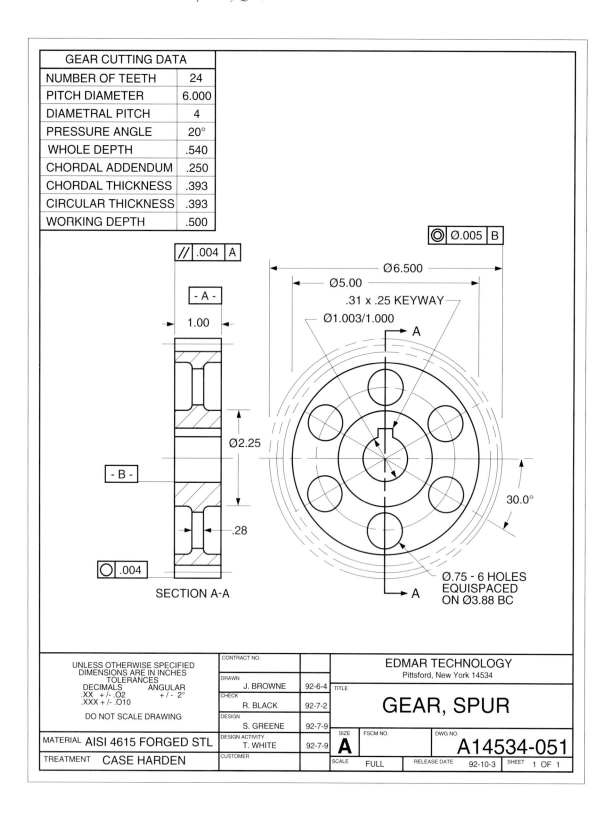

Section 11: *Power Transmission Elements*   **263**

Student _____   Date _____

PART C INTERPRETING GEARS

1. For the spur gear, drawing A14534-051 (see page 262), respond to the following questions.

    a. What is the pitch diameter of the gear? _____

    b. What is the outside diameter of the gear? _____

    c. This gear is the smaller of two gears in mesh. What is it called? _____

    d. What is the material of the gear? _____

    e. What is the tolerance for the bore? _____

    f. What is the tolerance for parallelism for the sides of the gear? _____

    g. What is the size of the keyway? _____

    h. What is the diameter of the six equally spaced holes? _____

    i. What is the working depth of the teeth? _____

    j. What is the thickness of the web? _____

    k. What is the general tolerance for two-place decimals? _____

    l. What is the diameter of the hub? _____

    m. What is the treatment for the gear? _____

©1995 West Publishing Company

264 Section 11: *Competency Quiz*

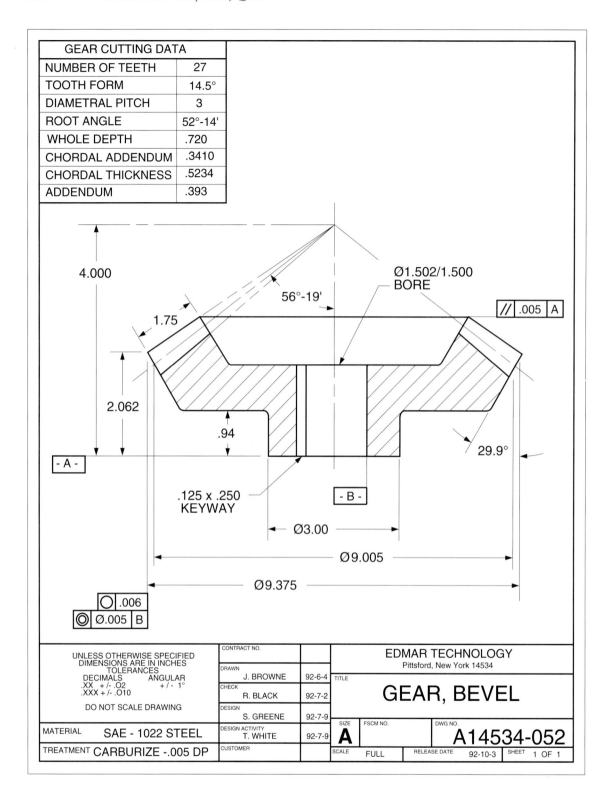

Section 11: *Power Transmission Elements*   **265**

Student _____ Date _____

2. For the bevel gear, drawing A14534-052 (see page 264), respond to the following questions.

    a. How many teeth are on the gear? _____

    b. What is the root angle? _____

    c. What is the pitch angle? _____

    d. What is the pitch diameter of the gear? _____

    e. What is the face width of the gear? _____

    f. What is the hub diameter? _____

    g. What does the 4.000-inch diameter represent? _____

    h. What is the upper limit of the bore? _____

    i. At what angle is the tooth form? _____

    j. Of what material is the gear made? _____

    k. What is the tolerance for angular dimensions? _____

    l. What is the meaning of the symbol $\boxed{\bigcirc\,.006}$? _____

    m. What is the depth of the addendum? _____

©1995 West Publishing Company

266 Section 11: *Competency Quiz*

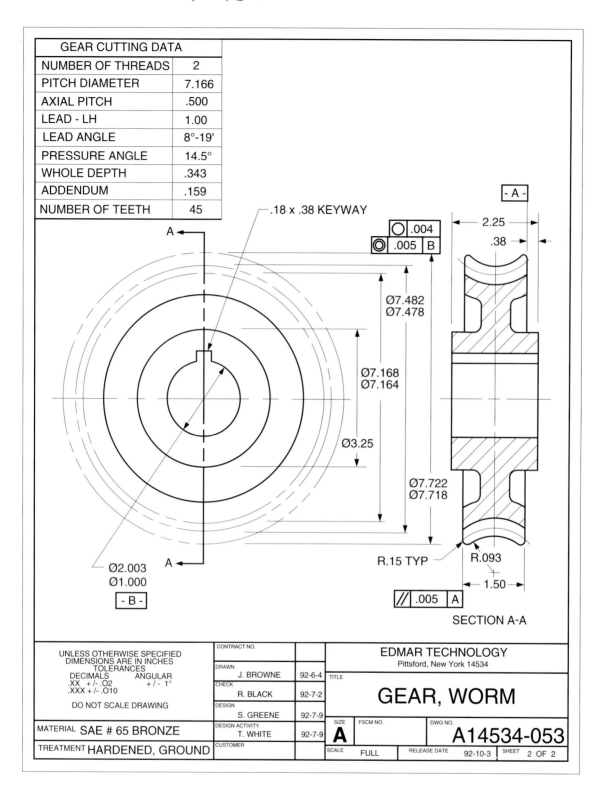

Section 11: *Power Transmission Elements*   **267**

Student _____   Date _____

3. For the worm gear, drawing A14534-053 (see page 266), respond to the following questions.

    a. What is the material for the gear? _____

    b. Are the teeth on the gear left hand or right hand? _____

    c. What are the upper and lower limits for the pitch diameter? _____

    d. How many teeth are required? _____

    e. What is the lead angle for the gear? _____

    f. On which dimension is parallelism to be held? _____

    g. What size is the keyway? _____

    h. What is the throat radius of the gear? _____

    i. What is the face width? _____

    j. What is the tolerance on the hub? _____

    k. What is the tolerance for roundness? _____

    l. What is the pressure angle? _____

    m. What kind of section is section A-A on the drawing? _____

©1995 West Publishing Company

268 Section 11: *Competency Quiz*

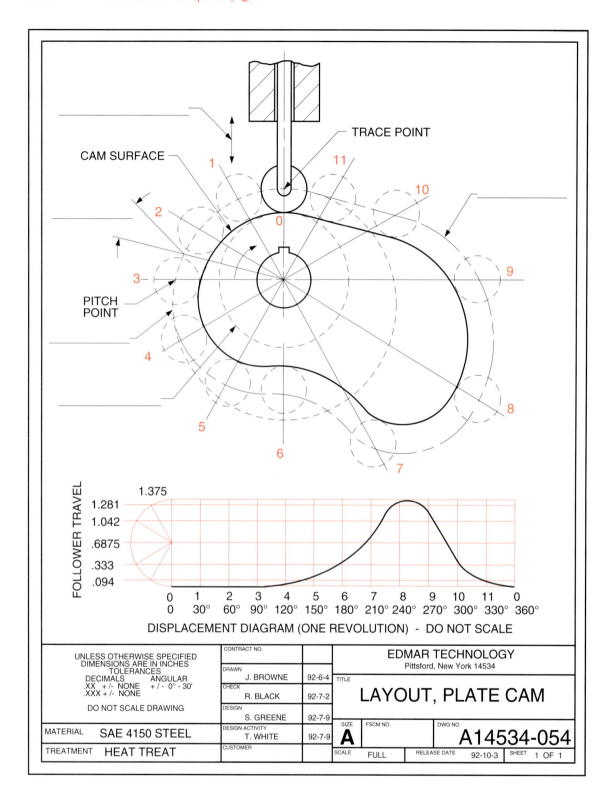

Section 11: *Power Transmission Elements*   **269**

Student _____   Date _____

PART D INTERPRETING CAMS, PULLEYS, AND SPROCKETS

1. For the plate cam layout, drawing A14534-054 (see page 268), respond to the following questions and exercises.

   a. Fill in each of the five blank spaces on the layout with the correct cam nomenclature.

   b. For approximately how long is the cam in a dwell period? _____ In a fall period? _____ Answer in degrees.

   c. What is the material specified for the cam? _____

   d. From the zero position, how much does the cam rise at the 180-degree point? _____

   e. Beginning at the zero position, it takes the cam 4 minutes to complete one revolution. Approximately how long will it take to reach position #9? _____ Position #6? _____ Position #2? _____

   f. What is the distance from the prime circle to the outside surface of the cam for the following positions? Position #2, 60 degrees _____ Position #7, 210 degrees _____ Position #11, 330 degrees _____

   g. What type of follower is specified on the cam layout? _____

   h. In what direction is the cam moving? _____

   i. What is the maximum follower travel distance? _____

©1995 West Publishing Company

270 Section 11: *Competency Quiz*

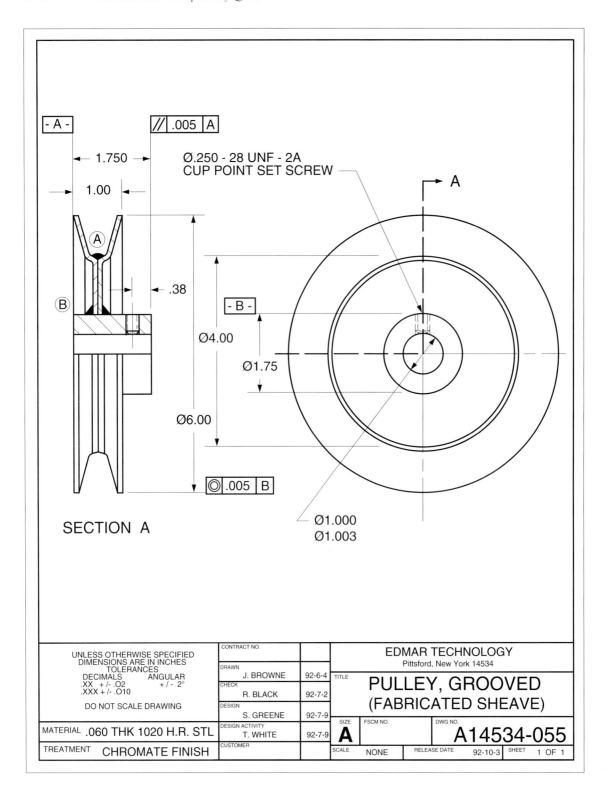

Section 11: *Power Transmission Elements* **271**

Student _____   Date _____

2. For the grooved pulley, drawing A14534-055 (see page 270), respond to the following questions and exercises.

a. What type of belt is required to accommodate the pulley? _____

b. From Figure 11.18 in the text, select the belt that would fit this pulley _____

c. What welding symbol is needed for the weld at Ⓐ? _____

Interpret the symbol. _____

d. What is the correct weld symbol for the weld at Ⓑ? _____

Interpret the symbol. _____

e. What is the length of the hub? _____

f. What is the size of the set screw? _____

g. How long a set screw should be used? _____

h. What is the lower limit for the bore? _____

i. What is the material for the pulley? _____

j. Which dimensions have two-place tolerances? _____

k. What is the general tolerance for the three-place decimal dimensions? _____

l. What is the scale for the drawing? _____

m. How is the pulley to be finished? _____

©1995 West Publishing Company

Section 11: Competency Quiz

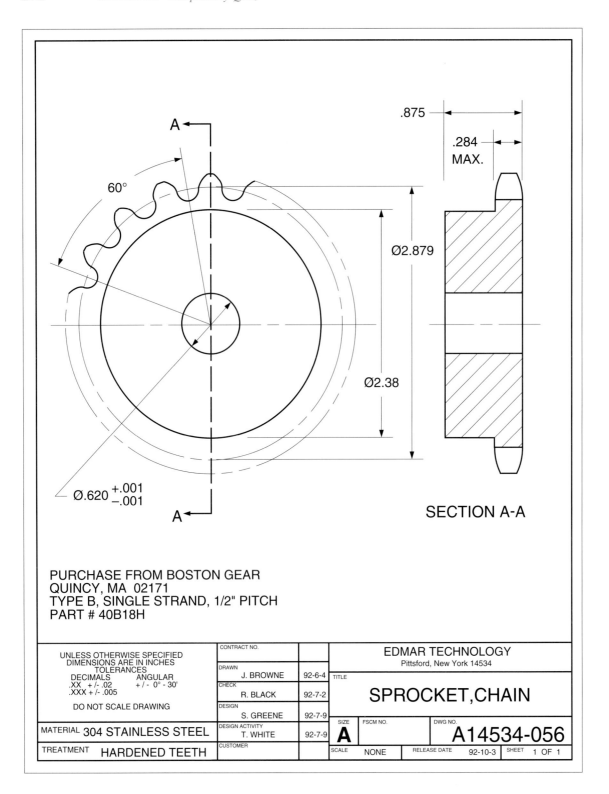

Section 11: *Power Transmission Elements*   **273**

Student _____   Date _____

3. For the chain sprocket, drawing A14534-056 (see page 272), respond to the following questions.

   a. On what type of drive mechanism is this sprocket to be used? _____

   b. How many teeth are on the sprocket? _____

   c. From what company is the sprocket purchased? _____

   d. What is the vendor part number and type of sprocket? _____

   e. What is the pitch diameter? _____

   f. From what material is the sprocket made? _____

   g. What is the basic size of the bore? _____

   h. What are the upper and lower limits for the bore? _____

   i. What is the diameter of the hub? _____

   j. What is the tolerance for the width of the teeth? _____

   k. How will the teeth be treated? _____

   l. What type of section view is shown? _____

   m. What is the angular tolerance for the teeth? _____

   n. What do the circular phantom lines indicate on the drawing? _____

©1995 West Publishing Company

PART E TECHNICAL MATHEMATICS

GEAR CALCULATIONS

In the solution of gear problems, calculations using formulas are necessary. The following basic formulas are used to solve gear problems. For the solution to the problems, use a separate sheet for calculations.

$$D = N/P \qquad\qquad DR = D - 2b$$
$$Do = (N + 2)/\pi \qquad\qquad a = 1/P$$
$$p = \pi D/N \qquad\qquad b = 1.157/P$$
$$\qquad\qquad\qquad\qquad WD = a + b$$

where

P	= diametral pitch		a	= addendum
N	= number of teeth		b	= dedendum
D	= pitch diameter (inches)		Do	= outside diameter
p	= circular pitch (inches)		WD	= whole depth
DR	= root diameter			

1. A spur gear has 32 teeth and a diametral pitch of 6. What is the pitch diameter of the gear?

2. A gear has 16 teeth and a circular pitch of .7854. What is the outside diameter of the gear?

3. Find the addendum of a gear that has an outside diameter of 6.000 inches and a pitch diameter of 5.500 inches.

4. Determine the root diameter of a spur gear that has a pitch diameter of 12.000 inches and 48 teeth.

5. Given a 14.5-degree involute spur gear, which has a pitch diameter of 5.000 inches and 40 teeth, calculate for its addendum, dedendum, outside diameter, circular pitch, and whole depth.

6. A 40-tooth gear has a diametral pitch of 2. Find the circular pitch, pitch diameter, and outside diameter.

Section 11: *Power Transmission Elements*   **275**

Student _____ Date _____

CENTER DISTANCE BETWEEN GEARS

The center distance (*CD*) between two gears is determined by adding the pitch diameters of the two gears and dividing by two.

Therefore: $CD = \dfrac{D1\ (\textbf{gear}) + D2\ (\textbf{pinion})}{2}$

7. A gear with a pitch diameter of 8.333 meshes with a pinion whose pitch diameter is 5.63. What is the center distance between the two?

8. A 20-tooth pinion has a diametral pitch of 6 and meshes with a 72-tooth gear. Determine the center distance.

9. A pinion has a pitch diameter of 3.750 inches and its mating gear has a pitch diameter of 12.000 inches. Find the center distance.

GEAR RATIO

The ratio of gears is the relationship between any two of the following:

 a. The pitch diameter of the gears
 b. The number of teeth of the gears
 c. The revolutions per minute (rpm) of the gears

Therefore: **R (gear ratio)** $= \dfrac{\textbf{larger value}}{\textbf{the corresponding smaller value}}$

10. A gear with a pitch diameter of 12.750 inches meshes with a pinion that has a pitch diameter of 3.1875 inches. What is the gear ratio?

11. A pinion on a spur gear drive has 32 teeth, and its gear has 96 teeth. What is the gear ratio?

©1995 West Publishing Company

276   **Section 11:** *Competency Quiz*

BELT DRIVE SPEEDS

Belt drive speeds are found by determining the relationship between rotational speed and pulley diameters by using the following formula:

$$\frac{D1}{D2} = \frac{N2}{N1}$$

where

$N1$ = **rotational speed of drive pulley (in revolutions per minute)**
$N2$ = **rotational speed of driven pulley (in revolutions per minute)**
$D1$ = **drive pulley diameter (in feet, inches, or millimeters)**
$D2$ = **driven pulley diameter (in feet, inches, or millimeters)**

12. If a driving pulley on a belt drive is 6.00 inches in diameter and the driven pulley is 12.00 inches in diameter, what is the speed of the driven pulley if the driver rotates at 200 rpm?

13. On a belt drive, the driven pulley is 17.00 inches in diameter and the drive pulley is 7.00 inches in diameter. The 7.00-inch pulley rotates at 350 rpm. What is the rotational speed of the 17.00-inch pulley?

14. A motor drives a compressor with a belt drive. If the motor speed is 1750 rpm and the compressor pulleys are 2.50 inches for the drive pulley and 8.00 inches for the driven, what is the speed of the driven pulley?

FLAT BELT LENGTH

When it is necessary to calculate the length of a flat belt in an open drive, the following equation may be used:

$$L = \frac{2C + 1.57\,(D + d) + (D - d)^2}{4C}$$

where

D = **diameter of the large pulley (inches or millimeters)**
d = **diameter of the small pulley (inches or millimeters)**
C = **center distance between pulleys (inches or millimeters)**
L = **length of belt**

15. Compute the length of a belt whose system includes pulleys that are 9 inches and 13 inches in diameter. Their center distance is 32 inches.

16. Two pulleys have diameters of 11.00 inches and 5.50 inches, respectively. The distance between their centers is 11.50 inches. What length of belt is required to drive this system?

SECTION 12

Surface Texture and Protective Coatings

LEARNER OUTCOMES

You will be able to:

- Distinguish between surface texture and protective coatings.
- Define twelve technical terms for surface texture.
- Name the component parts of the surface control system.
- Identify two major reasons for controlling surface texture.
- Explain the meaning of several lay and roughness symbols.
- Apply several surface texture symbols to an engineering drawing.
- List four reasons for the use of protective coatings.
- Identify three basic types of protective coatings.
- Explain the electroplating process.
- Define *chemical conversion*.
- Discuss the differences between various polymer coatings.
- Demonstrate your learning through the successful completion of competency exercises.

12.1 INTRODUCTION

Because of today's technological advances in machining, high-tech materials, demand for higher production, and tighter tolerances for items (including automotive parts, engines, and aerospace products), the need to apply standards to the texture of surfaces is a key quality-control issue. Part of total quality management (TQM) in industry is the accurate application and interpretation of finish symbols for machined surfaces. The responsibility lies with the engineering department to control surface finish on a drawing. If the wrong surface finish is specified on a drawing, it may cause a part to be too costly to be produced. If the specification for a finish is too rough for the requirement, the part may fail or may not function as intended.

For large multinational companies, parts and subassemblies may be engineered in one location, produced in another location or plant, and assembled in a third location. For uniformity in quality, it is imperative that surface texture standards be followed. Two major reasons for controlling surface texture are (1) to reduce friction, and (2) to control wear on parts or assemblies.

The American National Standards Institute Specifications ANSI B 46.1 and ANSI Y 14.36 are the standards used when specifying and working with surface texture. They deal with surface texture and its components: roughness, waviness, lay, and flaws. In addition, the specifications call for a set of standard symbols to be used on engineering drawings.

Protective coatings differ from surface texture. Surface texture deals with the texture of a produced surface for a mechanical component, while a protective coating refers to a film or base material applied to an item to protect it from corrosion, abrasion, erosion, and other forms of deterioration. Protective coatings are available in several forms, including chemical, organic, metallic, vitreous, and lubricant treatments. The protective coatings that will be addressed in this section include polymers, plating, and conversion coatings.

12.2 SURFACE TEXTURE DEFINITION AND NOMENCLATURE

The terms *surface texture*, *surface finish*, and *surface irregularities* are used interchangeably and may be defined as the repetitive or random deviation from the nominal surface that forms the three-dimensional topography of the **surface**. The following are basic terms a shop technician will encounter when interpreting engineering drawings on which surface texture is specified. Refer to Figure 12.1 and Figure 12.2 for a pictorial description of surface texture nomenclature.

Nomenclature related to the surfaces of solid materials includes the following:

- *Surface*—The surface of an object is the boundary that separates that object from another object, substance, or space. It is produced by such means as abrading, casting, coating, cutting, etching, plastic deformation, sintering, wear, erosion, and so forth.
- *Nominal surface*—The intended surface contour, the shape and extent of which is usually shown and dimensioned on a drawing or descriptive specification.
- *Measured surface*—A representation of the surface obtained by instrumentation.
- *Roughness*—The surface feature of random and repetitive spaced minute or finer deviations from the center line, having roughness height and width as included within a typical sampling length known as the *roughness-width cutoff*.
- *Waviness*—The surface feature of more widely spaced components of surface texture. Unless otherwise specified, waviness includes all irregularities whose spacing is greater than the

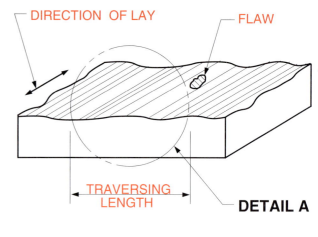
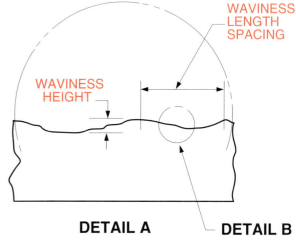
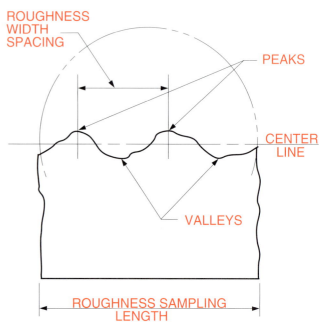

FIGURE 12.1
Surface texture nomenclature

FIGURE 12.2
Nominal and measured profiles

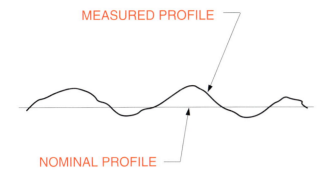

roughness sampling length and less than the waviness sampling length. Waviness may result from such factors as machine or work defects, vibration, chatter, heat treatment, or warping strains. Roughness may be considered superimposed on a wavy surface.

- *Lay*—The direction of the predominant surface pattern, determined by the production method used.
- *Flaws*—Unintentional irregularities that occur at one place as a direct result of a manufacturing process, including defects singularly or in groups, such as cracks, blow holes, checks, ridges, scratches, gouges, nicks, cuts, punctures, dimples, chemical erosion, and so forth.

Terms related to the measurement of surface texture include the following:

- **Profile**—The contour of a surface in a plane perpendicular to a surface, unless some other angle is specified.
- *Nominal profile*—A profile of the nominal surface; it is the intended profile.
- *Measured profile*—A representation of the profile obtained by instrumental or other means. When the measured profile is a graphical representation, it will usually be distorted through the use of different vertical and horizontal magnifications but will otherwise be as faithful to the profile as is technically feasible.
- **Center line (mean line)**—A theoretical line parallel to the general surface profile at a midpoint between the high and low measurements of surface features.
- **Peak**—The point of maximum height on that portion of a profile lying above the center line and between two intersections of the profile and the center line.
- **Valley**—The point of maximum depth on that portion of a profile lying below the center line and between two intersections of the profile and the center line.
- **Spacing**—The distance between specified points on the profile, measured parallel to the nominal profile.
- *Roughness width spacing*—The average spacing between adjacent peaks of the measured profile within the roughness sampling length.
- *Roughness sampling length*—The sampling length within which the roughness average is determined.
- *Waviness spacing*—The average spacing between adjacent peaks of the measured profile within the waviness sampling length.
- *Waviness height*—The peak-to-valley height of the modified profile, from which the roughness and flaws have been removed by filtering, smoothing, or some other means.
- **Sampling length**—The nominal spacing within which a surface characteristic is determined.
- *Traversing length*—The length of a profile that is traversed by a stylus (a measuring device with a sensitive tip) to establish a representative measurement.
- **Micrometer**—One millionth of a meter (.000001 meter). When used, its designation is μm.
- **Microinch**—One millionth of an inch (.000001 inch). When used, its designation is μin.

12.2.1 Surface Control System

The *surface control system* is used to indicate the limits of surface characteristics on an engineering drawing. It features a surface texture symbol that includes a check mark with a horizontal top extension as shown in Figure 12.3, which also indicates the proper size for the symbol as it appears on a drawing. The longer leg and its extension are to the right, as read on the drawing. The top extension is shown as long as it needs to be to accommodate all the necessary information. The symbol is modified, when necessary, to require or prohibit the removal of material.

280 Section 12: *Surface Texture and Protective Coatings*

FIGURE 12.3
Surface texture symbol size

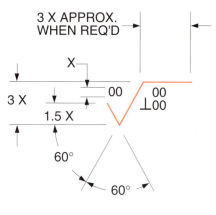

X = LETTER HEIGHT

When surface texture values other than roughness are specified, the symbol appears with additional features to indicate procedures and limits. The variations that are possible within this context are illustrated in Figure 12.4.

■ 12.2.2 Roughness

Roughness deals with the finer irregularities of a surface. Included in the irregularities are those which result from the mechanical action of the production process. Longitudinal and radial marks or ridges can be left on a mechanical surface by tools used when turning, grinding, milling, shaping, and so forth. Figure 12.1, detail B, shows an enlarged view of the roughness of a surface. It also illustrates the peaks and valleys of roughness as well as the center line (mean). Roughness measurements are read in millimeters (metric scale) or in micrometer (microinch) increments. There are many factors that help determine the specific surface roughness of a mechanical part: the material to be machined, the condition of the cutting tool, and the speed of the cutting, turning, grinding, or shaping operation, for example. General values for surface roughness produced by common production and machining methods are shown in Figure 12.5

FIGURE 12.4
Surface texture symbols variations

SYMBOL	PURPOSE	INTENT
 x.x ✓	BASIC SYMBOL: ROUGHNESS AVERAGE SPECIFIED	WHERE ROUGHNESS HEIGHT ONLY IS INDICATED, THE SURFACE MAY BE PRODUCED BY ANY METHOD.
 ✓	REMOVAL OF MATERIAL REQ'D TO PRODUCE PART	MATL REMOVAL BY MACHINING IS REQ'D. THE HORIZONTAL BAR INDICATES THAT MATL REMOVAL IS REQ'D TO PRODUCE THE SURFACE, AND MATL MUST BE PROVIDED FOR THAT PURPOSE.
 x.x ✓	REMOVAL OF MATERIAL REQ'D TO ACHIEVE FINISH	MATL REMOVAL ALLOWANCE. THE NUMBER INDICATES THE AMOUNT OF STOCK TO BE REMOVED BY MACHINING (MM OR INCHES). TOLERANCES MAY BE ADDED TO THE BASIC VALUE SHOWN OR BY NOTE .
 ✓	NO MATERIAL REMOVAL PERMITTED	MATL REMOVAL PROHIBITED. THE CIRCLE INDICATES THAT THE SURFACE MUST BE PRODUCED BY PROCESSES SUCH AS CASTING, FORGING, HOT OR COLD FINISHING, ETC. WITHOUT SUBSEQUENT MATL REMOVAL.
 ✓	SPECIAL SURFACE CHARACTERISTICS INDICATED	SURFACE TEXTURE SYMBOL. USED WHEN ANY SURFACE CHARACTERISTICS ARE SPECIFIED ABOVE THE HORIZONTAL LINE OR TO THE RIGHT OF THE SYMBOL. SURFACE MAY BE PRODUCED BY ANY METHOD EXCEPT WHEN THE BAR OR CIRCLE IS SPECIFIED.

Section 12: *Surface Texture and Protective Coatings* 281

■ 12.2.3 Waviness

Figure 12.1, detail A, illustrates examples of **waviness** height, which refers to the peaks and valleys of waves, and waviness length spacing, which is the average spacing between adjacent peaks of waves. Waviness height and width measurements are normally described in millimeters.

■ 12.2.4 Lay

The surface pattern on a part is a result of the production method used in its manufacture, subsequent machining, or both. In the case of a machined part, tool marks or grain lines will be visible. To illustrate **lay**, symbols are used to provide the drawing interpreter information through symbols. Figure 12.6 shows these symbols as well as their meaning.

FIGURE 12.5
Surface roughness produced by common production and machining methods

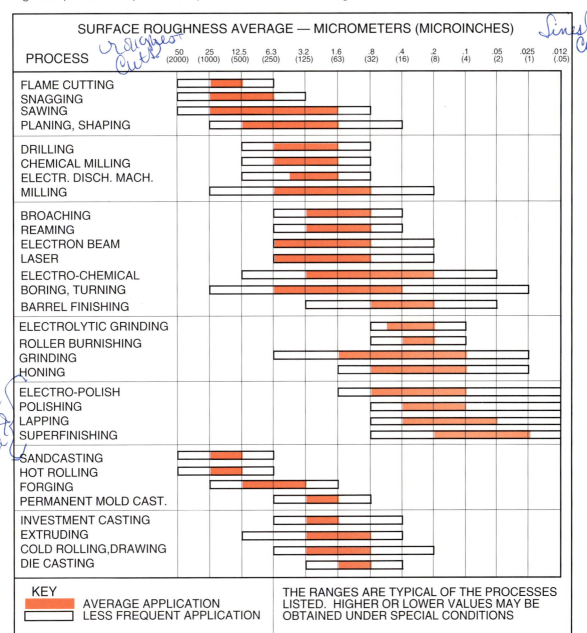

282   **Section 12:** *Surface Texture and Protective Coatings*

FIGURE 12.6
Lay symbols

LAY SYMBOL	MEANING	DIRECTION OF TOOL MARKS
—	LAY IS PARALLEL TO THE LINE THAT REPRESENTS THE SURFACE TO WHICH THE SYMBOL IS APPLIED	
⊥	LAY IS PERPENDICULAR TO THE LINE WHICH REPRESENTS THE SURFACE TO WHICH THE SYMBOL IS APPLIED	
X	LAY IS ANGULAR IN BOTH DIRECTIONS TO THE LINE REPRESENTING THE SURFACE TO WHICH THE SYMBOL IS APPLIED	
M	LAY IS MULTIDIRECTIONAL	
C	LAY IS CIRCULAR RELATIVE TO THE CENTER OF THE SURFACE TO WHICH THE SYMBOL IS APPLIED	
R	LAY IS RADIAL RELATIVE TO THE CENTER OF THE SURFACE TO WHICH THE SYMBOL IS APPLIED	
P	LAY IS NONDIRECTIONAL, OR PROTURBENT IN NATURE	

■ 12.2.5 Flaws

Flaws exist because of manufacturing defects, not as a result of a machining process.

The effect of flaws is not included in the measurement of roughness height unless otherwise specified. When flaws are to be considered a characteristic of a surface, a note stating specific requirements will normally appear on the drawing as follows:

⚠ SURFACES INDICATED TO BE FREE FROM SCRATCHES, DENTS, STEPS, OR FLAWS EXCEEDING [XXX] MICROMETERS IN DEPTH.

12.3 APPLICATION OF SURFACE TEXTURE SYMBOLS

The surface texture symbol denotes lay, roughness, and waviness characteristics on an engineering drawing. When interpreting a drawing on which surface texture symbols are used, the apex of the symbol is shown touching the surface that is to be controlled. When this is not practical, the symbol appears with the apex touching an extension line from the surface to which it applies. Symbols are shown in a horizontal reading position only, and not repeated in another section or view. When the control symbol is used with a dimension, it affects all surfaces related to that dimension. Dimensions and areas of transition, such as fillets and chamfers, conform to the roughest adjacent surface area unless otherwise noted. Figure 12.7 illustrates the application of surface texture symbols.

Representative examples of the application of surface roughness height values and texture symbols are illustrated in Figure 12.8.

12.4 PROTECTIVE COATINGS

Because of today's technological advances, there are many types of **protective coatings** from which to choose for various commercial, consumer, industrial, and military products. They exist, primarily, under three categories: metallic, nonmetallic, and polymer.

FIGURE 12.7
Application of surface texture symbols

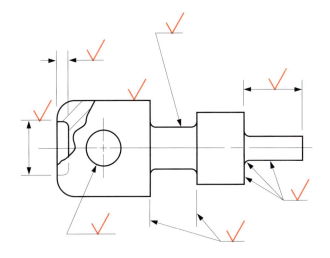

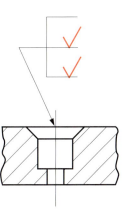

UNLESS OTHERWISE SPECIFIED
ALL SURFACES 63

There are several reasons for requiring coatings on a part or assembly. Some of the more prominent reasons include:

- to enhance the looks of a part, for eye appeal
- to improve resistance to corrosion
- to control friction and wear
- to alter dimensions

The automobile is an example in which individual components as well as the fully assembled automobile requires coatings for corrosion resistance, friction and wear, and decorative purposes. Coatings are used in engine parts and high-speed machinery to control friction. One of the more recent reasons for coatings is to alter the dimensions on a part for the purpose of adding or deleting material from out-of-tolerance dimensions.

Our discussions in this section will include the coatings identified in Figure 12.9. Each process will be explained, as well as examples of where each is used.

12.4.1 Metallic Coatings

Three types of *metallic coatings* are plating techniques: electroplating, electroless plating, and immersion plating. All include the conversion of metal ions in a solution to metal ions on a surface. The differences between the three are the properties of each and how the metal-reduction process occurs.

The **electroplating** process began soon after the advent of electricity. The basic process includes the use of two electrodes immersed in an *electrolyte* and connected to a power supply. The electrolyte material is usually a water solution of the salts of the metal to be plated. For example, copper sulfate is used as an electrolyte for copper plating. With the suitable combination of metals in a plating cell, ions from one electrode (A) will be attracted to ions of the other electrode (B), as illustrated in Figure 12.10. Thus, material from A will be deposited on B, completing the electroplating process.

Familiar electroplating processes include silver plating for waveguide applications for radar equipment, cadmium plating for hardware, and nickel and chromium plating for items that require a bright finish, such as toasters, interior automobile parts, and other consumer products.

Electroless plating was developed at the end of World War II and does not require any external electrical energy in its process. Objects are merely dipped in a plating bath, and an adherent coating of metal develops. The process includes the attraction of chemically induced ions. This process is frequently used as a pretreatment for nickel plating because it produces a hard, durable surface on items such as fixtures, machine parts, and base plates.

Immersion plating, also called galvanic deposition, is probably the simplest of all plating processes.

284   **Section 12:** *Surface Texture and Protective Coatings*

FIGURE 12.8

Application of surface roughness height values and symbols

ROUGHNESS HEIGHT		APPLICATION
μ m	μ in.	
25	1000	PRODUCED BY SAND CASTINGS, SAW AND TORCH CUTTING, AND ROUGH FORGING. THIS FINISH IS SUITABLE FOR ROUGH ITEMS THAT DO NOT REQUIRE A MACHINING OPERATION.
12.5	500	A SURFACE PRODUCED BY HEAVY CUTS AND COARSE FEEDS IN MILLING, TUIRNING, BORING, AND ROUGH FILING.
6.3	250	A SURFACE PRODUCED BY FAST FEEDS IN MILLING, BORING, SHAPING, PLANING, AND FILING.
3.2	125	A FINISH PRODUCED BY MACHINING OPERATIONS WITH SHARP TOOLS, HIGH SPEEDS, FINE FEEDS, AND LIGHT CUTS. PRODUCED ON MACHINES SUCH AS LATHES, MILLS, SHAPERS, AND GRINDERS.
1.6	63	SPECIFIED FOR CLOSE FITS. PRODUCED WITH SHARP CUTTERS, HIGH SPEEDS, AND FINE FEEDS. ALSO PRODUCED BY EXTRUDING, ROLLING, DIE CASTING, AND PERMANENT MOLD CASTINGS.
.8	32	THIS IS AN EXTREMELY HIGH QUALITY FINISH PRODUCED BY LATHES, MILLING MACHINES, AND CENTERLESS, CYLINDRICAL, AND SURFACE GRINDING.
.4	16	PRODUCED BY SMOOTH REAMING, COARSE GRINDING, CYLINDRICAL GRINDING, AND SURFACE GRINDING ON ROTATING SHAFTS AND HEAVILY LOADED BEARINGS.
.2	8	A SURFACE PRODUCED BY HONING, LAPPING, AND BUFFING REQUIRED FOR INTERIOR HONED SURFACES, PRECISION GAUGES, AND ROTATING SHAFTS AND BEARINGS.
.1	4	A FINELY HONED AND LAPPED SURFACE, USUALLY REQUIRED FOR INSTRUMENT AND GAUGE WORK .

FIGURE 12.9
Protective coating applications

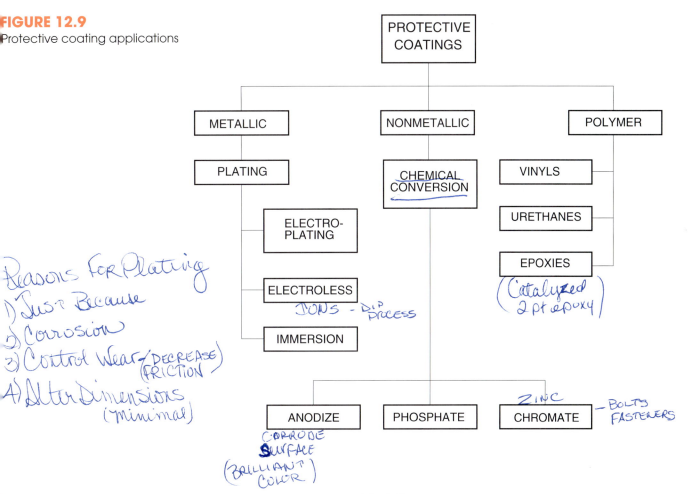

FIGURE 12.10
Electroplating cell

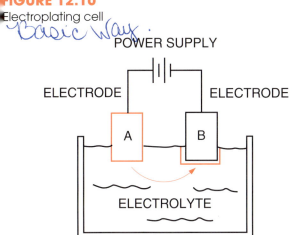

The deposition of the coating metal takes place by an electrolytic reaction between the substrate metal and the metal ion in solution. No external electrical power source or energy is used. This process is often a preliminary step in hard-to-plate metals, such as magnesium or aluminum. It produces a very thin coating that can be wiped off easily.

12.4.2 Nonmetallic Coatings

Chemical conversion is considered a nonmetallic coating process that occurs when a superficial layer of a compound is produced by the chemical or electrochemical treatment of a surface. This chemical treatment of a metal surface converts the surface into a corrosion-resistant, nonreactive form that becomes an integral part of the substrate. This type of coating provides an excellent surface for organic finishes.

Bauxite, an ore from which aluminum is extracted, is the most abundant of all metals on earth. Aluminum, because of its poor wear and weathering characteristics, requires some form of surface treatment. The most common chemical conversion treatment is **anodize**. The anodizing process includes an arrangement similar to electroplating in that a tank, solution, and a power source are required. In this process, aluminum is converted from metallic aluminum to aluminum oxide. In this conversion, a ceramic compound results and is deposited on the object to be plated, in the form of anodize. Anodize is available in many different colors and may be found

286 **Section 12:** *Surface Texture and Protective Coatings*

on storm windows and screens, decorative items, architectural shapes, and automotive trim pieces. Anodized parts are resistant to chipping and peeling and have excellent wear characteristics.

A phosphate coating is a thin coating produced by the chemical conversion of a metal surface to a phosphate compound. This type of coating is produced by the spraying or the immersion of an object in a heated solution of dilute phosphoric acid with additives. The resultant chemical concentration attacks the metal surface of the object. The pleasant black coating that is formed adheres to the object. The metal used most often for phosphate coatings is steel. This type of coating has good wear resistance.

Chromate coating appears on several automotive components, including the zinc-aluminum bodies of carburetors. This type of coating is characterized by its yellow-brownish color. Chromate coatings are similar to phosphate coatings except that the protective film is formed by the reaction of water solutions of chromic acid or chromium salts, such as potassium dichromate or sodium. The coating forms when the metal to be coated is attacked by the chromate solution. This treatment is applicable to zinc and aluminum castings and galvanized steel parts.

■ 12.4.3 Polymer Coatings

For hundreds of years, paints have been used for decorative purposes. Varnishes and enamels were given their film properties by natural resins, dyes, and pigments. Traditional paint is a suspension of pigment in a liquid that dries to form a solid film. The coatings that have replaced the natural resins, dyes, and pigments are polymers, sometimes referred to as synthetics. Polymer coatings are a compound of high molecular weight derived by the addition of many smaller molecules of the same kind. Some polymers are applied to parts by dipping, while other parts are sprayed with a plasticized polymer. In either case, surface preparation is important to the coating process. The major difference between the various polymer coating processes is the method of film formation and the type of polymer used. The following paragraphs discuss the more popular coatings for commercial and industrial use.

Vinyls, especially vinyl paints, are used for general chemical resistance. These paints combine the use of PVC (polyvinylchloride) and vinyl acetate or vinyl butyral. Industrial vinyl paints are used on outdoor structures, such as steel buildings, pipes, and tanks. Latex vinyl paints are used as house paints. Other uses include plasticized coatings for tool handles. Vinyl coatings are not particularly hard, nor are they abrasion resistant.

Urethanes, in particular polyurethane, were discovered in Germany during the middle of the nineteenth century. The curing process for urethanes can be accomplished in one of two ways: by absorption and chemical reaction with moisture in the air, or by oxidation in urethane oils and alkyds. For industrial and commercial applications, heavy coatings are used to resist abrasion. Urethanes' resistance to chemicals also makes them suitable for paints on machine elements and aircraft. One of the most common places one would find their use is on industrial plant floors, including shop floors. In addition, urethanes are water, weather, and corrosion resistant, so they are used widely for marine applications and exterior maintenance work.

Epoxies are produced by an ethoxyline resin, which is a thermosetting resin formed by the reaction of biphenol and lorohydrin. Properly formulated, surface coatings can be applied to parts made of zinc, aluminum, wood, and molded plastics. Epoxies are abrasion, solvent, and chemical resistant, and they possess excellent adherence qualities. They are widely used for machine enamels and as coatings on equipment in chemical plants. This type of coating ranges in hardness from a viscous liquid to a hard, brittle resin.

■ 12.4.4 Finish Block Entries on an Engineering Drawing

Any information relative to coatings, be they metallic, nonmetallic, chemical, or a polymer, will normally appear in the supplemental data block on an engineering drawing, which is usually located adjacent to the title block. Figure 12.11 shows where the coating (treatment and finish) information appears on the data block.

When space within the supplemental data block does not permit the entry of all the necessary data, or when it is more practical, a delta note will appear to specify the finish and treatment requirements. The delta note may appear as follows:

FIGURE 12.11
Protective coating information

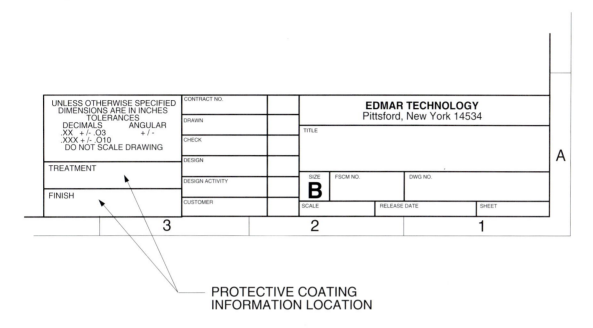

⚠ CHROMATE _____ , TYPE _____ , CLASS _____ .

or

⚠ PRIME PER _____ , 2 COATS

or

⚠ FINISH PER MIIL-C-4857, TYPE _____ , CLASS _____

12.5 SUMMARY

The need to apply standards to the texture of surfaces is a result of emerging technical advances that require higher production and tighter tolerances on products. The engineering department is responsible for controlling surface finish on a drawing. Key terms used in the control of surface texture include roughness, waviness, lay, and flaws, which were all covered in this section. In addition, surface texture symbols, including their characteristics and application, were explained in depth. Controlling the texture of a surface takes place during the machining process.

After individual parts have been machined and before they are placed in the stockroom, they probably will require some sort of protective coating, in the form of a metallic, nonmetallic, or polymer coating,
or a combination of two of them. Sometimes, final assemblies—such as large machinery, and certainly, automobiles—require a protective and decorative coating.

Several protective coatings were covered in this section, including metallic, nonmetallic, and polymer types. Reasons for protective coatings were outlined as being (1) to enhance the looks of a part, (2) to improve resistance to corrosion, (3) to control friction and wear, and (4) to alter dimensions. Three kinds of metallic coatings were detailed: electroplating, electroless plating, and immersion plating. Examples of situations in which each is successfully applied were given. Nonmetallic coatings that were covered fell under the category of chemical conversion, which includes anodize, phosphate, and chromate treatments. It was stated that chemical conversion occurs when a superficial layer of a compound is produced by a chemical or electrochemical treatment of a surface.

Although utilizing natural ingredients have been used for the production of paints and varnishes for many years, they have been for the most part replaced by polymers. The more common polymer coatings used for industrial and commercial products were referred to in this section. They include vinyls, urethanes, and epoxies. Each coating process was discussed in general, and an application of each was given.

288 **Section 12:** *Surface Texture and Protective Coatings*

Finally, finish block entries were presented. Examples showed the location and type of information included in the supplemental data block on an engineering drawing.

TECHNICAL TERMS FOR STUDY

anodize   The most common of chemical conversion treatments. It is available in many different colors and is resistant to chipping and peeling. It has excellent wear characteristics.

bauxite   An ore from which aluminum is extracted. It is the most abundant of all metals on earth.

center line (mean line)   A theoretical line parallel to the general profile at a midpoint between the high and low measurements of the surface features.

chemical conversion   A nonmetallic coating process that occurs when a superficial layer of a compound is produced by the chemical or electrochemical treatment of a surface.

chromate coating   A type of chemical conversion that has a yellow-brownish color. It is applicable to zinc and aluminum castings and galvanized parts.

electroless plating   A dipping process that does not require external energy. It is used as a pretreatment for nickel plating.

electroplating   A metallic coating that requires the use of two electrodes immersed in an electrolyte and connected to a power supply. It is an ion transfer process.

epoxies   Polymer-type coatings that are thermosetting resins. Epoxies are widely used for machine enamels and are abrasion, solvent, and chemical resistant, with excellent adherence qualities.

flaws   Manufacturing defects in objects. They are not included in the measurement of roughness. Flaws may include any of the following: scratches, dents, holes, voids, and so forth.

immersion plating   A coating process in which an object is immersed in a tank containing a plating bath. No external power source is required. This process produces a very thin coating that can be wiped off easily.

lay   The direction of the predominant surface pattern, determined by the production method.

microinch   A unit of measure used for measuring surface roughness. It equals one millionth (.000001) of an inch. Its designation is μin.

micrometer   The metric measurement for surface roughness. It equals one millionth (.000001) of a meter. Its designation is μm.

peak   The point of maximum height on that portion of a profile that lies above the center line and between two intersections of the profile and the center line.

phosphate coating   A thin coating that results from a chemical conversion of a metal surface to a phosphate compound. Produced by spraying or by immersion. Used primarily on steels; it is black in color.

polymer coatings   A compound of high molecular weight derived by the addition of smaller molecules of the same kind. Some polymers are applied to parts by dipping, while certain parts are sprayed with a plasticized polymer.

profile   The contour of a surface in a plane perpendicular to a surface, unless some other surface is specified.

protective coating   A film or base material applied to an item to protect it from corrosion, abrasion, erosion, or other forms of deterioration.

roughness   A surface feature of randomly and repetitively spaced minute or finer deviations from the center line, within a certain range of height and width.

sampling length   The nominal spacing within which a surface characteristic is determined.

spacing   The distance between specified points on the profile, measured parallel to the nominal profile.

surface   The boundary that separates one object from another object, substance, or space, produced by such means as abrading, casting, coating, cutting, etching, plastic deformation, sintering, wear, or erosion.

urethanes Polyurethanes with industrial and marine applications. They are resistant to chemicals, water, weather, and corrosion.

valley The point of maximum depth on that portion of a profile that lies below the center line and between two intersections of the profile of the center line.

vinyls Polymer coatings used as paints and produced by combining PVC and vinyl acetate or vinyl butyral. It is used on outdoor structures, such as steel buildings, pipes, and tanks. Vinyls are not particularly hard or abrasion resistant.

waviness The surface feature of more widely spaced components of surface texture. It may result from machine or work defects, including vibration, chatter, heat treatment, or warping strains.

Student _____ Date _____

Section 12: Competency Quiz

PART A COMPREHENSION

1. What are two major reasons for controlling surface texture? (4 pts)

2. What is the purpose of protective coatings? (4 pts)

3. What are the four components of surface texture? (4 pts)

4. Define the following surface texture terms: (4 pts)

   a. roughness

   b. waviness

   c. lay

   d. flaws

5. What is the purpose of the surface control system? (5 pts)

6. In the space below, draw freehand a basic surface texture symbol. (4 pts)

©1995 West Publishing Company

7. What do the following lay symbols mean? (6 pts)

   X =

   R =

   C =

8. Give four reasons for the use of protective coatings. (4 pts)

9. What are the three major categories of protective coatings? (6 pts)

10. By what production methods can the following roughness values be produced? (6 pts)

    63/▽

11. Name three types of protective coatings produced by chemical conversion. (6 pts)

Section 12: *Surface Texture and Protective Coatings*   **293**

Student _____   Date _____

12. Which of the three plating methods is accomplished through the use of two electrodes, an external power source, and an electrolyte? (5 pts)

13. What is bauxite? (5 pts)

14. What is the most common chemical conversion treatment for aluminum? (5 pts)

15. Which non-metallic surface treatment has a yellow-brownish color? (5 pts)

16. What polymer coating is suitable for industrial, commercial, and marine use? (5 pts)

17. Name three uses for vinyl paints. (6 pts)

18. What is the major difference between the various polymer coatings described in this section? (6 pts)

19. Which polymer varies in hardness from a viscous liquid to a hard, brittle resin? (5 pts)

20. Where on an engineering drawing should information relative to finish and treatments be placed? (5 pts)

©1995 West Publishing Company

294   Section 12: *Competency Quiz*

PART B TECHNICAL TERMS

For each definition, select the correct technical term from the list on the bottom of the page. (10 pts each)

1. _Chem. Conversion_ A nonmetallic coating process that occurs when a superficial layer of a compound is produced by the chemical or electrochemical treatment of a surface.

2. _Chromate_ A chemical conversion coat that appears on automotive components, including the zinc-aluminum bodies of carburetors.

3. _electrolyte_ A material in the electroplating process that consists of a water solution of the salts of the metal to be plated.

4. _electroplating_ A plating process that includes the use of two electrodes immersed in an electrolyte and connected to a power supply.

5. _spacing_ The boundary that separates an object from another object, substance, or space, produced by such means as abrading, casting, coating, cutting, etching, plastic deformation, sintering, wear, or erosion.

6. _____ The location, on an engineering drawing, where treatment and finish entries are placed.

7. _profile_ The contour of a surface in a plane perpendicular to the surface, unless some other angle is specified.

8. _Protective Coating_ A film or base material applied to an item to protect it from corrosion, abrasion, erosion, and other forms of deterioration.

9. _____ On a surface, the distance between specified points on the profile, measured parallel to the nominal profile.

10. _Surface Control_ This item features a surface texture symbol that includes a check mark with a horizontal top extension. _system_

A. chemical conversion
B. chromate plating
C. electrolyte
D. electroplating
E. immersion plating
F. peak
G. profile
H. protective coating
I. roughness
J. spacing
K. supplemental data block
L. surface
M. surface control system
N. valley

Student _____   Date _____

PART C APPLICATION OF SURFACE TEXTURE SYMBOLS

1. Convert the following surface roughness measurements to the metric (μm) form.

63 = _____ μm   125 = _____ μm

8 = _____ μm   500 = _____ μm

Competency Quiz continues on next page.

Section 12: *Competency Quiz*

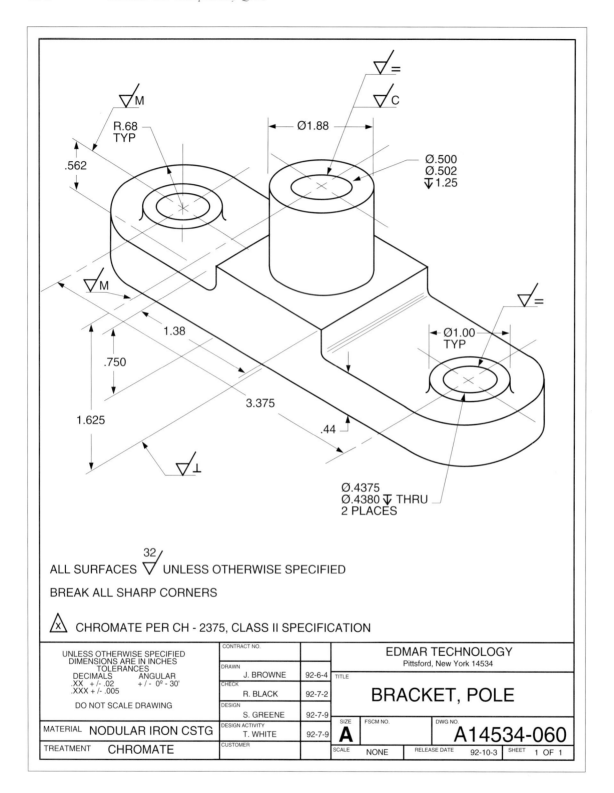

Section 12: *Surface Texture and Protective Coatings* **297**

Student _____   Date _____

2. For the pole bracket, drawing A14534-060 (see page 296), respond to the following questions and exercises.

    a. If the surfaces for the ∅1.00 bosses are machined by milling, what would be the range of surface roughness average, in micrometers? _____

    b. If the bottom of the Pole Bracket is machined by grinding, what would be the mean surface roughness average, in microinches? _____

    c. From what material is the pole bracket made? _____

    d. What surface roughness is required for all areas not otherwise specified? _____

    e. What surface treatment is specified on the drawing? _____

    f. For the bored hole ∅.500 / ∅.502, 1.25 DP, why are there two surface texture symbols shown? What would be the mean average roughness, in micrometers? _____

    g. The ∅.4375/∅.4380 holes are drilled. What would the required average surface roughness range for the holes, in micrometers and microinches? _____

    h. The end of the ∅1.88 extension is also milled. What would be the mean surface roughness, in microinches? _____

    i. Interpret the following lay symbols shown at various locations on the drawing:

        C = _____   M = _____

        ⊥ = _____   ═ = _____

    j. Why are the ⚠ symbol and note needed on the drawing? _____

    k. What is the general tolerance for two-place decimals? _____

    l. What is the dimensional height of the two ∅1.00-inch raised surfaces (called bosses)? _____

    m. What is the overall length of the object? _____

    n. What is the overall width of the object? _____

©1995 West Publishing Company

SECTION 13

Detail and Assembly Drawings

LEARNER OUTCOMES

You will be able to:

- Identify two groups of engineering drawings.
- Name two types of detail drawings and explain their differences.
- List twelve types of drawings in use today by business and industry.
- Describe the purpose of the design layout drawing.
- Identify six items of information normally found on an assembly drawing.
- Explain the difference between an inseparable and separable assembly drawing.
- Define the term *expanded assembly drawing*.
- Demonstrate your learning through the successful completion of competency exercises.

13.1 INTRODUCTION

The industrial sector requires many types of drawings and diagrams for the various applications and uses of consumer, commercial, and military products produced throughout the world. Some types of drawings are required more often than others, depending on the nature of an industry's product or product line and on the size and structure of the company. This section is devoted to a discussion of detail and assembly drawings.

13.2 TYPES OF DETAIL AND ASSEMBLY DRAWINGS

The following list gives an indication of the multitude of different types of detail and assembly drawings used by industry.

- The assembly drawing
- The specification control drawing
- The source control drawing
- The envelope drawing
- The design layout drawing
- The detail drawing
- The erection drawing
- The expanded assembly drawing
- The installation drawing
- The kit drawing
- The mechanical schematic drawing
- The modification drawing

- The numerical control drawing
- The piping diagram
- The proposal drawing
- The tabulated assembly drawing
- The tube bend drawing

Included in the preceding list are those drawings used in the electrical and electronics fields as well as those for use in the areas of building and construction, oil and gas lines, and field service. Because the primary users of this worktext are those who are or will be working in the machine trades, four types of drawings will be featured: the design layout drawing, the detail drawing, the assembly drawing, and the expanded assembly drawing.

Detail and assembly drawings are the basis of manufacturing throughout the world. For this reason, interpreting them accurately and understanding their intent is of extreme importance in the production of quality goods. The title of this worktext refers to the interpretation of engineering drawings. Engineering drawings are produced in an engineering department for the purpose of becoming data in print or in an electronically produced visual form that shop personnel can work from. They are called **working drawings**, or **production drawings**. Whatever they are called, they are technical in nature and present information, ideas, and instructions in pictorial and text form.

Engineering drawings are classified into two groups: **detail drawings** and **assembly drawings**. Since the parts in these drawings may be manufactured in one facility and assembled in another location, the drawings must be consistent with some standard practice. In the United States, the standard used by most medium and large corporations is one that has been developed by the American National Standards Institute (ANSI). ANSI is a professional organization that proposes, develops, modifies, approves, and publishes drafting and manufacturing standards. For the production of military products, the Department of Defense (DOD) has standards to which the companies doing work for the United States government must conform.

While a detail drawing is a basic drawing that represents all the necessary information to fabricate an item—including the shape description, the size description, and the specifications of the part—an assembly drawing depicts a product in its completed form. It is a combination of detail parts that are joined to form a subassembly or a complete unit. The assembly drawing consists of at least two parts and may include hundreds, depending on the requirements.

■ 13.2.1 The Design Layout Drawing

In an engineering department, before detail and assembly drawings are produced, there needs to be a **design layout drawing** for anything more complex than the very simplest of designs. The development of the design layout drawing is usually the responsibility of the designer or the design engineer. The design layout drawing graphically illustrates a design concept and is the origin or basis for all detail and assembly drawings. The design layout is not for the purchasing of parts, for inspection, or for manufacturing purposes, but rather provides sufficient information to the detail drafting group so that a complete set of engineering drawings may be completed with a minimum amount of consultation with the designer. Layouts are drawn accurately and usually at full scale so that detail drawings may be prepared by scaling the layout drawing.

In the process of producing the design layout drawing, the designer makes certain that sufficient information is included. At times, information such as pertinent dimensions, critical tolerances, clearances, materials, finishes, important manufacturing processes, and specific notes may be required. In other instances, the detailer is responsible for taking all the information off the layout. Figure 13.1 provides an illustration of a design layout drawing of a hydraulic piston, while Figure 13.2 shows a layout of a motor yacht rudder weldment. More often than not, the design layout drawing is produced on polyester film for purposes of accuracy and longevity.

■ 13.2.2 The Detail Drawing

After the design layout drawing is complete and approved, the drafting group is called on to produce the details. This is accomplished by producing a drawing of each individual part on the layout. Two types of detail drawings are used by industry. One is referred to as the **mono-detail drawing**, and the other is the **multi-detail drawing**. *Mono* implies that only one

FIGURE 13.1
Design layout drawing

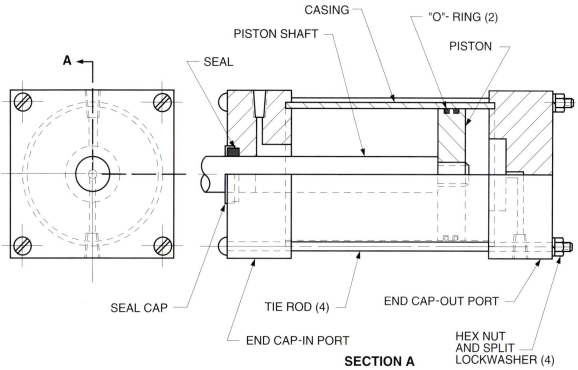

NOTES:
1. FIT BETWEEN INSIDE OF CASING AND PISTON IS A RUNNING FIT
2. IN AND OUT PORTS ARE .250 NPT
3. UNC-2 THREAD IN PISTON

DESIGN LAYOUT
DRAWING # HPX-101
ASSY, HYDRAULIC PISTON

DRAWN BY: EAM
DATE: 30 JUNE 1993
SCALE: FULL

detail drawing is produced for each part. *Multi* means that two or more detail parts may appear on a single drawing sheet of drawing medium. Figures 13.3a, an adjusting screw, and 13.3b, a machine screw jack, clearly illustrate the difference between the mono-detail and multi-detail drawings.

13.2.3 The Assembly Drawing

Functional groupings of components for machines and mechanisms for various products are composed of numerous parts. The drawing that depicts a product in its completed form is called an assembly drawing. An assembly drawing represents the assembled relationship of (1) two or more parts; (2) a combination of detail parts that are joined to form a subassembly or a complete unit; or (3) a group of assemblies required to form an assembly of higher order. Assembly drawings differ from one another depending on their complexity and the information required, but they usually contain most of the following:

- A sufficient number of views to clearly show the relationships among parts.
- Sectional views that show internal features, function, and assembly.
- Enlarged views to identify necessary details.

FIGURE 13.2
Design layout drawing

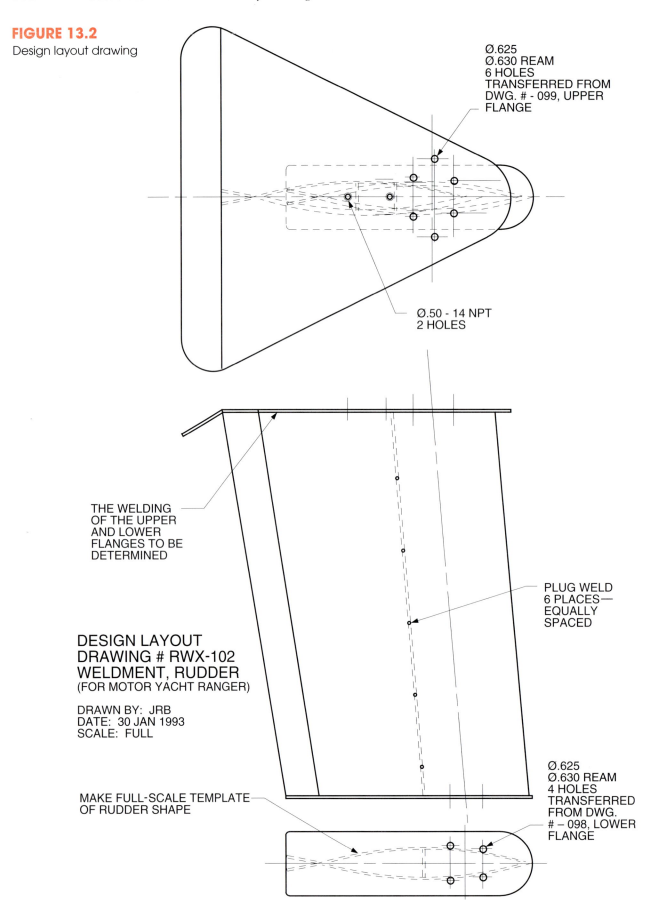

FIGURE 13.3a
Mono-detail drawing

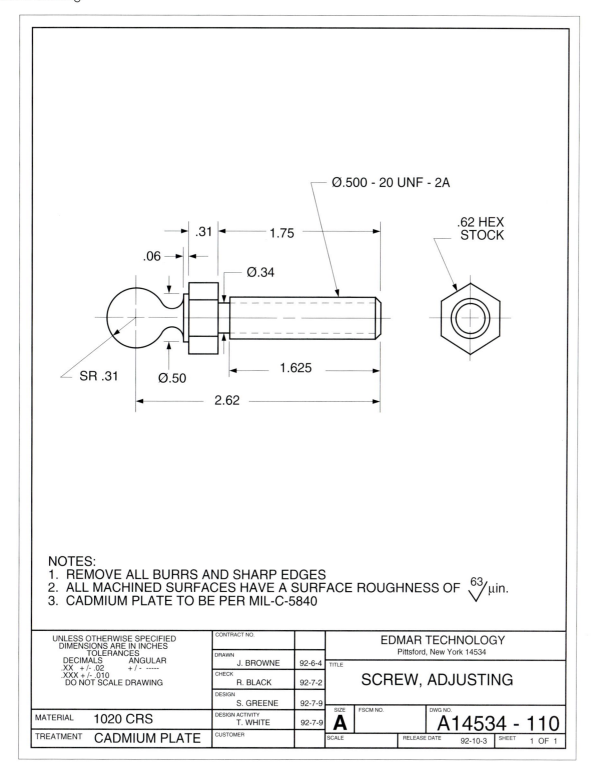

FIGURE 13.3b
Multi-detail drawing

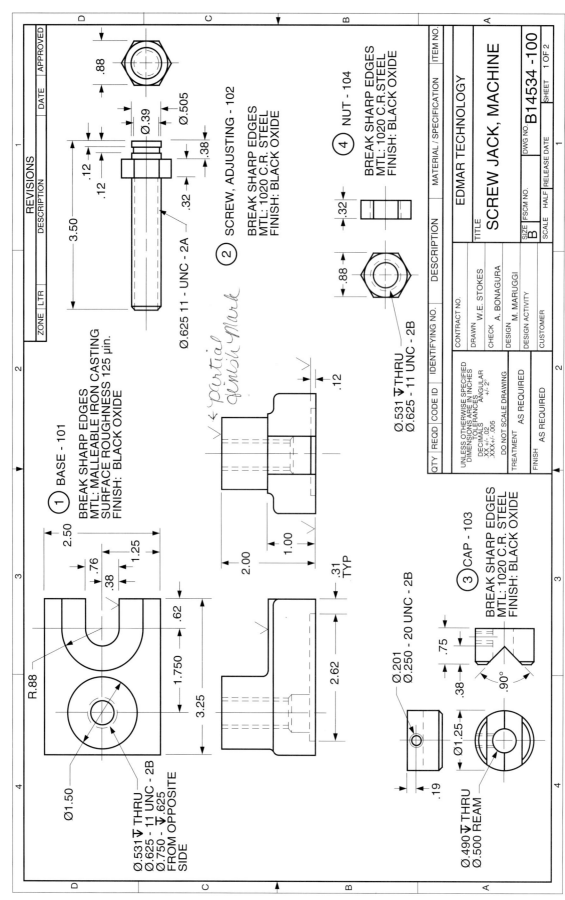

- Principal reference dimensions or the dimensions necessary for assembly.
- A parts list, including balloon number identifiers.
- Any affected manufacturing processes required for assembly.

13.2.3.1 Types of Assembly Drawings

Of the several different types of assembly drawings, the four most common include the separable, inseparable, detail, and expanded assembly drawings. A **separable assembly drawing** shows the assembled relationship between two or more items in which at least one of the items can be disassembled for servicing or replacement without causing damage or destruction to any of the other parts. This type of drawing is also referred to as a **general assembly drawing**, and it is, by far, the most common. Figure 13.4 provides an example of a separable assembly drawing.

An **inseparable assembly drawing** represents the assembled relationship between two or more parts separately fabricated but permanently joined together by manufacturing processes like brazing, riveting, cementing, welding, or soldering. This type of assembly is considered a single, stand-alone item. The clip assembly in Figure 13.5 is an inseparable assembly.

Figure 13.6 illustrates a **detail assembly drawing**. This type of drawing has several dimensions because, while it identifies the relationship between assembled items, it also details one or more parts of the assembly. Individual drawings are not required for drawings so illustrated, but fabrication details are an important consideration. All the detail parts are made from this single drawing.

An **expanded assembly drawing** is also referred to as an **exploded view drawing**. It graphically describes an assembly/disassembly relationship between parts in isometric or perspective form. The parts on this type of drawing appear to explode along the center lines. One will find this type of drawing in vendors' catalogs as well as in instruction booklets, maintenance manuals, or service bulletins for consumer, industrial, and military products. An example of a simple expanded assembly drawing of a parallel clamp assembly appears in Figure 13.7. Figure 13.8 is a more complex illustration of a vibration mounting assembly. Note how easy it is to determine how the parts are assembled.

13.3 SUMMARY

The production of goods throughout the world requires many different kinds of drawings to serve the many purposes of business and industry. Of the several drawings identified in this section, four commonly used by the industrial community were covered in depth: the design layout drawing, the detail drawing, the assembly drawing, and the expanded assembly drawing. The design layout is usually the first drawing produced on a project of any magnitude, and it is the drawing in which detail drawings are initiated. It was pointed out that the design layout drawing illustrates a design concept and is the basis for all detail and assembly drawings. It is an engineering drawing that later translates into a working drawing, also known as a production drawing.

Mono-detail and multi-detail drawings were illustrated, and examples were provided. The assembly drawing was defined as being one that represents the relationship of (1) two or more parts; (2) a combination of detail parts that are joined to form a subassembly or a complete unit; or (3) a group of assemblies. Several types of assembly drawings were covered, including the separable, inseparable, detail, and expanded. Illustrations of each were provided.

TECHNICAL TERMS FOR STUDY

assembly drawing A drawing that represents the assembled relationship of (1) two or more parts; (2) a combination of detail parts that are joined to form a subassembly or a complete unit; or (3) a group of assemblies required to form an assembly of higher order.

design layout drawing A drawing that graphically illustrates a design concept and is the origin or basis for all detail and assembly drawings.

Section 13: *Detail and Assembly Drawings*

detail assembly drawing   A drawing identifying the relationship between assembled items while detailing one or more parts of an assembly. It usually requires dimensions.

detail drawing   A drawing that represents all the necessary information to fabricate an item, including the shape description, size description, and specifications of the part.

expanded assembly drawing   Also referred to as an exploded view drawing. This drawing graphically describes the assembly/disassembly relationship between parts in isometric or perspective form.

exploded view drawing   *See* **expanded view drawing**.

general assembly drawing   Also referred to as a separable assembly drawing. It shows the relationship between two or more items in which at least one of the items can be disassembled for servicing or replacement without causing damage or destruction to any of the other parts.

inseparable assembly drawing   A drawing that represents the assembled relationship between two or more parts separately fabricated but permanently joined together by processes like brazing, riveting, cementing, welding, or soldering.

mono-detail drawing   A drawing in which only one part or item is displayed.

multi-detail drawing   A drawing in which two or more parts or items appear on a single drawing sheet or drawing medium.

production drawing   A drawing with sufficient detail and information so that it is capable of being manufactured by shop personnel. Same as a working drawing.

separable assembly drawing   *See* **general assembly drawing**.

working drawing   *See* **production drawing**.

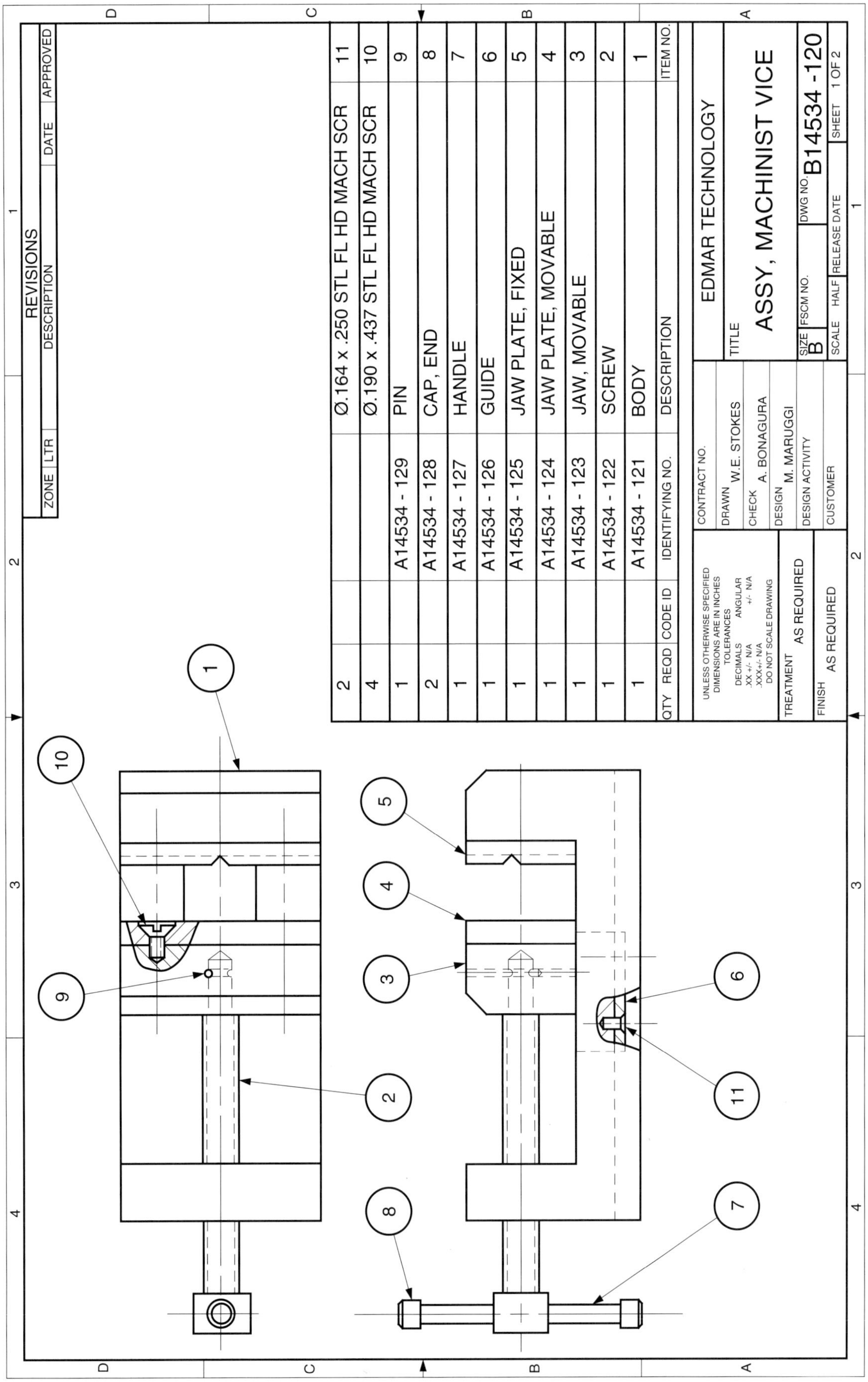

FIGURE 13.4

Separable assembly drawing

REVISIONS

ZONE	LTR	DESCRIPTION	DATE	APPROVED

Parts List

QTY REQD	CODE ID	IDENTIFYING NO.	DESCRIPTION	ITEM NO.
1		A14534 - 121	BODY	1
1		A14534 - 122	SCREW	2
1		A14534 - 123	JAW, MOVABLE	3
1		A14534 - 124	JAW PLATE, MOVABLE	4
1		A14534 - 125	JAW PLATE, FIXED	5
1		A14534 - 126	GUIDE	6
1		A14534 - 127	HANDLE	7
2		A14534 - 128	CAP, END	8
1		A14534 - 129	PIN	9
4			Ø.190 x .437 STL FL HD MACH SCR	10
2			Ø.164 x .250 STL FL HD MACH SCR	11

CONTRACT NO.

DRAWN	W.E. STOKES
CHECK	A. BONAGURA
DESIGN	M. MARUGGI
DESIGN ACTIVITY	
CUSTOMER	

UNLESS OTHERWISE SPECIFIED
DIMENSIONS ARE IN INCHES
TOLERANCES
DECIMALS   ANGULAR
.XX +/- N/A   +/- N/A
.XXX +/- N/A
DO NOT SCALE DRAWING

TREATMENT   AS REQUIRED

FINISH   AS REQUIRED

EDMAR TECHNOLOGY

TITLE

ASSY, MACHINIST VICE

SIZE **B**	FSCM NO.	DWG NO. **B14534 -120**
SCALE HALF	RELEASE DATE	SHEET 1 OF 2

Section 13: *Detail and Assembly Drawings*

FIGURE 13.5
Inseparable assembly drawing

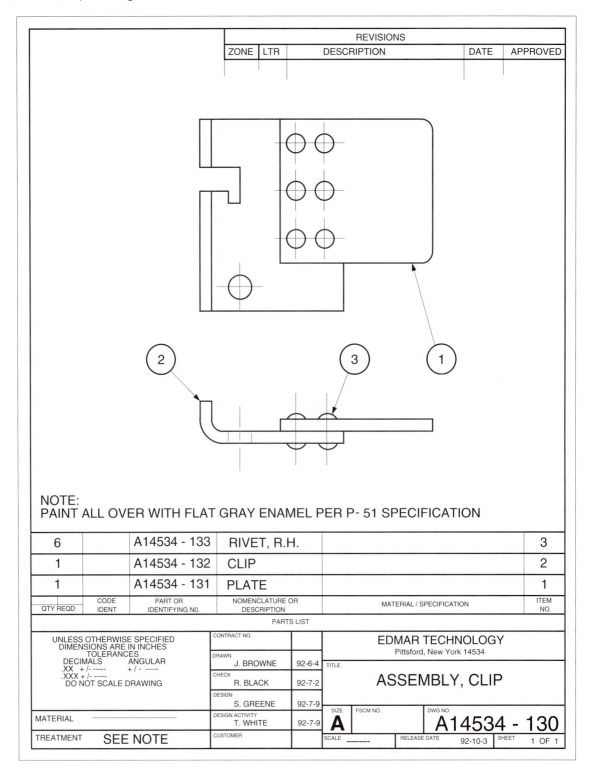

FIGURE 13.6
Detail assembly drawing

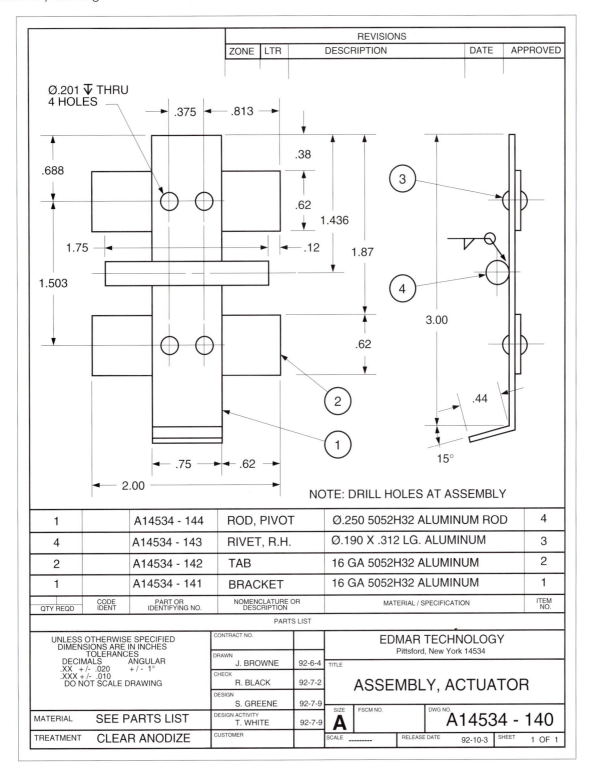

Section 13: *Detail and Assembly Drawings*

FIGURE 13.7
Expanded assembly drawing

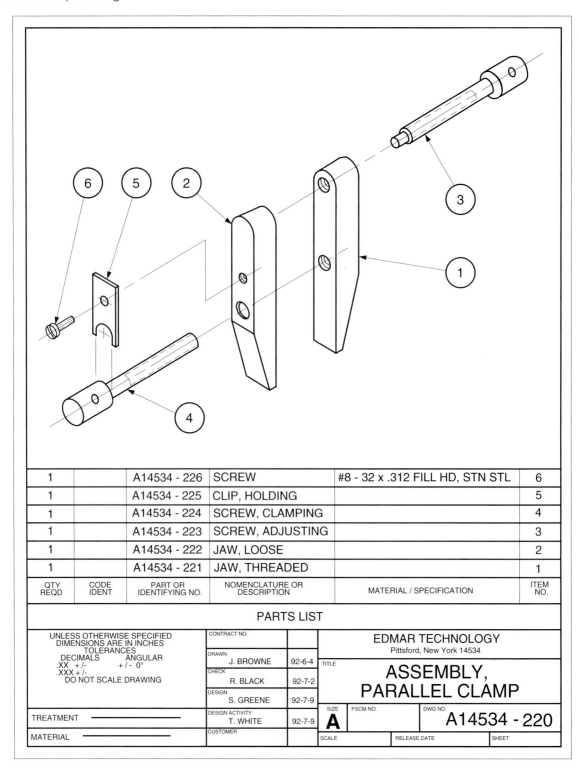

Section 13: *Detail and Assembly Drawings* **311**

FIGURE 13.8
Expanded assembly drawing

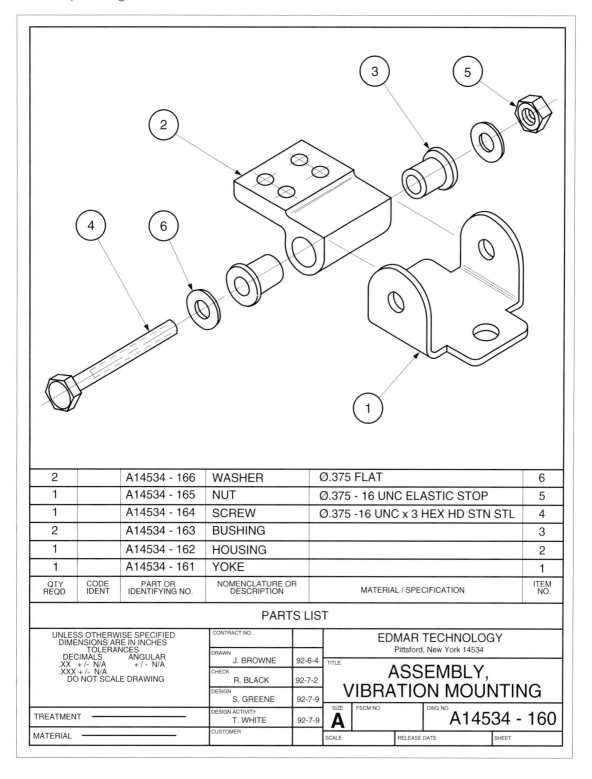

Section 13: *Detail and Assembly Drawings* 313

Student _____   Date _____

Section 13: Competency Quiz

PART A COMPREHENSION

1. Define the term *design layout drawing*. (8 pts)

Graphically illustrate a design concept to develop all design & assembly drawings

2. What types of information are visually described on a detail drawing? (10 pts)

Shape description, size & spec of part all info to fabricate the item

3. Explain the difference between a mono-detail and multi-detail drawing. (10 pts)

Mono - 1 detail drawing for each part
Multi - more than 1

4. What does an assembly drawing represent? (8 pts)

Shows a part in completed form

5. Identify two types of engineering drawings. (4 pts)

detail
assembly

6. List twelve types of drawings in use today. (12 pts)

1 assembly   5) detail   9) electronic schematic
2 spec control   6) installation   10) tooling
3 source control   7) kit   11) proposal
4 design layout   8) modification   12) tabulated assembly

7. Identify six items of information normally found on an assembly drawing. (12 pts)

1) sufficient top views   5) parts list w/ balloon identifiers
2) sectional views
3) enlarged view   6) affected manufacturing
4) principal ref. dimension   processes for assembly

©1995 West Publishing Company

314   **Section 13:** *Competency Quiz*

8. Explain the difference between an inseparable and separable assembly drawing. (8 pts)

9. What is another term for *working drawing*? (8 pts)

10. Describe an expanded assembly drawing. (5 pts) By what other name is it also called? (5 pts)

11. List four places where the expanded assembly drawing would most likely be found. (8 pts)

Section 13: *Detail and Assembly Drawings*   **315**

Student _____   Date _____

PART B TECHNICAL TERMS

For each definition, select the correct technical term from the list on the bottom of the page. (10 pts each)

1. _____ A drawing that graphically illustrates a design concept and is the origin or basis for all detail and assembly drawings.

2. _____ A drawing in which two or more parts appear on a single drawing sheet or drawing medium.

3. _____ A drawing that graphically describes the assembly/disassembly between parts in isometric or perspective form.

4. _____ A drawing that represents all the necessary information to fabricate an item, including the shape description, size description, and specifications of the part.

5. _____ A drawing showing the relationship between two or more items in which at least one of the items can be disassembled for servicing or replacement without causing damage to any of the other parts.

6. _____ A drawing identifying the relationship between assembled items while detailing one or more parts of an assembly.

7. _____ A drawing that represents the assembled relationship between two or more parts separately fabricated but permanently joined together by processes like brazing, riveting, cementing, welding, or soldering..

8. _____ A drawing with sufficient detail and information so that it is capable of being manufactured by shop personnel.

9. _____ A drawing on which only one part or item is displayed.

10. _____ Another term used for *working drawing*.

A. design layout drawing
B. detail assembly drawing
C. detail drawing
D. expanded assembly drawing
E. general assembly drawing
F. inseparable assembly drawing
G. installation drawing
H. kit drawing
I. mono-detail drawing
J. multi-detail drawing
K. numerical control drawing
L. production drawing
M. proposal drawing
N. source control drawing

©1995 West Publishing Company

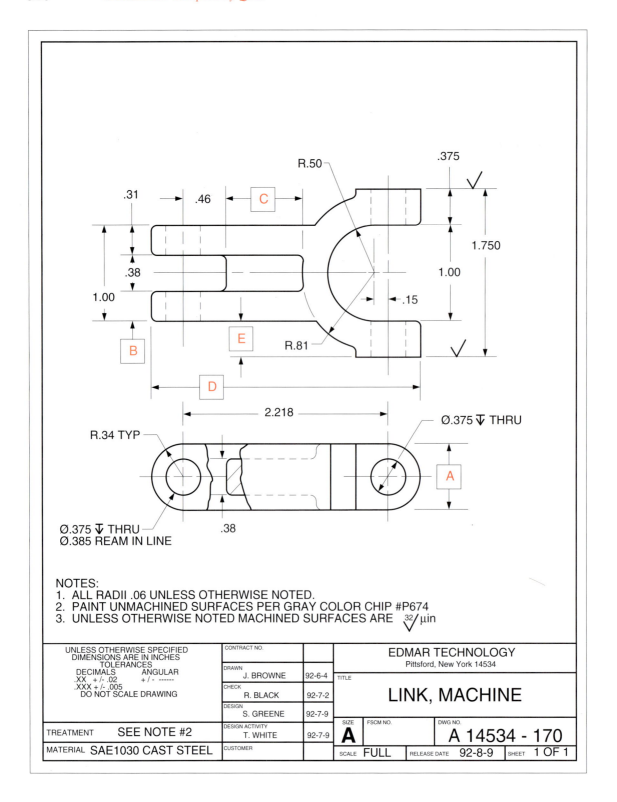

Section 13: *Detail and Assembly Drawings*   **317**

Student _____   Date _____

PART C INTERPRETING DETAIL DRAWINGS

1. For the machine link, drawing A14534-170 (see page 316), respond to the following questions.

   a. What specific type of drawing is A14534-170? _____

   b. Of what material is the part made? _____

   c. What is the upper limit and lower limit of the 2.218 dimension? _____

   d. What is the tolerance for the 1.750 dimension? _____

   e. What type of section is shown on the front view? _____

   f. What is height A of the part, shown in the lower view? _____

   g. What is the finished, machined, surface texture of the part? _____

   h. Which areas are painted on the part? _____

   i. What is the size of the reamed holes? _____

   j. How many views are called for? _____
      Name them.

   k. What are the dimensional limits for the R .81 dimension? _____

   l. What is the overall length of part D ? _____

   m. What is the dimension at B ? _____

   n. How many dimensioned radii are there on the drawing? _____

   o. What does the .15 dimension represent? _____

   p. What is the tolerance for all two-place decimals? _____

   q. What is the drawing scale? _____

   r. What is distance C ? _____

   s. What is the dimension at E ? _____

©1995 West Publishing Company

Section 13: Competency Quiz

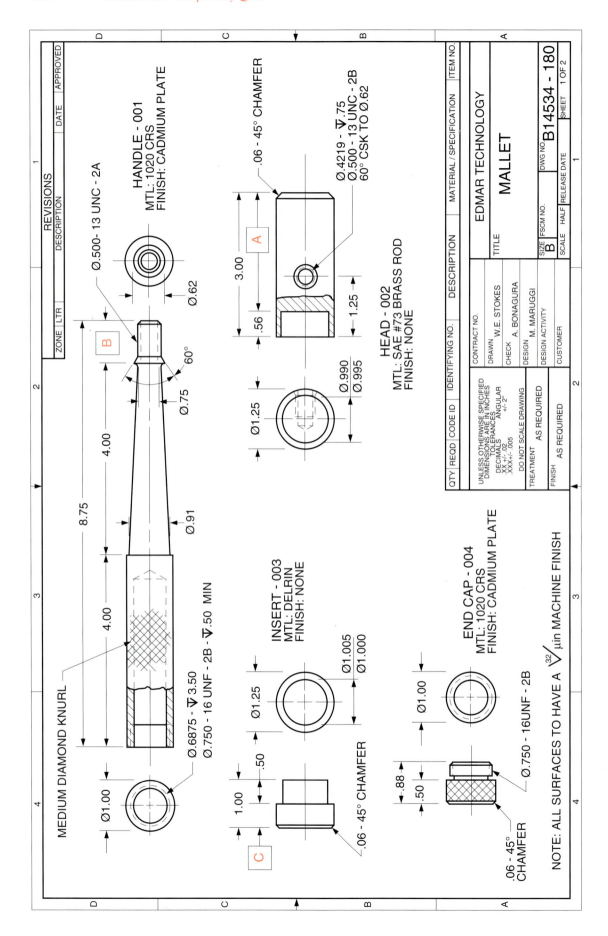

Section 13: *Detail and Assembly Drawings* **319**

Student _____ Date _____

2. For the mallet, drawing B14534-180 (see page 318), respond to the following questions.

   a. How many different detail parts are shown on the drawing? _____

   b. What specific name is given to a drawing of this type? _____ *multi detail* _____

   c. What is the material for the Head-002? _____ *SAE #773 Blass Red* _____

   d. What protective coating is required on the Insert-003? _____ *None* _____

   e. In the assembly of parts, the insert is fitted into the end of the head. What is the maximum interference between the two parts? _____ *.02* _____

   f. What is the thread size and series for the internal threads shown on the Handle-001? *⌀.750 - 16 UNF*

   g. What does CRS signify in a material specification? _____ *Cold Rolled Steel* _____

   h. What is the general tolerance for all two-place decimals, unless otherwise noted? _____ *+/- .02* _____

   i. What is the tap drill size for the ⌀.500-13 UNC-2B thread shown on the Head-002? _____
   What does UNC mean? *Unified threads*

   j. What is the minimum depth for the ⌀.750-16 UNF-2B threads shown on the Handle-001? *⌀.700*

   k. What is the finish requirement for the End Cap-004? _____

   l. What are the upper and lower limits for the overall length of the Handle-001? _____

   m. What is the dimension for ☐A on the Head-002? _____ *2.44* _____

   n. What is the dimension for ☐B on the Handle-001? _____ *.75* _____

   o. What is the dimension for ☐C on the Insert-003? _____ *.50* _____

   p. To what scale was the drawing produced? _____

   q. To what diameter is the countersink produced on the Head-002? _____ *⌀.62* _____

©1995 West Publishing Company

320   Section 13: *Competency Quiz*

r. What size chamfer is called for on (a) the Head-002, (b) Insert-003, and (c) End Cap-004?

a. _.06 45°_   **b.** _.06 × 45°_   **c.** _.06 -45°_

s. What is the angular tolerance for the 60-degree dimension on the Handle-001? _±2°_

t. What is the machine finish requirement for all the parts? _32 µin_

Section 13: *Detail and Assembly Drawings* **321**

Student _____   Date _____

PART D INTERPRETING ASSEMBLY DRAWINGS

Competency Quiz continues on next page.

©1995 West Publishing Company

Section 13: Competency Quiz

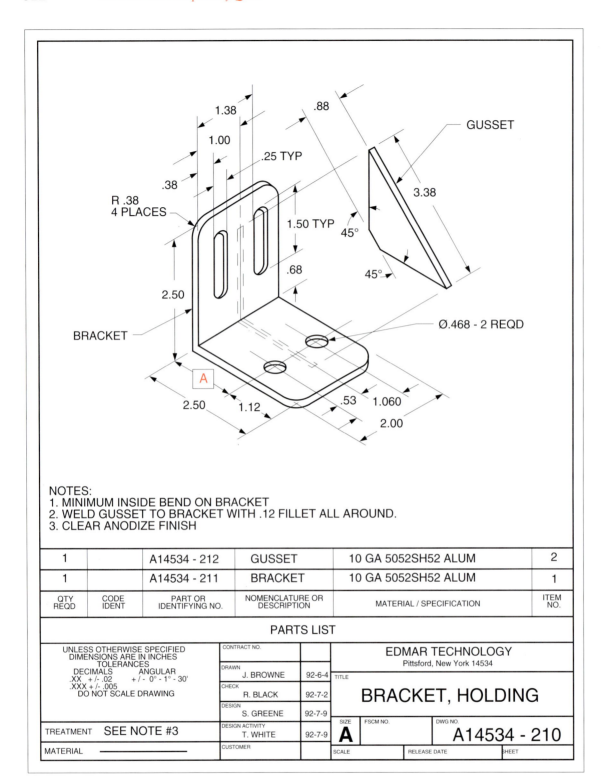

Section 13: *Detail and Assembly Drawings*   **323**

Student _____   Date _____

1. For the holding bracket, drawing A144534-210 (see page 322), respond to the following questions.

   a. What specific type of drawing is shown? _detail dwg._

   b. How many parts comprise the assembly? _2_

   c. What chemical conversion treatment is required to protect the parts? _alodine_

   d. What is the angular tolerance required on this drawing? _+/- 0 -1° -30_

   e. What is the height of the part? _2.50_

   f. What is the material for both the bracket and the gusset? _10 GA 5052.29752 Alum_

   g. With what type of weld are the parts joined? _.12 fillet_

   h. What radius is required on the corners of the bracket? _.38 4 places_

   i. What are the length and width dimensions for the slots on the bracket? _1.50 TYP .25 TYP_

   j. What dimension shows the location for the gusset on the bracket? _____

   k. What are the length and width dimensions for the gusset? _3.38 .88_

   l. What is the part number for the gusset? _A144534 - 212_

   m. What are the upper and lower limits for the Ø.468 holes? _.473 / .463_

   n. What is the dimension at [A] on the bracket? _1.30_

   o. How many gussets are required for the assembly? _2_

   p. What are the limits to the 3.38 dimension? _+/- .02_

   q. What is the fractional equivalent for the .25 dimension? _1/4_

   r. What is the size of the drawing? _A_

   s. What is the decimal equivalent for #10 gage aluminum? _____

©1995 West Publishing Company

Section 13: *Competency Quiz*

REVISIONS

ZONE	LTR	DESCRIPTION	DATE	APPROVED

ITEM	IDENTIFYING NO.	DESCRIPTION	MATERIAL / SPECIFICATION	QTY REQD	CODE ID
12	A14534 - 202	RING, RETAINING	EXTERNAL - #E375	6	
11	A14534 - 201	PIN		3	
10	A14534 - 200	RING, RETAINING	EXTERNAL - #E625	2	
9	A14534 - 199	PIN		1	
8	A14534 - 198	LOCKWASHER	Ø.500 STEEL	2	
7	A14534 - 197	NUT, HEX	Ø.500-13 UNC-2A STL	2	
6	A14534 - 196	BELT		1	
5	A14534 - 195	BUSHING		3	
4	A14534 - 194	STUD		1	
3	A14534 - 193	PULLEY		1	
2	B14534 - 192	PLATE, MOUNTING		2	
1	A14534 - 191	HOOK		1	

EDMAR TECHNOLOGY

ASSEMBLY, HOIST

CONTRACT NO.

DRAWN W.E. STOKES
CHECK A. BONAGURA
DESIGN M. MARUGGI
DESIGN ACTIVITY
CUSTOMER

UNLESS OTHERWISE SPECIFIED
DIMENSIONS ARE IN INCHES
TOLERANCES
DECIMALS   ANGULAR
XX +/-
XXX +/-
DO NOT SCALE DRAWING

TREATMENT NONE
FINISH NONE

SIZE B   FSCM NO.   DWG NO. B14534 - 190
SCALE HALF   RELEASE DATE   SHEET 1 OF 13

Section 13: *Detail and Assembly Drawings*   **325**

Student _____   Date _____

2. For the hoist assembly, drawing B14534-190 (see page 324), respond to the following questions and exercises.

    a. What total number of parts comprise the assembly? _____ *25* _____

    b. How is item ①, the hook, held in place? _____ *Hex Nut* _____

    c. What is the physical size of the detail drawing for the mounting plate? _____ *half* _____

    d. What is the thread size for the hex nut, A14534-197? _____ *0.50-13UNC* _____

    e. How many different-size retaining rings are required on the assembly? _____ *2* _____
        What are their part numbers? _____ *A14534-200* _____ / _____ *A14534-202* _____

    f. How many different-size pins are required on the assembly? _____ *2* _____

    g. What is the purpose of the pulley? _____ *to thread the belt* _____

    h. Identify part numbers and part names for parts other than items of hardware (nuts, bolts, pins, etc.).

PART NUMBER	PART NAME
A14534-191	*Hook*
B14534-192	*Mounting Plate*
A14534-193	*Pulley*
A14534-196	*Belt*

    i. What is item ④? _____ *Stud* _____
        What is its part number? _____ *A14534-194* _____

    j. How many of item ⑤ are required? _____ *3* _____
        What is their purpose? _____ *as a spacer* _____

©1995 West Publishing Company

Section 13: *Competency Quiz*

k. Why are there no tolerance requirements for the assembly? _____ all are tightened full_____

l. What is the material for the lockwasher? _____ Steel _____

m. Why are there different part numbers for the retaining rings? _____ different size _____

n. Are there multiple sheets to this drawing? _____ yes _____
   How many? _____ 3 _____

o. What type of lines are represented by [A], [B], [C], and [D]?

   A _hidden_____   B _object_____

   C _extension_____   D _dimension_____

SECTION 14

Development and Sheet Metal Drawings

LEARNER OUTCOMES

You will be able to:

- Identify three major types of development drawings.

- Describe the characteristics of the parallel line development, the radial line development, and the triangulation development.

- Explain the purpose of the fold line, the baseline, the seam line, and tabs in the layout of development drawings.

- State the purpose of a development drawing.

- Describe the intent of a transition piece.

- List eight methods of producing sheet metal parts.

- Identify six military, commercial, or consumer products that require the use of sheet metal components.

- Define ten technical terms related to sheet metal products.

- Demonstrate your learning through the successful completion of competency exercises.

14.1 INTRODUCTION

In industry, there is a very close relationship between development drawings and sheet metal drawings. Quite often, a **development drawing** is the basis for a **sheet metal drawing**. For example, a development drawing is usually a design layout drawing of a pattern or of a template. It is drawn and developed in a single plane by the drafter in preparation for folding, rolling, or bending to form some predetermined shape. These flat patterns or shapes, in turn, facilitate the cutting of a desired **blank** from sheet metal.

The purpose of a development drawing is to produce a model of an item with a particular shape or form, regardless of whether one or hundreds of thousands of the same item are to be produced. Examples of items that require a development drawing include heating and air conditioning ductwork, furnace parts, mailboxes, and metal containers and tins of all sizes and shapes. In addition, a one-of-a-kind item could be a special plenum, or transition piece, that attaches to the top of a furnace. Such a piece is illustrated in Figure 14.1. Note that the bottom of the plenum is square and the top is round; hence the name *transition piece*. It is made of thin-gage sheet metal.

Another item for which a development drawing is required is the paper cup. Paper containers are produced in hundreds of round, conical, square, rectangular, and oval sizes and shapes, as shown in Figure 14.2.

Before a paper container assumes its final form, there needs to be a design layout of the container. The development drawing of the sidewall of a typical paper cup, when laid flat in a single plane, has the

328 Section 14: *Development and Sheet Metal Drawings*

FIGURE 14.1
Furnace plenum, or transition piece

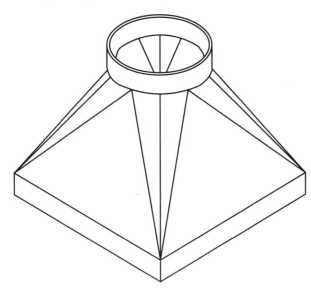

shape illustrated in Figure 14.3a. Its bottom blank is shown in Figure 14.3b.

Figure 14.4 illustrates how valuable a development drawing can be as it sequences the assembly of a paper cup.

Sheet metal drawings are required in many areas of industry, including the automotive, construction, and aircraft industries, but they are particularly useful in the electronics industry, where equipment enclosures and mountings for mechanical and electromechanical components can and do take many forms. Sheet metal parts are usually produced by *shearing, bending, forming, drawing, piercing, rolling, beading, spinning, trimming, punching, perforating, blanking,* or any combination of these processes.

FIGURE 14.2
Paper cups and containers

Photo courtesy of Paper Machinery Corp.

FIGURE 14.3
Sidewall and bottom blanks of a paper cup

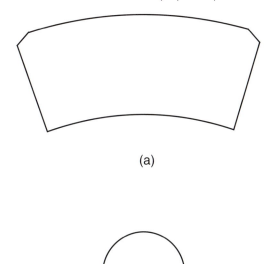

(a)

(b)

FIGURE 14.4
Sequence of operations for the assembly of a paper cup

14.2 TYPES OF DEVELOPMENT DRAWINGS

A surface that can be unfolded or rolled out without distortion is considered *developable*, which means that it can be laid out in a flat, single plane. Objects that consist of single-curved surfaces are said to be developable. Examples of single-curved surfaces are presented in Figure 14.5.

Double-curved, distorted, or warped surfaces, similar to ones shown in Figure 14.6, are termed *non-developable*. Surfaces of this type can be developed only by approximation and will appear on a development drawing as though the material has been stretched. For example, a sphere, such as a ball, cannot be wrapped smoothly because of its curvature; however, if it is to be covered with a flexible, pliable, somewhat elastic material, the covering can be stretched to fit. Figure 14.6 provides examples of distorted, double-curved, or warped surfaces.

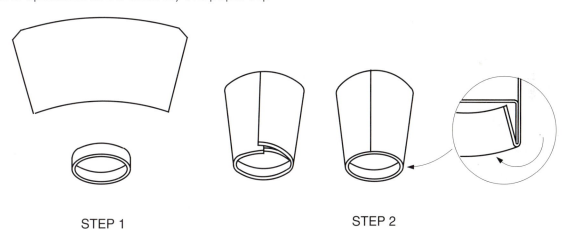

STEP 1

STEP 2

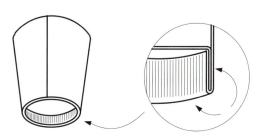

STEP 3

COMPLETED CUP

330 **Section 14:** *Development and Sheet Metal Drawings*

FIGURE 14.5
Developable, single-curved surfaces

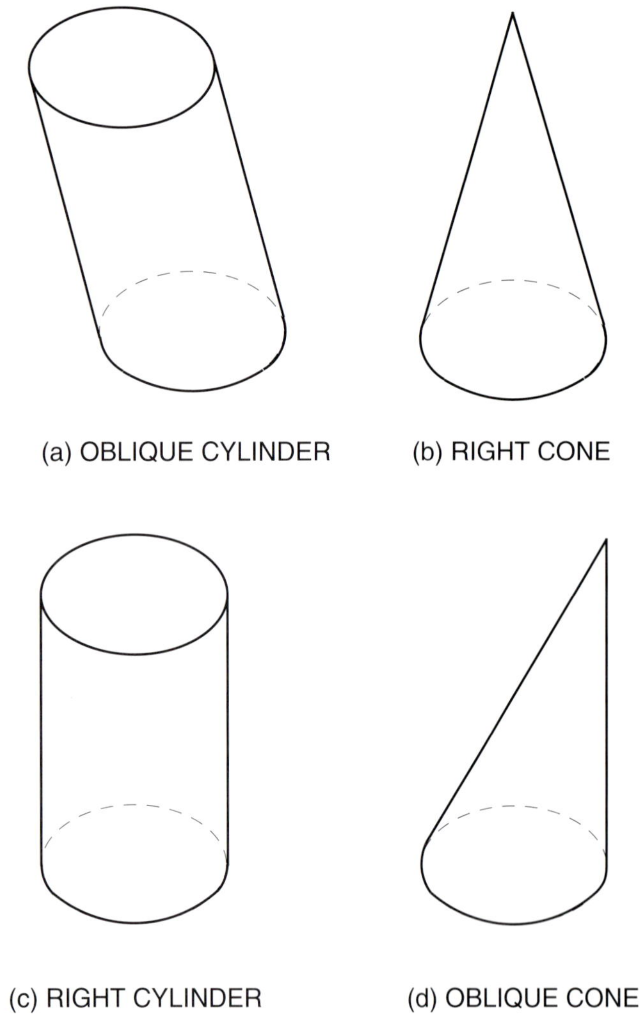

(a) OBLIQUE CYLINDER

(b) RIGHT CONE

(c) RIGHT CYLINDER

(d) OBLIQUE CONE

FIGURE 14.6
Distorted, double-curved, or warped surfaces

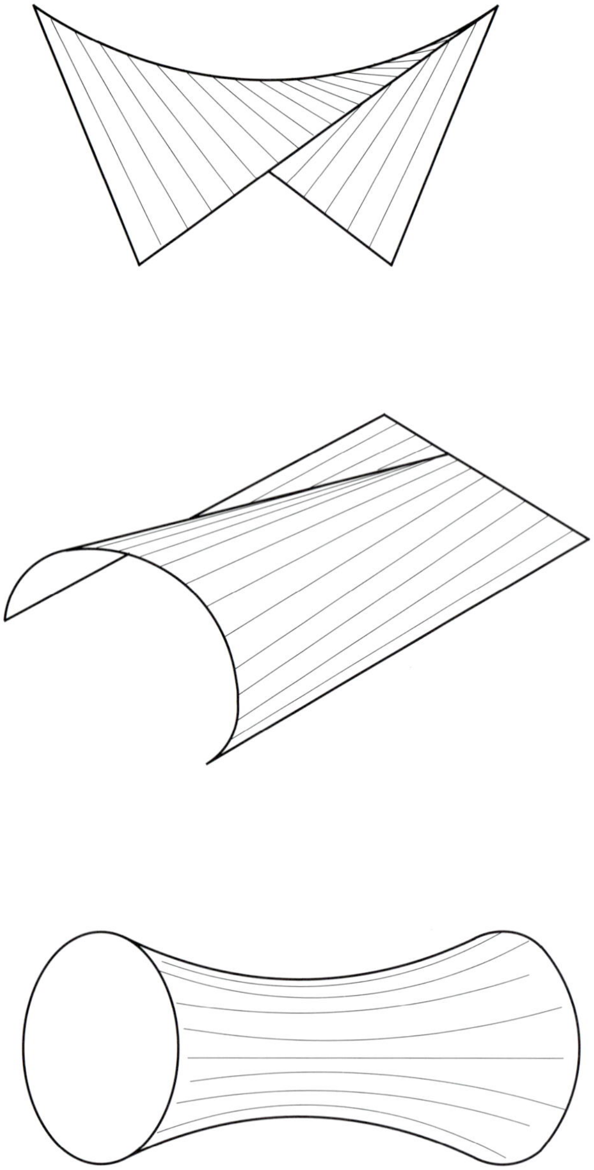

There are three major types of single-curved surface developments used in industry. They include the **parallel line development**, the **radial line development**, and the **triangulation development**.

■ 14.2.1 Parallel Line Development

Prisms and rectangular, cubical, and cylindrical shapes are those for which the parallel line development is used. Parallel line development implies that the pattern is drawn full size using parallel lines for the height, width, and depth features of an object. Although a pictorial view is important for the visualization of an object, the development drawing requires the orthographic projection of the views and surfaces involved. Figure 14.7 shows a pictorial view of a file as well as an orthographic projection of the file,

which is a simple rectangular box used as a wall-hung basket whose material is thin-gage sheet metal. Note that the lateral surfaces are parallel. Figure 14.7c shows the object in an unfolding position. Figure 14.7d represents the file in the flat, opened position. Because the object cannot be completely closed or sealed as laid out, it needs to have tabs added to several of its surfaces to be usable, as illustrated in Figure 14.7e. At this point, it is considered to be a complete development of the file. Note the addition of two holes for fastening it to a wall.

Section 14: *Development and Sheet Metal Drawings* 331

FIGURE 14.7
Parallel line development

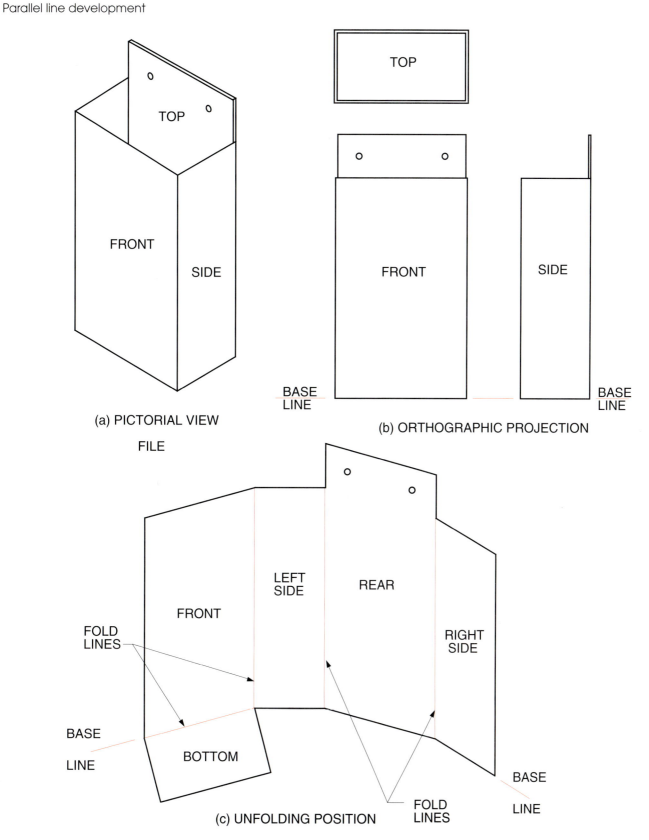

Section 14: *Development and Sheet Metal Drawings*

FIGURE 14.7
Continued

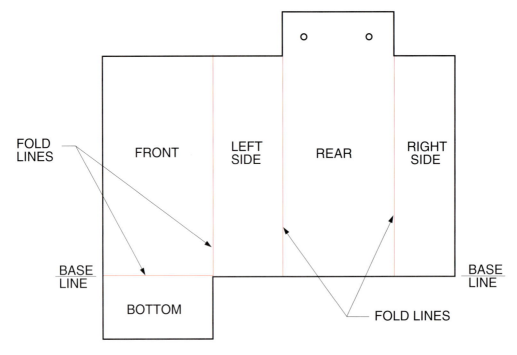

(d) FLAT, OPENED POSITION

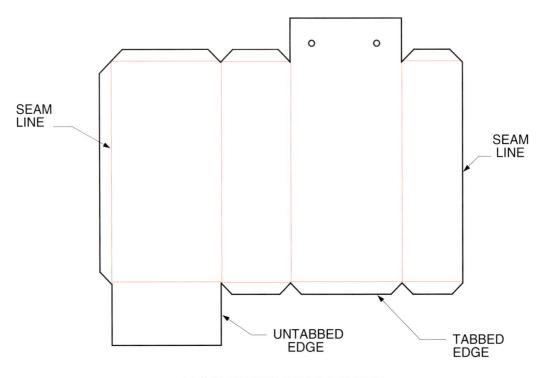

(e) COMPLETED DEVELOPMENT

Section 14: *Development and Sheet Metal Drawings* **333**

FIGURE 14.8
Parallel line development of a truncated cylinder

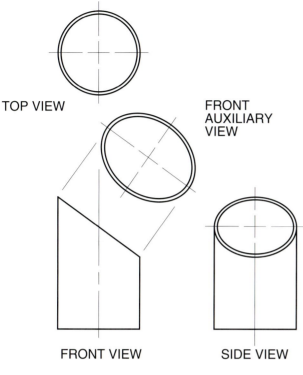

Important terms in this type of development include **baseline**, which is the point from where the layout is developed; **fold lines**, which are lines parallel to each other and perpendicular to the baseline; **seam line**, the point or line that, when joined, allows the object to take form; and **tabs**, which are extensions to the main body of an object and are required to fasten the object together to complete its boxlike shape.

If an object is cylindrical in shape and is cut at an angle (truncated), as shown in Figure 14.8a, the pattern is developed by dividing a circle into an equal number of segments (usually twelve) and laying it out in the flat, as Figure 14.8 illustrates.

■ 14.2.2 Radial Line Development

The radial line development differs from the parallel line development in that all fold lines converge on a single point. Therefore, fold lines are not parallel, as in parallel line development. The pyramid shown in Figure 14.9 is an example of an object whose development is produced using the radial line method. Note that all four edges and sides meet at a single point. Also, in this type of pattern drawing, the true length of each line (edge) must be determined through some manipulation of the object. The resulting flat pattern is illustrated in Figure 14.9c. The addition of tabs will make the pattern complete.

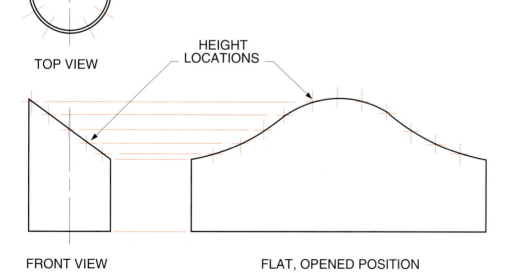

334 Section 14: *Development and Sheet Metal Drawings*

FIGURE 14.9
Radial line development of a pyramid

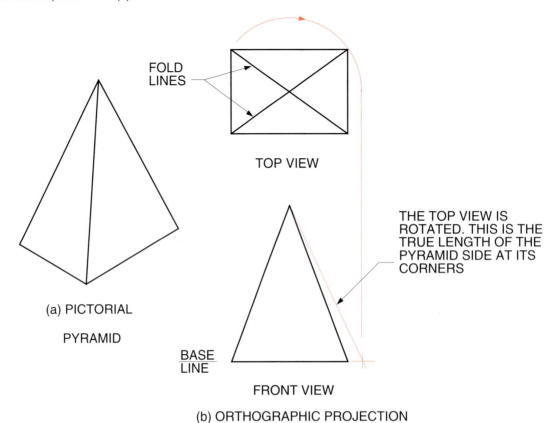

14.2.3 Triangulation Development

The term *triangulation* is derived from the word *triangle*. Triangulation development is a method of dividing a surface into a number of triangles and then transferring the true sizes and forms of the triangles into a development or pattern drawing. An example of an object for which a triangulation development can be used is a transition piece, such as the offset hopper assembly shown in Figure 14.10a. Transition pieces are used whenever it is necessary to connect the openings of different geometric shapes. The offset hopper assembly is an example of the development of rectangular-to-round transition. Figure 14.10b shows the flat, developed result of the offset hopper without its collar.

14.3 SHEET METAL DRAWINGS

With an increase in the number of applications for thin-gage metal products for such items as instrument cases, gages, equipment enclosures for mechanical and electromechanical components, copier parts, medical equipment components, and office furniture, the sheet metal fabrication business is a large one in this country. A sampling of thin-gage fabricated parts is pictured in Figure 14.11.

One of the major areas of sheet metal fabrication is the commercial and military electronics field. Equipment enclosures and mountings for electronics, electromechanical, and electro-optics packaging can take many forms. Quite often, components are mounted on or in equipment such as a rack, chassis, cabinet, or console. For the most part, these mountings are produced from thin-gage sheet metal. This type of fabrication includes practices that are quite different from those used for machined parts. As a result, there are several terms and fabrication techniques with which shop personnel who interpret drawings should be familiar. They include the following:

- **Rack**—A thin-gage metal structure that includes long vertical members (angles). It is normally made of steel or aluminum and houses electronic, electromechanical, or mechanical components, either permanently mounted or mounted

FIGURE 14.10a
Offset hopper assembly

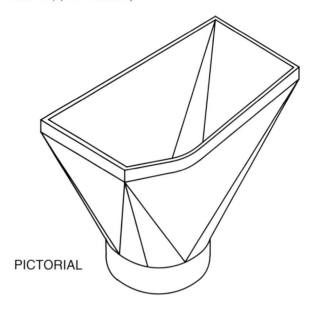

PICTORIAL

HOPPER ASSEMBLY, OFFSET

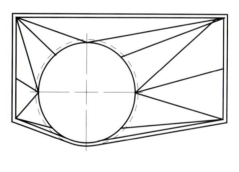

TOP VIEW

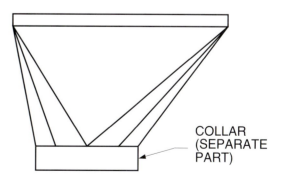

COLLAR (SEPARATE PART)

FRONT VIEW

336 Section 14: *Development and Sheet Metal Drawings*

FIGURE 14.10b
Triangulation development of the offset hopper

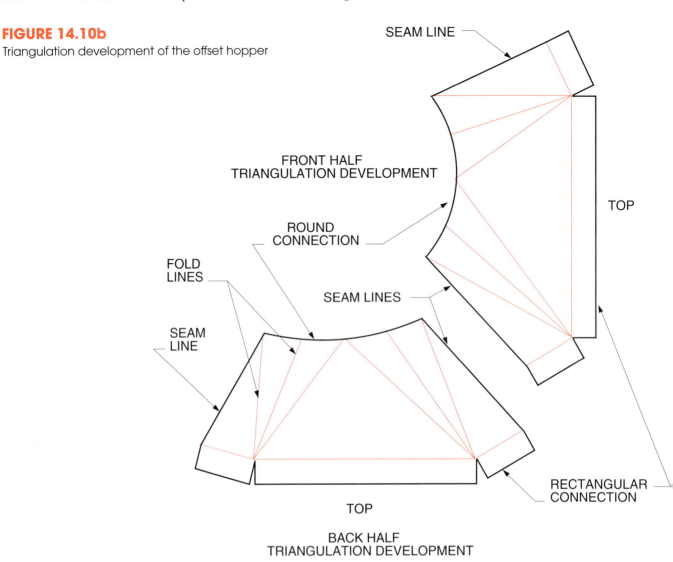

FIGURE 14.11
Thin-gage metal parts

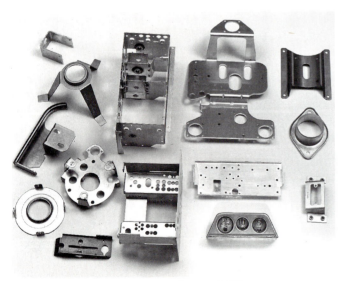

Photograph courtesy of Boker's, Inc., Minneapolis, MN

in drawers with slides. They are available in various heights, most being approximately six feet high. A standard rack is illustrated in Figure 14.12.

- **Chassis**—A thin-gage metal structure or base that is designed to support mechanical or electronic components. It may be of any size or shape and is capable of being mounted on a standard rack panel or used separately. It may or may not have a cover or be enclosed. Examples of a U-type, box type, and rack-mounted chassis appear in Figure 14.13.

- **Cabinet**—This is normally a unit made of aluminum or steel of either a standard or a special shape, depending on the application. The horizontal mountings are designed to accept a standard rack-mounted panel. A cabinet with a desk-

FIGURE 14.12
Standard equipment rack

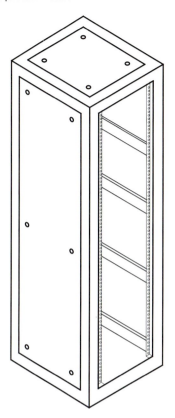

FIGURE 14.13
Common chassis types

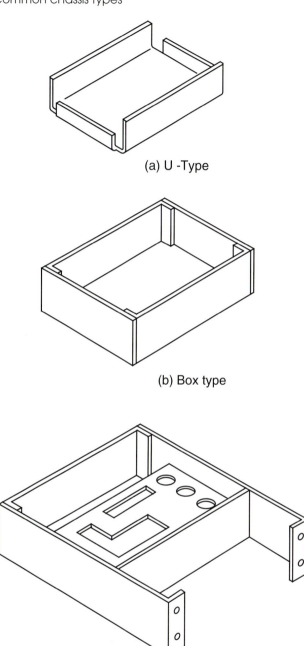

(a) U-Type

(b) Box type

(c) Rack mounted type

top surface may also be referred to as a **console**. Figure 14.14 shows a console.

Additional terms used throughout metal-fabricating industries are centered around laying out parts produced through forming, bending, and stamping. These terms are as follows:

- **Bend allowance**—The length of material around a bend from bend line to bend line, as represented in Figure 14.15.
- **Bend line**—The line of tangency where a bend changes to a flat surface, as depicted in Figure 14.15. Note that each bend has two bend lines, one on each surface.
- **Bend angle**—The angle to which a piece of sheet metal is bent. It is measured from the flat through the bend to the finished angle after bending, as pictured in Figure 14.16.
- **Leg**—The flat or straight section of a part after bending or forming. Two legs are shown in Figures 14.15 and 14.16. Legs are also referred to as *straight lengths*.

■ 14.3.1 The Developed Length

One of the most common operations performed on sheet metal parts in industry is the forming or bending operation. This operation usually takes place after all holes, slots, and cutouts are produced on the blank, or when the part is in the flat. Bending is performed with a power-operated machine called a **brake**. There

FIGURE 14.14
Console

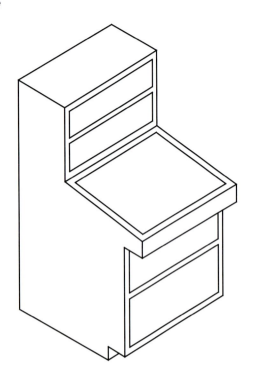

FIGURE 14.15
Sheet metal terms

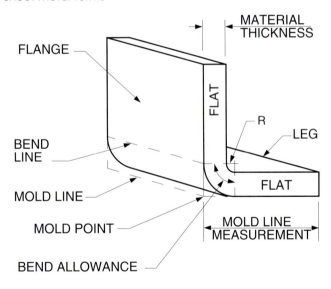

FIGURE 14.16
Bend angle

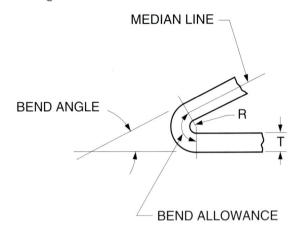

are small brakes and large brakes, which are capable of bending to a near zero bend radius. The recommended radius of a bend is dependent on the type and thickness of the material used. If too small a radius is specified, the material may develop a crack or actually fracture at the bend. Appendix G can be used for reference in determining the appropriate bend radius for a given application. It can also be used to solve problems that require the calculation of developed lengths with 90-degree or greater bends.

■ 14.3.2 Details of Sheet Metal Parts

The cost of sheet metal parts varies with the quantity to be produced and the design of the part. The per-unit cost for a small quantity of piece parts is generally higher than that for large quantities because the cost of tooling must be spread over the total number of parts produced. Also, the type and number of mechanical operations to be performed on each unit affects the final cost.

When interpreting sheet metal part drawings, there are peculiarities that occur only when dealing with thin-gage parts. The following identifies several design criteria and features that are most often used with sheet metal part drawings:

- *Progressive or datum dimensions*—This dimension locates part features with respect to datum lines. It was previously described in Section 8.4, "Application of Dimensions." Figure 14.17 illustrates progressive dimensioning of a plate, which is a thin-gage, flat-stamped part with a series of holes and slots.
- *Dimensions for flanged or formed parts*—Dimensions will normally be located from an unformed edge, usually beginning from a lower-left-side edge, as shown on the bracket in Figure 14.18.

Section 14: *Development and Sheet Metal Drawings* **339**

FIGURE 14.17
Progressive dimensioning of a flat sheet metal part

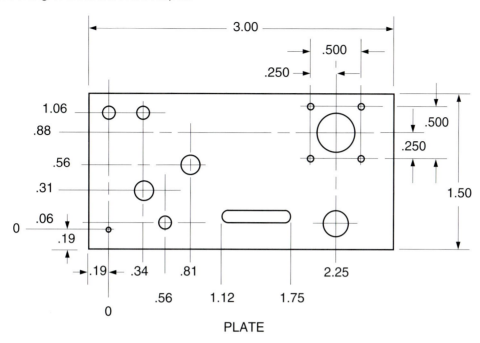

PLATE

FIGURE 14.18
Dimensions on a formed part

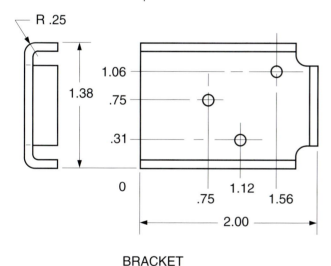

BRACKET

FIGURE 14.19
Noncircular openings

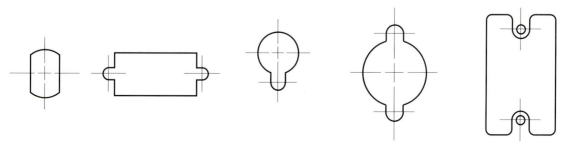

- *Noncircular openings*—Many odd-shaped cutouts or noncircular openings will be found on sheet metal parts because of the objects that need to be mounted in or on them. Figure 14.19 provides representative examples of this feature.
- **Bend relief**—When sheet metal bends intersect, such as at a corner, material is removed from the intersection area to prevent the wrinkling or tearing of the material during the forming operation. There are several configurations for a bend relief at corners, three of which are shown as top and side views in Figure 14.20.
- **Beads**—Raised or depressed areas in sheet metal parts are called beads, which add stiffness to parts. The top view of the bead shown in Figure 14.21 illustrates the outline of the bead at the mold line. In addition, breakout sections usu-

FIGURE 14.20
Corner bend reliefs

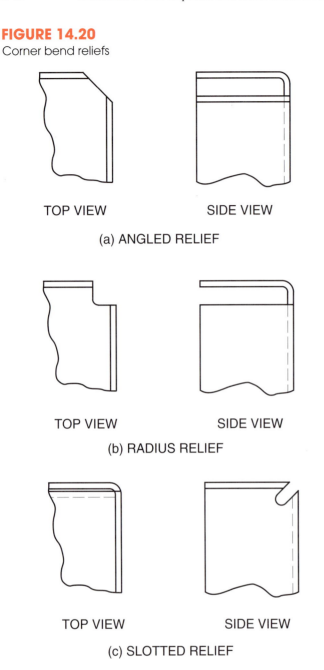

FIGURE 14.21
Beads

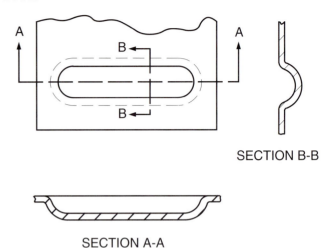

14.4 SUMMARY

The ability to interpret development drawings is a skill that shop personnel should possess. In this section, three types of development drawings were discussed: the parallel line development, the radial line development, and the triangulation development. The purpose of the development of pattern drawing was established, and examples of items that require development drawings were presented. Developable and undevelopable objects were illustrated. The text material covered the parallel line development of a wall-hung file and a truncated cylinder, the radial line development of a pyramid, and the triangulation development of an offset hopper assembly.

Sheet metal drawings were also covered in this section. Sheet metal drawings are used in several different industries, including the automotive, construction, and aircraft industries, and in the production of military electronics. Sheet metal parts are usually produced by one or several of the following processes: shearing, bending, forming, drawing, piercing, rolling, beading, spinning, trimming, perforating, and blanking. Sheet metal fabrication includes practices that are quite different than those usually employed for machined parts. As a result, several new technical terms were introduced. In addition, detailed information that one would expect to encounter when interpreting sheet metal drawings was explained.

ally appear on such a drawing to define the bead area.

- **Drawn parts**—Parts that require the stretching or compressing of material are formed through the process of drawing. Dimensions for drawn parts are shown as inside dimensions or outside dimensions, depending on the intended use for the drawn part. Figure 14.22 illustrates both conditions. Note that regardless of whether dimensions are internal or external, radii always appear as internal radii.

Section 14: *Development and Sheet Metal Drawings*   **341**

FIGURE 14.22
Drawn parts

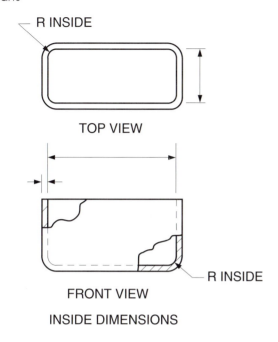

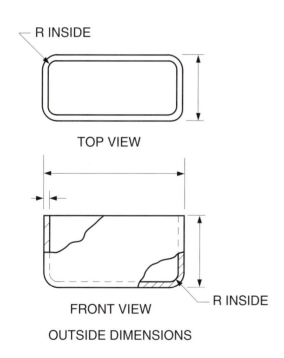

bead   A raised or depressed area in sheet metal parts, often used to strengthen or to provide a stiffener section to the part.

bend allowance   The length of material around a bend, from bend line to bend line.

bend angle   The angle to which a piece of metal is bent; the section of metal that is measured from the flat through the bend to the finished angle after bending.

bend line   The line of tangency where a bend changes to a flat surface.

bend relief   Material that is removed at a corner or intersection area to prevent the wrinkling or tearing of the material during the forming operation.

blank   A part, usually shaped by a punching operation, that forms the part in-the-flat, before bending.

brake   A manual or power-operated machine used to bend sheet metal.

cabinet   An enclosure made of thin-gage aluminum or steel of either a standard or special shape. Usually, horizontal equipment mountings are designed to accept a standard rack-mounted panel.

chassis   A thin-gage metal structure designed to house mechanical or electronic components. It may or may not have a cover or be enclosed.

console   A **cabinet** with a desktop surface.

development drawing   Usually a design layout drawing of a pattern or a template. It is developed in a single plane in preparation for folding, rolling, or bending to form some predetermined shape.

drawn parts   Parts that are formed through the stretching or compressing of material.

fold line   Lines where bending occurs; these lines are parallel to each other and perpendicular to the baseline in the development process.

leg   The flat or straight section of a part after bending or forming. Also referred to as a straight length.

parallel line development   A development drawing that requires the use of parallel lines for the height, width, and depth features of an object.

TECHNICAL TERMS FOR STUDY

baseline   The point or line on a development drawing where the layout begins.

rack   A thin-gage metal structure that includes long vertical members. It is normally made of aluminum or steel. Its purpose is to support or house electronic, electromechanical, or mechanical equipment.

radial line development   A development drawing in which all fold lines converge on a single point; the true length of this development's sides are determined through some manipulation of the object.

seam line   The point or line that, when joined, allows the object to take form.

sheet metal drawing   A drawing of thin-gage metal parts, requiring the use of one or more of the following processes to complete a finished part: shearing, bending, forming, piercing, rolling, beading, spinning, trimming, punching, perforating, or blanking.

tabs   Extensions to the main body of an object for the purpose of fastening the object together to complete its intended form.

triangulation development   A development method that requires dividing a surface into a number of triangles and then transferring the true sizes and forms of the triangles into a development or pattern drawing.

Section 14: *Development and Sheet Metal Drawings* 343

Student _____ Date _____

Section 14: Competency Quiz

PART A COMPREHENSION

1. What is the purpose of a development drawing? (8 pts)

2. Identify three major types of development drawings. (6 pts)

3. Describe the main characteristics of the following: (12 pts)

    a. parallel line development

    b. radial line development

    c. triangulation development

4. Explain the purpose of the following in development drawings: (8 pts)

    a. fold lines

    b. baseline

    c. seam lines

    d. tabs

©1995 West Publishing Company

344   **Section 14:** *Competency Quiz*

5. Describe the intent of the transition piece. (6 pts)

6. Identify eight military, commercial, or consumer products for which sheet metal is used. (16 pts)

7. List ten different fabrication processes that may be used in producing sheet metal products. (20 pts)

8. List three chassis types used in industry. (9 pts)

9. What is the purpose of a corner bend relief? (5 pts)

10. Describe the difference between a surface that is considered to be *developable* versus one that is *non-developable*. (10 pts)

Section 14: *Development and Sheet Metal Drawings*   **345**

Student _____   Date _____

PART B TECHNICAL TERMS

For each definition, select the correct technical term from the list on the bottom of the page. (10 pts each)

1. _____ The length of material around a bend, from bend line to bend line.

2. _____ The line of tangency where a bend changes to a flat surface.

3. _____ A manual or power-operated machine used to perform a bending operation.

4. _____ A thin-gage metal structure designed to support mechanical or electronic components. It may or may not have a cover.

5. _____ A design layout drawing of a pattern or a template.

6. _____ The flat or straight section of a part after bending or forming.

7. _____ A type of development drawing that uses parallel lines for the height, width, and depth features of an object.

8. _____ A type of development drawing in which all fold lines converge on a single point and the true length of each line is determined through some manipulation of the object.

9. _____ Extensions to the main body of an object, needed to complete the form of the object.

10. _____ A development method that requires the dividing of a surface into a series of triangles.

A. bend allowance
B. bend line
C. bend relief
D. brake
E. cabinet
F. chassis
G. console
H. development drawing
I. leg
J. parallel line development
K. rack
L. radial line development
M. tabs
N. triangulation development

©1995 West Publishing Company

346 Section 14: *Competency Quiz*

PART C DEVELOPMENT

1. Given the development of a pattern for a rectangular bin, respond to the following questions and exercises:

    a. What type of development is it? _____

    b. What type of line is Ⓐ ? _____
       How many are shown?

    c. The line at Ⓑ is what type? _____
       How many are shown?

    d. How many untabbed edges are identified? _____

    e. What type of line is Ⓒ? _____
       How many are shown?

    f. Add missing tabs.

    g. Cut out the pattern, apply glue or tape to the tabs, and enclose the bin.

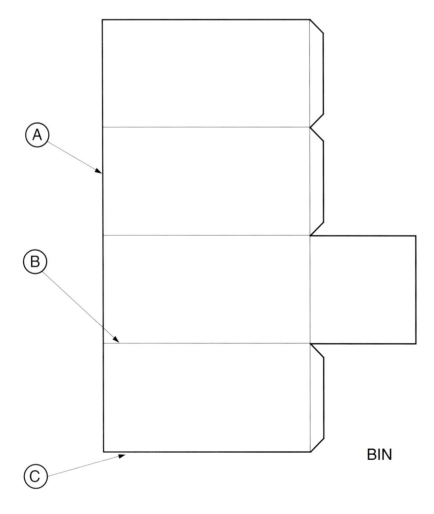

BIN

Section 14: *Development and Sheet Metal Drawings* **347**

Student _____   Date _____

Competency Quiz continues on next page.

©1995 West Publishing Company

Section 14: *Competency Quiz*

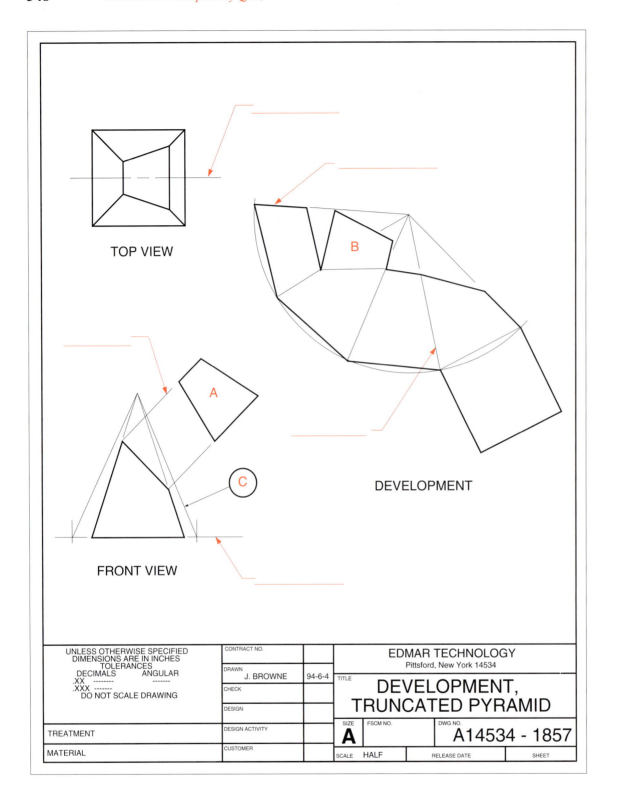

Section 14: *Development and Sheet Metal Drawings*   **349**

Student _____   Date _____

2. For the development of the truncated pyramid, A14534-1857 (see page 348), respond to the following questions and exercises.

   a. Name this type of development. _____

   b. Fill in the blank spaces on the drawing to identify the different types of lines.

   c. What type of view is Ⓐ ? _____

   d. How many sides does the pyramid have? _____

   e. Identify the bottom of the pyramid on the drawing, with the letter D.

   f. How many tabbed edges would be required to complete the development? _____

   g. What does the letter B represent on the drawing? _____

   h. How many views are shown? _____
      Identify them.

   i. Why is the pyramid referred to as being *truncated*? _____

   j. What does line Ⓒ represent? _____

   k. What is the drawing size? _____

   l. To what scale was the drawing produced? _____

   m. Why is the object called a pyramid? _____

©1995 West Publishing Company

PART D SHEET METAL DRAWING

1. For the chassis shown below, respond to the following question and exercises:

   a. What type of chassis is the object? _____

   b. What type of bend relief is required at the corners? _____

   c. What is the material? _____ The finish? _____

   d. In the grid area, develop a blank form (pattern) for the chassis. Include bend reliefs, holes, and cutouts. No dimensions are necessary. Draw freehand.

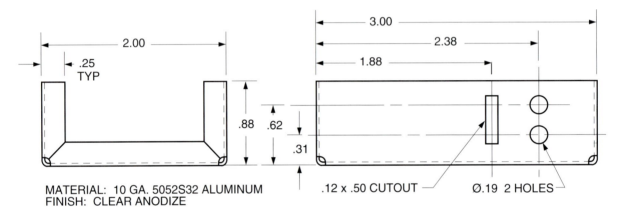

MATERIAL: 10 GA. 5052S32 ALUMINUM
FINISH: CLEAR ANODIZE

CHASSIS

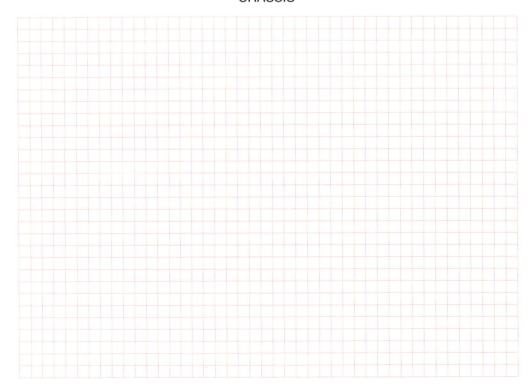

Section 14: *Development and Sheet Metal Drawings* **351**

Student _____ Date _____

2. Calculate the total developed length for the mounting bracket shown here. Use Appendix G and H for reference. Show all calculations.

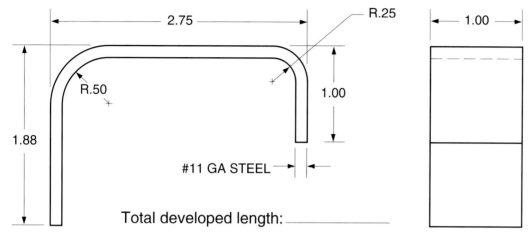

Total developed length: _____

BRACKET, MOUNTING

3. Calculate the total developed length and width for the cover shown here. (Inside bend radius equals .094 inch.)

a. Total developed width: _____

b. Total developed length: _____

c. Describe the meaning of the weld symbol. _____

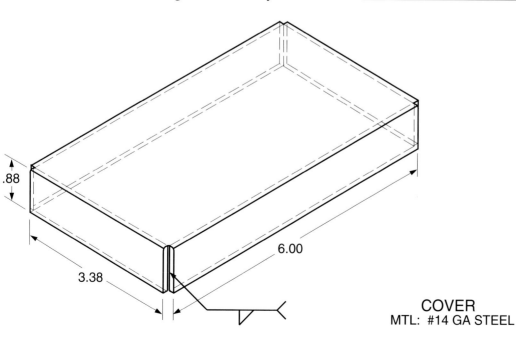

COVER
MTL: #14 GA STEEL

©1995 West Publishing Company

SECTION 15

Computer-Generated Drawings

LEARNER OUTCOMES

You will be able to:

- State the purpose of a computer-aided drafting (CAD) system.

- Explain the difference between hardware and software.

- List the major hardware components of a CAD system.

- Give several examples of an input device, output device, and processing device.

- Identify four types of CAD systems.

- Define several technical terms associated with computer-generated drawings.

- List three different types of printers.

- Identify the process for producing a simple engineering drawing using the coordinate drawing system.

- Explain the purpose of layers for engineering drawings.

- Demonstrate your learning through the successful completion of competency exercises.

15.1 INTRODUCTION

An integral part of the mechanism for producing goods and services for consumers throughout the world is the engineering drawing. In this process, information in the form of a drawing is communicated to production and manufacturing facilities through approved drafting standards and procedures. As illustrated in Section 1, the engineering drawing may be prepared by manual means or by the use of a computer. With the proliferation of computer-aided drafting, design, and engineering equipment as well as the new CAD software being produced almost daily, is it time to throw away the pencil? The answer may be yes. *Paperless systems* for producing parts and assemblies are now being used by a segment of the industrial community.

The paperless process begins with an engineering or production drawing being prepared by an engineering department using a computer-aided drafting (CAD) system. When a part for the drawing needs to be produced, the machine operator on the shop floor calls up the drawing on his or her computer, and it appears on the monitor. Then the part is produced by the machinist and the part is either routed to the stockroom or directly to the assembly line for insertion into an assembly. In the assembly area, the assembly drawing is called up on the computer by production line personnel, and parts are assembled according to the requirements of the drawing. The drawing appears on the computer monitor. When the assembly is complete, it may be sent to stock, or it may be readied for shipment to the customer. If the assembly is part of an order, the assembly drawing along with other pertinent information is forwarded to the customer through telephone lines, using a fax/modem. So, the assembly is drawn, manufactured, assembled, and sent to the customer without paper; a paperless system has been born.

Computer-generated drawings have been produced by government agencies and by industry for several years, providing new and imaginative ways of producing drawings. The primary purpose of producing computer-generated drawings is to increase productivity by enabling the drafter and designer to produce neater, cleaner, more accurate drawings in less time, and to allow parts and assemblies to be manufactured directly from graphic part information. A CAD system that produces computer-generated graphics is an automated tool that replaces traditional equipment. The key to any CAD system, however, is the operator. Computer-generated drafting systems are being used to some degree in all engineering fields. One field where it is used extensively is in the mechanical manufacturing and fabrication areas.

An *acronym* is a word formed by using the first letters of a title or a phrase. Acronyms are used throughout the computer-generated graphics field by computer users and by the companies that manufacture computer graphics systems. Some of the more common acronyms that describe these systems include the following:

- *CAD*—computer-aided design/drafting; computer-assisted design/drafting
- *CADD*—computer-aided design and drafting; computer-assisted design and drafting
- *CAE*—computer-aided engineering
- *CAA*—computer-aided artwork
- *CAG*—computer-aided graphics
- *ADS*—automated drafting system
- *EGS*—engineering graphics system

15.2 THE BASIC CAD CONCEPT

When one is entering the field of CAD for the very first time, it is important that the basic components of the system be understood. As just stated, the key to any CAD system is the operator. In any CAD system, the computer receives information (**input**) from the operator via the *keyboard* or a *digitizer*; the computer then processes the information and displays the result (**output**) on a *monitor* or *plotter*. In our case, the output is an engineering drawing. Figure 15.1 illustrates the basic CAD concept.

FIGURE 15.1
Basic CAD concept

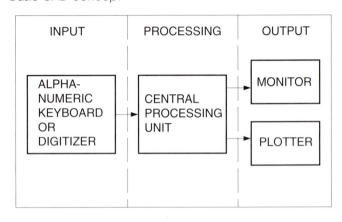

CAD systems are generally interactive, which means that an interaction or communication takes place among the various components of the system as well as with the operator. The types of mechanical engineering drawings that can be produced accurately and rapidly were identified in Section 1.1, "Rationale for Interpreting Drawings."

Although one must endure a time-consuming learning process before becoming proficient in the use of CAD system hardware and software, it is claimed that the productivity of engineering personnel can be increased by a factor of seven to one over traditional drafting and design methods, depending on the type of CAD system and the number of peripheral options used.

15.3 SYSTEM COMPONENTS

As discussed in Section 1.3, "Current Practices," computer-aided drafting systems consist of two basic elements: hardware and software. **Hardware** may be defined as the physical components of the system, including electrical, electronic, electromechanical, magnetic, and mechanical devices. **Software** consists of sets of procedures, programs, and related documentation that direct the operation of the system to produce graphics and related text material. It also provides support functions, such as input/output control, editing, storage assignment, and data management. **Application programs** allow the computer to perform desired tasks, for example, designing, drafting, and desktop publishing. *Operating system software* is required for operating the computer system.

Hardware for a computer always includes a *central processing unit (CPU)*, a *memory device*, and a *storage facility* for programs and files. A typical CAD workstation may also consist of a *digitizer* and a *digitizing device*, a *graphics tablet* with *stylus*, a function *keyboard*, a *graphics display* (monitor), and a *plotter* for producing graphical output. The system depicted in Figure 15.2 is an example of a complete interactive CAD system. Examples of software include information contained in memory, on either *floppy disks* or a *hard disk*, in files located in the CPU, and in storage.

■ 15.3.1 The Input Device

The *input device* provides the computer with data and instructions. Every input device takes data from the operator and prepares it for processing by the CPU of the CAD system. CAD systems use various means and types of input devices for processing information. Several in general use today are identified here.

- **Alphanumeric keyboard**—The alphanumeric keyboard is similar to a typewriter but contains special function keys, as well. This device allows the operator to input letter (*alpha*) and number (*numeric*) instructions to the CPU. It may also be used as a programming device or as a word-processing tool. An illustration of an alphanumeric keyboard is given in Figure 15.3
- **Cursor**—This is not an input device in the truest sense, but it assists in providing input. Its function is to determine the placement of the next

FIGURE 15.2
Interactive CAD system

FIGURE 15.3.
Alphanumeric keyboard

character on the screen, in the form of a bright marker. It may be shaped as a square, rectangle, check mark, or crosshair. Examples of cursors are shown in Figure 15.4.

FIGURE 15.4
On-screen cursors

- **Graphics tablet**—This device has a flat surface on which work is performed. In conjunction with the tablet, a stylus, puck, or mouse may be used to provide graphical data entry, and information is transmitted to the *cathode-ray tube (CRT)* by means of an electronically controlled grid beneath the table's surface. Figure 15.5 shows a typical graphics tablet.

FIGURE 15.5
Graphics tablet

- **Menu tablet**—A menu tablet has a flat surface on which functions are selected by the user with a stylus or puck. This device is also known as a menu pad. A typical menu tablet is displayed in Figure 15.6.

FIGURE 15.6
Menu tablet

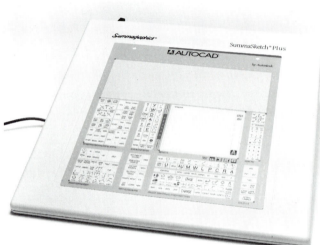

- **Modem**—A modem is both an input and an output device because it is capable of sending (inputting) and receiving (outputting) information. The vehicle for the transmission of data is a regular telephone line. Modems are available in various *baud* (speed) rates. A *fax/modem* is also used to further enhance communication transmission. Figure 15.7 illustrates a typical fax/modem.

FIGURE 15.7
Fax and fax/modem

Section 15: *Computer-Generated Drawings* **357**

- **Mouse**—A mouse is operated by moving it around a flat surface while crosshairs track its movement on the screen. To activate the point or menu item at which the crosshairs are positioned, the user presses a button on the mouse. Figure 15.8 shows an example of a mouse input device.

FIGURE 15.9
Examples of prompts

FIGURE 15.8
Mouse

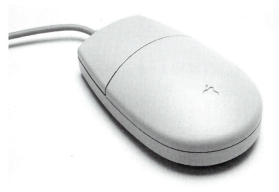

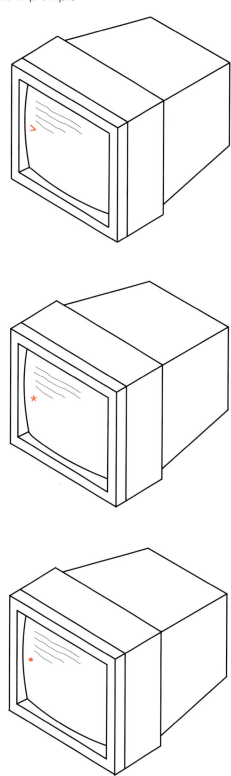

- **Prompt**—The prompt is usually an arrowhead (>), an asterisk (*), or a dot (•) at the beginning of a line on the screen, signifying that the system is ready for the user to input a command. Examples of prompts can be found in Figure 15.9.
- **Puck**—A puck is a handheld device that is moved on a graphics tablet. It has essentially the same effect as a mouse. A puck is illustrated in Figure 15.10
- **Screen menu**—On some CAD systems, a menu is displayed on the graphics screen. The menu allows the input commands to be entered by simply pointing to a command on the display screen. An example of a screen menu may be seen in Figure 15.11.
- **Touch pen**—This device works much like a light pen, but it is operated by touching a special panel on the face of the monitor. The user selects the point or menu item by sliding the pen tip along the panel until the crosshairs are positioned at the desired point, then lifting the pen from the screen. An example of this device is presented in Figure 15.12.

358 Section 15: *Computer-Generated Drawings*

FIGURE 15.10
Puck

FIGURE 15.12
Touch pen

- **Voice-data entry**—This system is activated by verbal instructions by the user. It is in wider use today than at any time. It allows the operator to keep his or her eyes and hands free to perform other tasks, which may result in fewer errors.

15.3.2 The Output Device

The reason for inputting information and data into a computer is to receive output. The output takes the form of either a **visual copy** or a **hard copy**. Visual copy is observed by the user's eye on the graphics screen. Hard copy is produced by a printer or plotter on paper or film. The *output device* is the vehicle that produces output. The major output devices in use today include the following:

- **CRT**—A CRT (cathode-ray tube), also called a *monitor*, projects electrons onto a fluorescent screen, which glows when the electron beams are excited and, as a result, produces a graphics display on a video screen. CRTs are available in several different sizes. A popular size is one in which two full pages of $8\frac{1}{2}$-by-11 sheets can be displayed. An example of a CRT is shown in Figure 15.13.

FIGURE 15.11
Screen menu

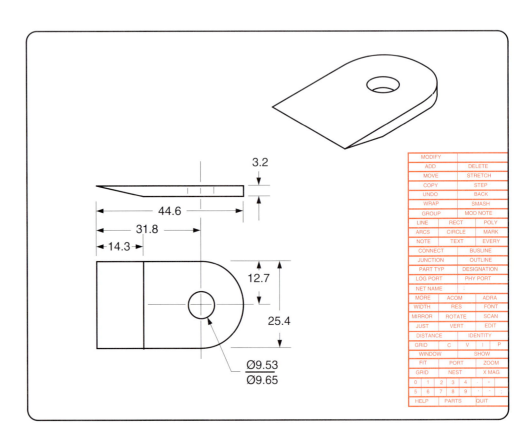

Section 15: *Computer-Generated Drawings* **359**

FIGURE 15.13
CRT

- **Plotter**—This output device is a piece of equipment that produces hard copy either mechanically (with a pen) or electrostatically on paper, vellum, or polyester film. Commonly used plotters for CAD work include *pen plotters* and *electrostatic plotters*. An example of each is illustrated in Figure 15.14.
- **Printer**—A printer is an output device capable of reproducing an image of a graphics display. One example is an alphanumeric unit similar to a typewriter. Because printers do not generally produce extremely high-quality output, their use is limited to the production of text and preliminary or check print copies rather than permanent graphic output. Devices of this type include the *dot-matrix*, *ink jet*, and *laser* printers (see Figure 15.15). All the mechanical drawings for this worktext were produced on an Hewlett-Packard IIIP LaserJet Printer.

FIGURE 15.14
Plotters: (a) pen and (b) electrostatic types

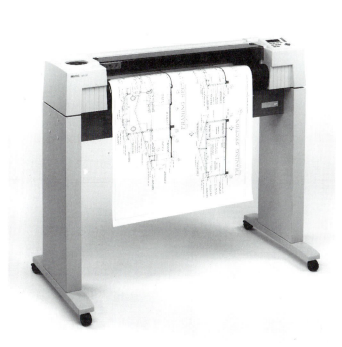

Photo courtesy of Hewlitt-Packard Company

(a)

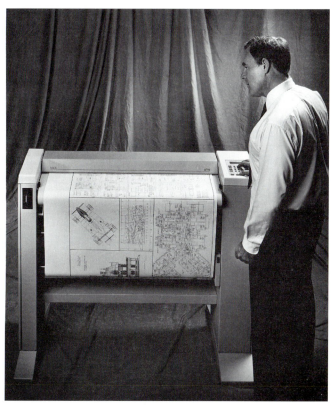

VERSATEC is a registered trademark of Xerox Corporation

(b)

FIGURE 15.15
Dot-matrix, ink jet, and laser printers

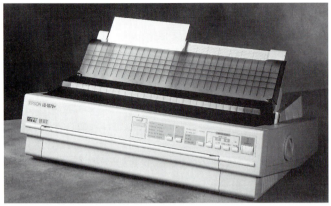
Courtesy of EPSON

Courtesy of EPSON

Reprinted with permission of Canon Computer Systems, Inc. All rights reserved.

■ 15.3.3 The Processing Device

As indicated in Figure 15.1, the three major areas of every CAD system are input, output, and processing. Processing devices are those units that process input information, which includes the storage of data (memory). After input is provided to the system, the data is processed—that is, it is manipulated before it appears as output. Processing devices not previously covered under input and/or output (I/O) devices are identified here:

- **Central processing unit (CPU)**—The CPU is considered the brains of any computing system. Integrated circuits form an important part of the microprocessor portion of the computer, which accomplishes the logical processing of data. The CPU contains arithmetic, logic, and control circuits, and it incorporates memory and the storage of files. Figure 15.16 shows the physical appearance of a typical CPU.

FIGURE 15.16
CPU

- **Memory**—The storage of data is a function of memory. The size of a computer's memory is determined by the amount of information it is capable of storing. The smallest unit of memory is the **bit**. The bit is designated by the binary number 0 or 1, and it may or may not contain an electrical charge. A **byte** is equal to eight bits. A CAD system may contain several thousand (*K*) bytes to several million (*mega*) bytes of memory.
- **Disk drive**—A disk drive retains, reads, and writes onto disks. Most systems use two disk drive units, a *floppy disk drive* and a *hard disk drive*. A disk drive is the device that operates the various software packages needed to perform computer operations. Figure 15.17 depicts an external floppy disk drive package for a microcomputer-based system. The **floppy disk** that is used in the floppy disk drive is a portable, flat, black, or brown circular plate made of thin vinyl

with a magnetic coating housed in a square-shaped envelope that is inserted and removed from a floppy disk drive unit. Floppy disks can contain entire software packages or they can be used to store data. They are available in standard square sizes of 3.5, 5.25, and 8.0 inches.

FIGURE 15.17
External floppy disk drive

A **hard disk** is a nonremovable disk with a large capacity for storing information in the form of files. Its size depends on the amount of memory that is desired to operate the system successfully. Because computers are interactive, data is easily transferred between floppy and hard disks.

15.4 TYPES OF COMPUTER SYSTEMS

Computer systems are of two general types, *digital* and *analog*. Systems, as they relate to CAD, are digital in nature. This means that they continuously count by digits. Examples of everyday digital items include the wrist watch, the clock radio, and the calculator. Analog systems are those used for other functions and operations in business and industry, such as for problem solving and for research. Both types of systems have gone through tremendous changes over the years. During the computer's formative years, it occupied racks and racks of equipment and was generally slow operating. Today, with miniaturization, super-fast computer chips, and price reductions, the average American can afford to buy a powerful computer.

■ 15.4.1 The Mainframe-Based System

The physical configuration of a system that produces computer-generated graphics can take many forms. The most versatile, powerful, and expensive form is the *mainframe-based system*. This system is capable of performing numerous tasks and functions across a wide range of complex engineering areas. The term mainframe means that this system consists of a CPU with a large capacity for design and analytical tasks that can accept the workloads of several workstations (or *terminals*) simultaneously. In this arrangement, each workstation has input and output capability, with the CPU being at a remote location. In most situations, several terminals operate off one main or *host computer*. This is called **networking** or *timesharing*. Figure 15.18 is a photographic representation of a mainframe-based computer system.

FIGURE 15.18
Mainframe-based computer system

International Business Machine Corporation

■ 15.4.2 The Minicomputer-Based System

The minicomputer-based system, at this point in time, may be considered a down-scaled mainframe-based system or an up-scaled microcomputer-based system. It is referred to as a *workstation*. It is only *mini*, or small, in terms of physical size. It is an example of the computer industry's emphasis on producing smaller but more powerful systems for CAD use. There are several of this type of workstation being produced by the leading computer manufacturers as well as by emerging computer companies. These systems are a family of network servers and modular components. These components include items such as a selection of processors with different performance levels, display systems, memory boards, interface cards, operating systems, and peripherals. Figure 15.19 shows several types of component options that may be used for a typical minicomputer-based computer system.

FIGURE 15.19
Minicomputer-based computer system options

International Business Machine Corporation

■ 15.4.3 The Microcomputer-Based System

Several microcomputer-based computer systems are available in the marketplace. They are not quite as powerful as (and thus cannot perform the large number of engineering functions of) either the mainframe-based or minicomputer-based systems. However, since the advent of the 286, 386, 486, and now 586 chip (integrated circuit), competition has been intense in the field of 16-bit and 32-bit systems produced by leading computer manufacturers. These are *desktop systems* that are capable of performing simulation, design, and CAD functions, and they are making an impact on education and industry because of their low cost and ever increasing high-technology capability. Various software programs for producing computer-generated drawings are in existence today, with three-dimensional packages available for the more powerful units. Because of the increased capability of the most recent systems, software programs are being produced at a rapid rate. Example of a system available today is shown in Figure 15.20.

FIGURE 15.20
Microcomputer-based computer system

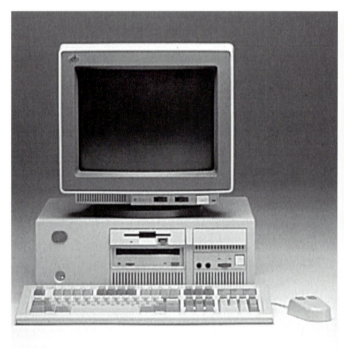

International Business Machine Corporation

■ 15.4.4 The Laptop Computer

Laptop or *portable computers* can also be used to produce engineering drawings. They are extremely useful because of their portability. As a result, work can be accomplished while riding in an automobile, in airport passenger areas, on trains, in libraries, and so forth. These machines are available with large hard disks as well as with floppy disks. Some have very high memory capability, with a built-in mouse and a modem feature. This type of hardware is probably the fastest growing area of the computer industry. Examples of laptop computers are shown in Figure 15.21.

FIGURE 15.21
Laptop computers

Reprinted with permission of Canon Computer Systems, Inc. All rights reserved.

15.5 CAD-BASED SOFTWARE FOR PRODUCING ENGINEERING DRAWINGS

The development of software packages for producing computer-generated drawings has increased at an unprecedented rate. For this reason, producers of existing software must continuously update and revise their programs to remain current with new products. Some software packages have been revised between five and twelve times to meet the ever-changing requirements of users. There are many high-quality CAD software packages available to the consumer, at a wide range of prices and with a variety of features. For graphics purposes, the software packages identified in Table 15.1 represent areas of capability that include engineering design, product design, technical illustration, tutorial options, wireframe and solid modeling features, 2-D and 3-D drawing capability, keyboard shortcuts, on-line help commands or menu drives, automatic dimensioning features, and technical support.

TABLE 15.1
CAD-based software

PRODUCT	MANUFACTURER
Ashlar Vellum	NESI
AutoCAD	Autodesk
CADAPPLE	T & W Systems
CADRA	ADRA
CADvance	Calcomp Systems
ClarisCAD	Claris
Conceptstation	Aries Technology
Euclid	Matra Datavision
Generic CAD	Generic Software
Intergraph EMS	Intergraph
Microstation	Intergraph
PRO/Engineer	Parametric Technology
Unigraphics II	EDS
VersaCAD	T & W Systems

15.6 CAD APPLICATIONS

Performing CAD applications is within the average person's capability. Before attempting to develop technical graphics on a CAD system, however, the operator must have a firm grasp of the hardware and software being used. To become familiar with the basics of a system, it is necessary to learn about its operation through reading the manuals provided, self-learning, or formal classroom instruction. At a minimum, a novice should understand how to select commands through input devices, store and retrieve data, and plot data.

■ 15.6.1 Preparing for Graphics Production

The very first step in producing graphics on a CAD system is to **boot the system**, which means, simply,

364   **Section 15:** *Computer-Generated Drawings*

to turn it on. *It* refers to the CAD system hardware, such as the CPU, monitor, and peripheral devices, including the plotter, menu tablet, and tape drive, if appropriate. From portables to mainframes, each system has its own start-up procedures. If a floppy disk is used in the CAD graphics process, **formatting** the disk may be necessary at this point.

Immediately after start-up, certain *messages* about the system configuration may be displayed, or certain information may be requested by the computer relative to the user's name and a confidential password. The user is now in the process of **logging on** the system. These procedures may also include a main or basic menu display on the monitor to guide the operator to the next step. Input performed by the user, at this point, may be through the alphanumeric keyboard or through a mouse. Typical information requested by the computer when logging on is illustrated in Figure 15.22.

FIGURE 15.22
Logging-on requirements

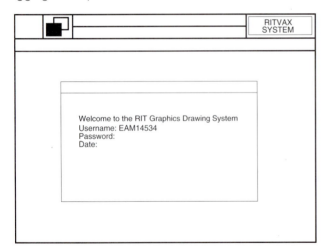

15.6.2 The Menu and Command System

The method of inputting CAD data to the computer varies from manufacturer to manufacturer of CAD equipment, but it usually consists of some form of screen display, including the drawing field, a sidebar or pull-down menu, and possibly an information menu, as illustrated in Figure 15.23.

In addition, the system will likely have several viewports. A **viewport** is a rectangular working area on the display screen. Viewports clearly define the many different views required to complete a drawing. Many CAD systems have four to eight viewports for the traditional views (front, top, side) and an isometric view, plus others, as shown in Figure 15.24.

FIGURE 15.23
Typical screen information when preparing to draw

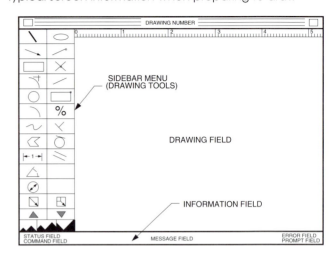

FIGURE 15.24
Examples of viewports

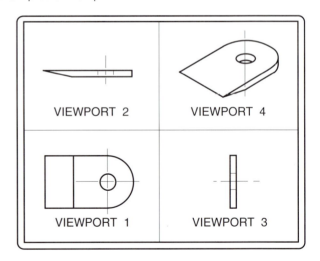

The menu is a listing of basic commands that may be used. There are also subcommands from which more detailed inputs can be made. Commands may be activated through an input device, such as a mouse, puck, or light pen, with or without a graphics tablet. Examples of basic command menus appear in Figure 15.25.

FIGURE 15.25
Basic CAD command menus

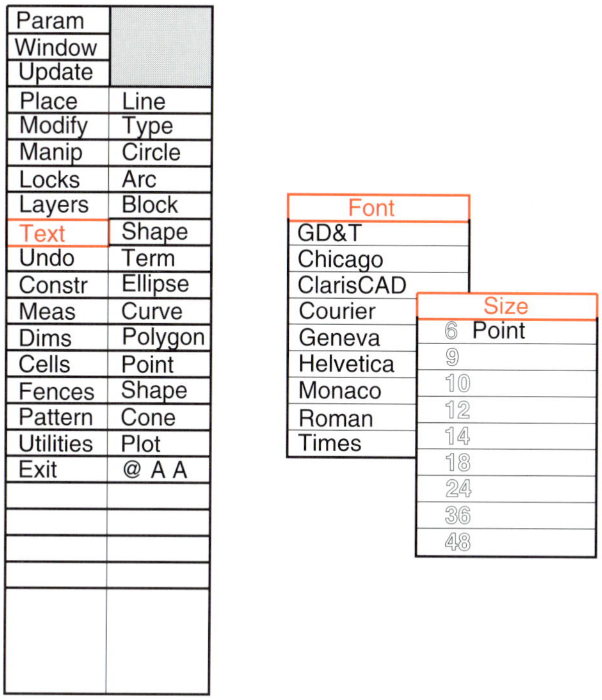

Param	
Window	
Update	
Place	Line
Modify	Type
Manip	Circle
Locks	Arc
Layers	Block
Text	Shape
Undo	Term
Constr	Ellipse
Meas	Curve
Dims	Polygon
Cells	Point
Fences	Shape
Pattern	Cone
Utilities	Plot
Exit	@ A A

Font
GD&T
Chicago
ClarisCAD
Courier
Geneva
Helvetica
Monaco
Roman
Times

Size	
6	Point
9	
10	
12	
14	
18	
24	
36	
48	

The screen cursor is controlled by the input device or by keyboard input. The location of the cursor assists the operator in visualizing important screen data, such as end points of lines and hole centers, as well as numerous graphics functions.

■ 15.6.3 Creating a Design File

Creating a **design file** on a CAD system is the same as finding a place in a four-drawer file cabinet at home or at the office to store data. When filing, one usually arranges data in some order and gives an appropriate name or number to the file. The same is true with a CAD design file. The CAD design file contains all the necessary technical data, including graphics and parts lists, for a complete project. When establishing this file, the user also needs to set up the computer for the desired parameters for the particular project or drawing (scale, dimensioning system, decimal or metric system, grid scale, drawing size, etc.).

■ 15.6.4 Drawing Line Forms

The CAD user has the option of selecting various line styles, line widths, and line colors for producing graphics. A line may be produced by selecting the LINE, PLACE LINE, or DRAW LINE command as

well as a subcommand for selecting the desired type and weight of line. Figure 15.26 shows a typical LINE command pull-down menu with its associated subheadings.

FIGURE 15.26
Typical line menu

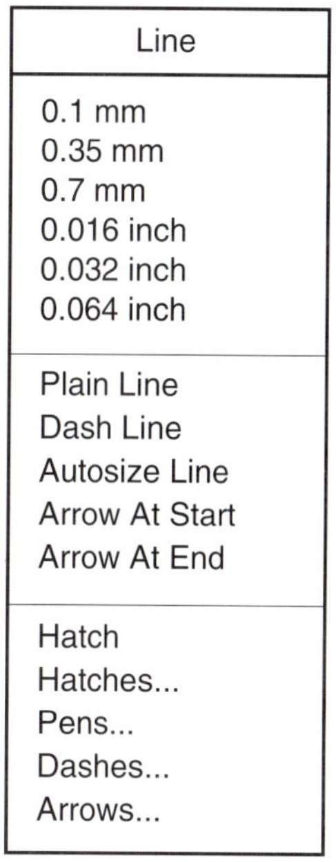

Line
0.1 mm
0.35 mm
0.7 mm
0.016 inch
0.032 inch
0.064 inch
Plain Line
Dash Line
Autosize Line
Arrow At Start
Arrow At End
Hatch
Hatches...
Pens...
Dashes...
Arrows...

Input for a line may be placed through the keyboard or through the screen's menu. If accomplished using the keyboard, the Cartesian, or Delta, coordinate system (X, Y) is often used. In this system, one must remember that the X coordinate is always on the horizontal axis, and the Y coordinate is on the vertical axis. The rectangle with its associated coordinates illustrated in Figure 15.27a is the result of using the coordinate system of input to draw vertical and horizontal lines. Note that the origin, or starting point, of the rectangle is 0, 0 (zero, zero). The command sequence or program necessary to produce the rectangle is shown in Figure 15.27b.

In most CAD software systems, the operator uses the mouse in conjunction with on-screen tools (in the form of a menu) as the input device for drawing lines. Systems such as AutoCAD, CADkey, and fastCAD, use these methods for line input. Typical CAD on-screen tools for drawing straight lines, circles, arcs, radii, angles, and ellipses (including automatic dimensioning) is shown on the drawing menu in Figure 15.23.

366 Section 15: *Computer-Generated Drawings*

FIGURE 15.27
Producing a rectangle using the coordinate system

FIGURE 15.28
Menu for text options

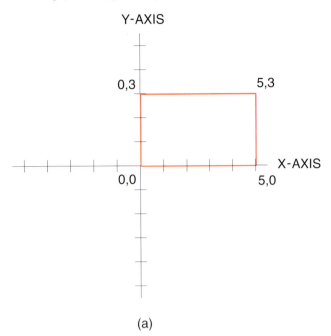

15.6.5 Creating Text

The need to create text on a CAD drawing is merely an exercise in word processing. Text on a CAD system consists of alphanumeric characters and punctuations in a variety of fonts. Most software manufacturers provide several fonts from which to choose. The TEXT or PLACE TEXT command or an on-screen tool signifying the letter *A* is often designated when the insertion of text is required. The operator is responsible for establishing text parameters that affect how text will look on the completed drawing. Parameters that can be controlled by the user include text height, color, font type, style, and justification. Figure 15.28 illustrates a typical pull-down menu for text options.

15.6.6 Establishing Layers

A **layer** is also referred to as a *level* or an *overlay* and is used to specify the level on which elements of a drawing need to appear. Layers are much like transparencies stacked (overlaid) on top of each other. The design area on most CAD systems includes up to 250 identically sized layers and literally thousands of color combinations.

Having different elements of a drawing assigned to a specific layer is convenient if drawing changes are necessary; it also allows the drafter the flexibility to move, add, and delete parts of a drawing or to plot drawings. Any combination of layers can be displayed at the same time. The layer on which the drafter is currently working is called the *active layer*. All the drawings displaying color in this worktext represent two or more layers. An example of layers for engineering drawings is portrayed in Figure 15.29, which identifies three different layers. In this instance, layer 1 and layer 2 combine to produce layer 3.

15.6.7 Plotting

Plotting is the output of the process of producing a paper or film copy of a drawing in the design file. This is called a *hard copy*. Plotting is not synonymous with printing. Printing is primarily the output of word processing. It is also used for creating a check copy or a screen dump copy (preliminary copy) of a

FIGURE 15.29a

Layer 1

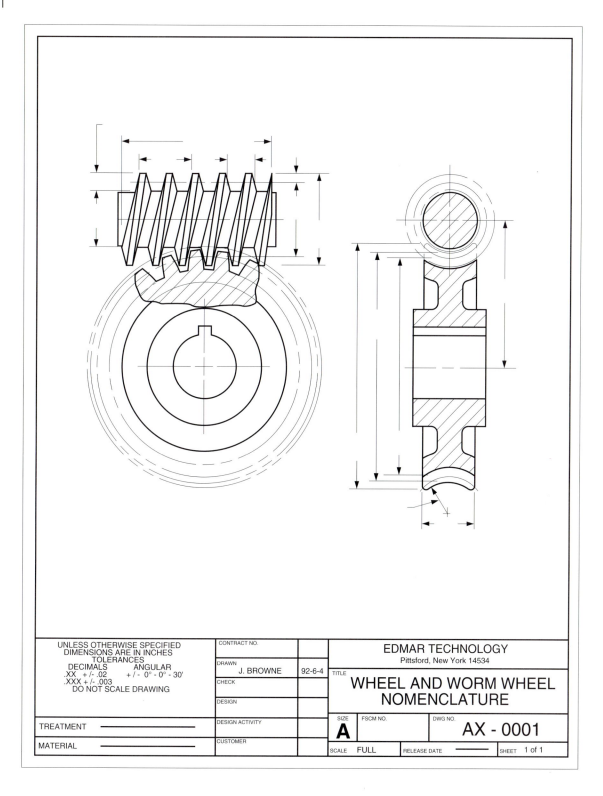

368   **Section 15:** *Computer-Generated Drawings*

FIGURE 15.29b
Layer 2

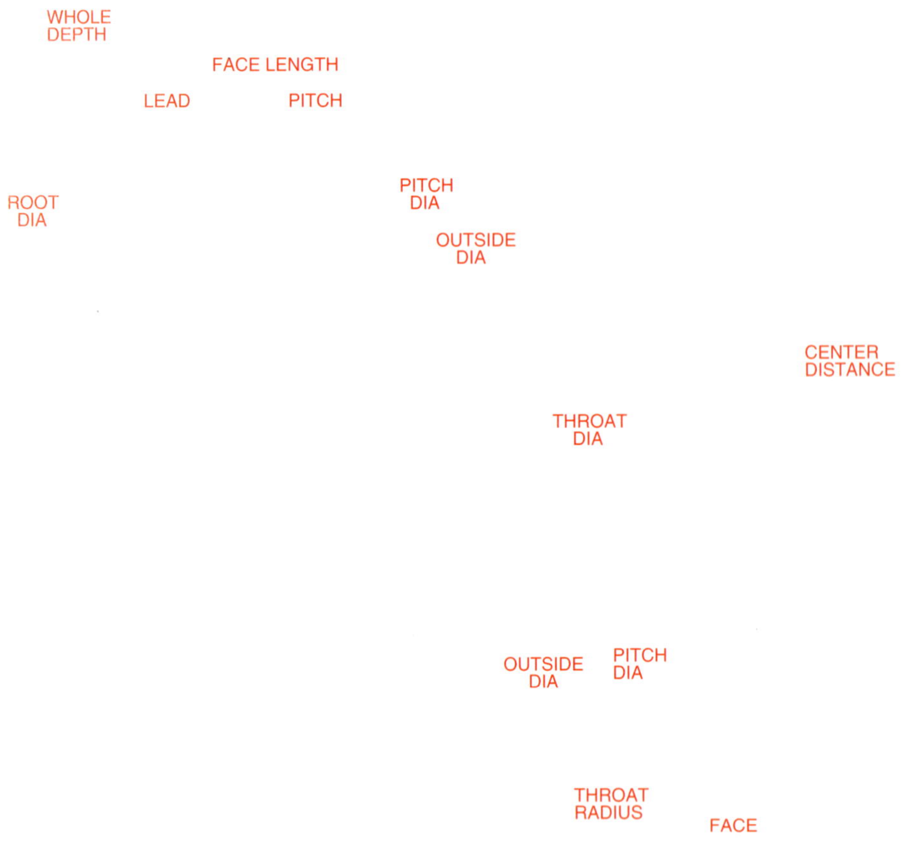

WHOLE
DEPTH

FACE LENGTH

LEAD   PITCH

ROOT
DIA

PITCH
DIA

OUTSIDE
DIA

CENTER
DISTANCE

THROAT
DIA

OUTSIDE   PITCH
DIA   DIA

THROAT
RADIUS   FACE

FIGURE 15.29c
Layer 3

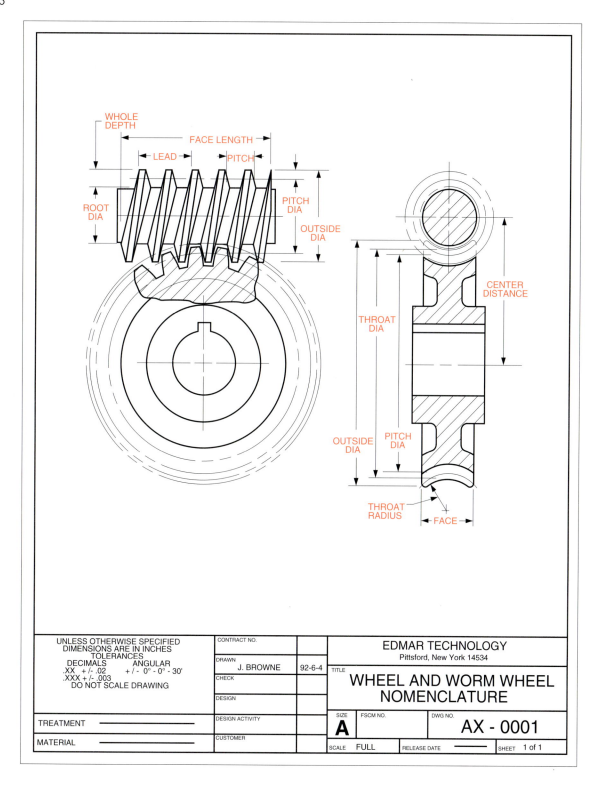

370   **Section 15:** *Computer-Generated Drawings*

CAD drawing. Plotting produces a final, ready-to-be released copy of a completed CAD drawing. On most CAD systems, there are three stages of manipulations that need to occur to produce a plot. They include the following:

1. Setting up the system. This means determining one or more of the following: name of the file, drawing orientation and scale, and pen arrangement.
2. Converting the file into plot data. This normally refers to changing the software elements into vector data that the plotter can understand.
3. Running the plot on the plotter.

Usually, the key words PLOT, CREATE PLOT, MAKE PLOT, or PRODUCE PLOT are the commands for initiating a plot.

15.7   SUMMARY

Producing drawings with the aid of a computer has had a tremendous impact on the way business is conducted in engineering departments, particularly as they relate to the design and drafting functions. The potential for the field of engineering is overwhelming. The next generation of computers may very well be optical computers that will operate a thousand times faster than anything currently in production—systems that simulate human brain functions. One of the greatest advances has been in the production of engineering drawings. Through the use of CAD, productivity has increased, and at the same time, cleaner, neater, and more accurate drawings are being produced. Equipment ranging from stand-alone mainframe systems to minicomputer, microcomputer, and laptop computer hardware and software is available. These systems are generally within the financial reach of most educational institutions and can be valuable tools for student and staff development, instruction, and training. Mainframe, minicomputer, microcomputer, and laptop computer systems were covered, as were input, output, and processing devices. Representative CAD software packages, which are being developed rapidly to keep up with the requirements of the user for faster, more powerful systems, were also discussed. Finally, a brief accounting of CAD applications, including the preparation for and process of producing engineering drawings, was outlined.

TECHNICAL TERMS FOR STUDY

The technical terms for study defined here are in addition to those already defined in Section 15.3.1, 15.3.2, and 15.3.3.

application program   Software that is the link between specific system use and its related tasks.

bit   A signal of 0 or 1, either of which is a binary number; taken from the term *binary digit*.

boot the system   To turn on or to start up a CAD system.

byte   A character of memory used as the basis of comparison of various systems and manufacturers. It equals eight binary digits (bits).

design file   A storage area within a CAD system's memory for saving drawings.

floppy disk   A software package for storing data, housed in standard square sizes.

formatting   A method of preparing a floppy disk for use on the computer.

hard copy   Data produced by a plotter or printer on paper or film.

hard disk   A nonremovable disk with a large capacity for storing information in the form of files. Most often built into a computer.

hardware   The physical components of a computer system, including electrical, electronic, electromechanical, magnetic, and mechanical devices.

input   To providing data and instructions to the computer through one or more input devices.

layer   An overlay or level on which elements of a drawing need to appear.

logging-on   Following a set of start-up procedures that allow the user to use the CAD system.

networking Interconnecting two or more workstations to perform computing functions, with the CPU being at a remote location.

output The result of input. It is the data that the computer produces.

plotting A process for producing a paper or film copy of a drawing in the design file. The result of plotting is a hard copy.

RAM Random-access memory; temporary memory, provides storage locations for entries by any input device.

ROM Read-only memory; permanent memory, or on-line storage. ROM cannot be changed by an operator.

software Sets of procedures, programs, and related documentation that direct the operation of the system to produce graphics and related text material.

viewport A rectangular working area on the monitor (display screen) that allows the user to view the several views required of a drawing.

visual copy Data that is observed by the user on a graphics screen.

Section 15: *Computer-Generated Drawings*   **373**

Student _____   Date _____

Section 15: Competency Quiz

PART A COMPREHENSION

1. Why is a CAD system valuable for producing engineering drawings? (5 pts)

2. Draw freehand a block diagram of the basic CAD concept. (5 pts)

3. What does the term *interactive* mean? (3 pts)

4. What three types of systems are available for producing computer-generated drawings? (6 pts)

5. What is menu tablet? (3 pts)

6. What piece of CAD equipment produces a hard copy? (3 pts)

7. What is the purpose for the alphanumeric keyboard? (3 pts)

8. What is a floppy disk? How is it used? (6 pts)

9. What is the function of a cursor? (3 pts)

©1995 West Publishing Company

374   **Section 15:** *Competency Quiz*

10. Why is a CPU important to a computer system? (3 pts)

11. What is an input device? Give five examples. (6 pts)

12. What is an output device? Give three examples. (6 pts)

13. What kind of device is a modem? How is it used? (6 pts)

14. List six different computer-based CAD software packages. (12 pts)

15. What is the difference between a printer and a plotter? (3 pts)

16. Which CAD system is the most powerful as well as the most expensive? (3 pts)

17. What is bit? How many bits are in a byte? (6 pts)

18. Of what use is a prompt? (3 pts)

Section 15: *Computer-Generated Drawings* **375**

Student _____   Date _____

19. Describe the paperless system for producing parts and assemblies in industry. (8 pts)

20. Why are laptop computers valuable and popular? (4 pts)

21. Identify three areas of engineering capability for CAD software packages. (3 pts)

©1995 West Publishing Company

PART B TECHNICAL TERMS

For each definition, select the correct technical term from the list at the bottom of the page. (5 pts each)

1. _____ The physical components of a computer system.

2. _____ Considered the brains of a computing system. It houses the arithmetic, logic, and control circuits, and often incorporates memory storage.

3. _____ Similar to a typewriter, allows the user to input alpha and numeric instructions to the computer.

4. _____ Sets of procedures, programs, and related documentation that direct the operation of the computer system.

5. _____ An output device that produces a hard copy.

6. _____ An output device that reproduces a visual copy.

7. _____ A bright marker used to determine the next character on a screen.

8. _____ An electronic tracing table or board that is used to enter existing drawings into a CAD system by means of a pointing device and a menu.

9. _____ A device that has a flat surface on which work is performed. Used in conjunction with a stylus, puck, or mouse.

10. _____ Consists of lever with a weighted base where the movement of the lever controls the screen's cursor.

11. _____ A device that produces a graphics display on a video screen; also referred to as a monitor.

12. _____ An input and an output device that involves the sending and reception of information through standard telephone lines.

13. _____ Either an arrowhead, crosshair, or dot signifying that the system is ready for input.

14. _____ A device that inputs commands merely by pointing to a command on the screen.

15. _____ A system that is activated by verbal instructions from the operator.

Section 15: *Computer-Generated Drawings*   **377**

Student _____   Date _____

16. _____ The amount of information that a computer system is capable of storing.

17. _____ A portable black or brown circular plate made of vinyl with a magnetic coating housed in a square-shaped envelope.

18. _____ Several workstations working off the same, remotely located, host computer.

19. _____ A computer system that is smaller in physical size than a mainframe or minicomputer but is capable of performing drawing functions.

20. _____ A nonremovable disk with a large capacity for storing information in the form of files.

A. alphanumeric keyboard
B. CPU
C. CRT
D. cursor
E. digitizer
F. floppy disk
G. graphics tablet
H. hard copy
 I. hard disk
J. hardware
K. joystick
L. light pen
M. memory
N. microcomputer
O. modem
P. mouse
Q. networking
R. plotter
S. printer
T. prompt
U. screen menu
V. software
W. touch pen
X. visual copy
Y. voice-entry data

©1995 West Publishing Company

SECTION 16

Supplementary Competency Exercises

LEARNER OUTCOMES

You will be able to:

- Identify various views of an object.
- Differentiate between different types of tolerancing methods.
- Explain the different symbols used in dimensioning practices.
- Determine tap drill sizes.
- Identify technical term abbreviations.
- Recognize zoning on drawings.
- Explain engineering changes to a drawing.
- Identify items and quantities of parts on a material list.
- Calculate dimensions on an engineering drawing.
- Evaluate a drawing for accuracy.
- Determine the producibility of engineering drawings.

16.1 INTRODUCTION

This section offers supplementary competency exercises for students so that they will have the opportunity to interpret actual industrial-type engineering drawings. The drawings selected represent two types: detail and subassembly drawings. The intent is for students to interpret various aspects of the drawings and to evaluate them as to their producibility in the shop.

Section 16: *Supplementary Competency Exercises* **381**

Student _____   Date _____

Section 16: Supplementary Competency Exercises

PART A DETAIL DRAWING INTERPRETATION

1. For the industrial-type detail drawing of the ring, drawing 335682-EP3 (see first foldout following page 384), respond to the following questions and exercises:.

   a. How many views are represented by the drawing? _____
   Name them. _____

   b. Is there an indication of where the section view was taken? _____

   c. What type of section view is shown? _____

   d. How many changes have been made on this drawing? _____

   e. How many reamed holes are there? _____

   f. How many tapped holes are required? _____

   g. What type of tolerance is given for the ∅1.0000 dimension? _____

   h. What is the vertical dimension between the two 8-32 tapped holes, on the far side of the part (G-3, 4)? _____

   i. To what does G-3, 4, refer in question h? _____

   j. Is there a material specified for the part? _____

   k. What is the drawing size? _____

   l. What is the distance between the two ∅.25-20 threaded holes? _____

   m. What heat treatment specification is required? _____

   n. What is the tolerance for four place decimals? _____

   o. What is the finish for the part? _____

©1995 West Publishing Company

382   Section 16: *Competency Quiz*

p. What size of chamfer is required? _____

q. What size of tap drill is needed for the 8-32 thread? _____
   For the Ø.25-20 thread? _____

r. How deep is the Ø1.125 shouldered bore? _____

s. Has the drawing been checked or approved? _____

t. Is there sufficient information on the ring drawing to produce the part? _____

2. For the industrial-type detail drawing of the block, drawing 457-637-Z (see second foldout following page 384), respond to the following questions and exercises:

   a. On the drawing, draw a cutting plane line (A-A) on the view that represents the section view shown. Designate the section view as Section A-A.

   b. On the drawing, what do you believe the following abbreviations mean?

      PLCS _____

      DWL _____

      THD _____

      CHAM _____

   c. What surface finish is required for the bottom surface? _____
      For the top surface? _____

   d. How deep are the Ø.625 holes? _____

   e. What is the tolerance for the R .6090 dimension? _____

   f. What is the horizontal distance between the Ø.41 holes? _____
      What is the tolerance for this dimension? _____

   g. From what material is the part made? _____

   h. What is the overall height of the part? _____

   i. What are the upper and lower limits for the height dimension? _____

Section 16: *Supplementary Competency Exercises*   **383**

Student _____   Date _____

j. How many ∅.375 dowel holes are required? _____

k. Explain what is meant by ∅.125, ⊔.25, ⤓ 1.00, far side, and 4 PLCS in zone C-7, 8.
_____

l. How many four-place decimals are on the drawing? _____

m. What are the overall height, width, and length dimensions of the block? _____

n. What is the tolerance for the .993 dimension (D-4, 5)? _____
This tolerance is called a _____ tolerance.

o. What are the minimum and maximum dimensions for the 2.906 ±.005 dimension? _____

p. The part is to be heat treated per what specification? _____

q. Is there sufficient information on the block drawing to produce the part? _____

©1995 West Publishing Company

384   Section 16: *Competency Quiz*

PART B ASSEMBLY DRAWING INTERPRETATION

1. For the industrial-type press subassembly, drawing 068209-579 (see third foldout following page 384), respond to the following questions and exercises. Note that there are several balloons appearing with two numbers; the upper number is the actual item number, while the lower number is the sheet number on which the part appears. In addition, there are no part numbers on the drawing. There is, however, a material list.

   a. How many items are called for on the drawing? _____

   b. In what zone is the cam motion diagram located? _____

   c. The letters S.H.C.S. appear several times on the material list. What do they signify? _____

   d. There are two items on the material list that are springs. What item numbers are they?
   _____ From what material are they made? _____

   e. What are the overall dimensions of the bearing block? _____

   f. How many $5/16$-inch ID lockwashers are required? _____

   g. What kind of pins are items 27 and 37? _____

   h. To what scale was the drawing produced? _____

   i. Is the line weight for items 24, 26, and 30 correct? _____
   If not, why? _____

   j. Has the drawing been checked or approved? _____

   k. What material is specified for item #14? _____

   l. What are the dimensions for the vertical slide rod? _____

   m. How many sheets in this particular set of drawings? _____

   n. Are there any dimensions on the drawing? _____

   o. Should there be any tolerances in the tolerance block? _____

   p. What is the total number of 10-32 screws on the material list? _____

   q. What is the total number of parts on the subassembly? _____

   r. The slide is item _____. On what sheet does the detail drawing
   appear? _____

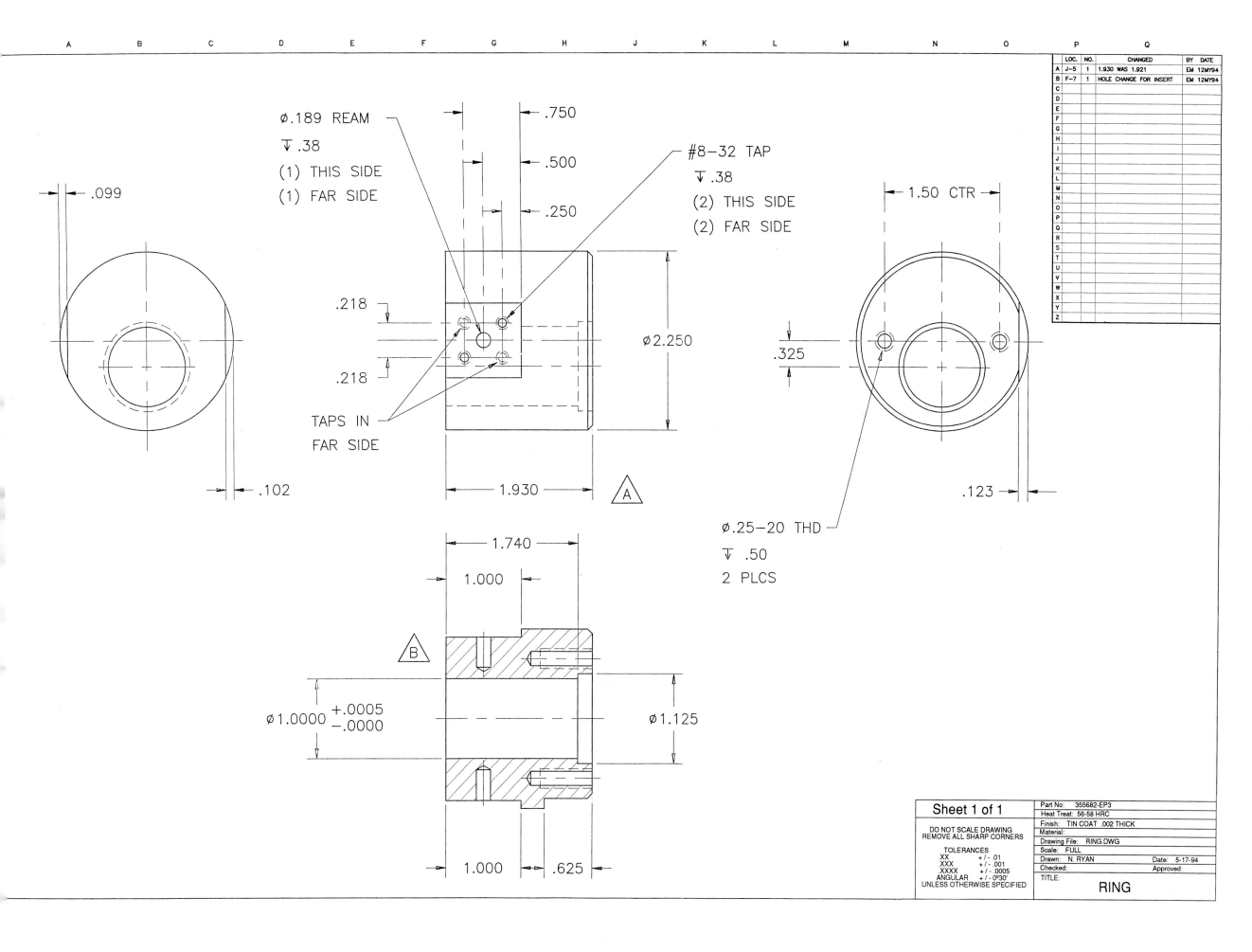

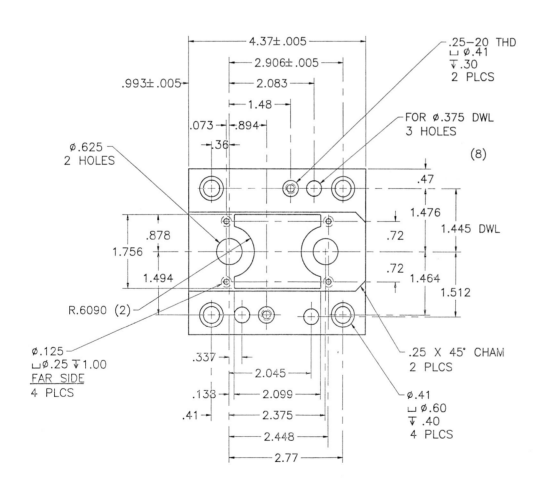
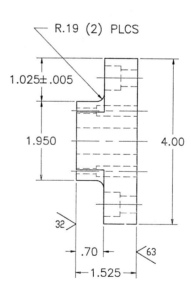

Part No:	457-637-Z
Heat Treat:	56-60 HRC
Finish:	ELECTROLESS NICKEL .002 THICK
Material:	
Drawing File:	BLOCK.DWG
Scale:	1:2
Drawn:	N. RYAN Date: 4-17-94
Checked:	Approved:
TITLE:	BLOCK

Sheet 1 of 1

DO NOT SCALE DRAWING
REMOVE ALL SHARP CORNERS

TOLERANCES
.XX   +/- .01
.XXX   +/- .001
.XXXX   +/- .0005
ANGULAR +/- 0°30'
UNLESS OTHERWISE SPECIFIED

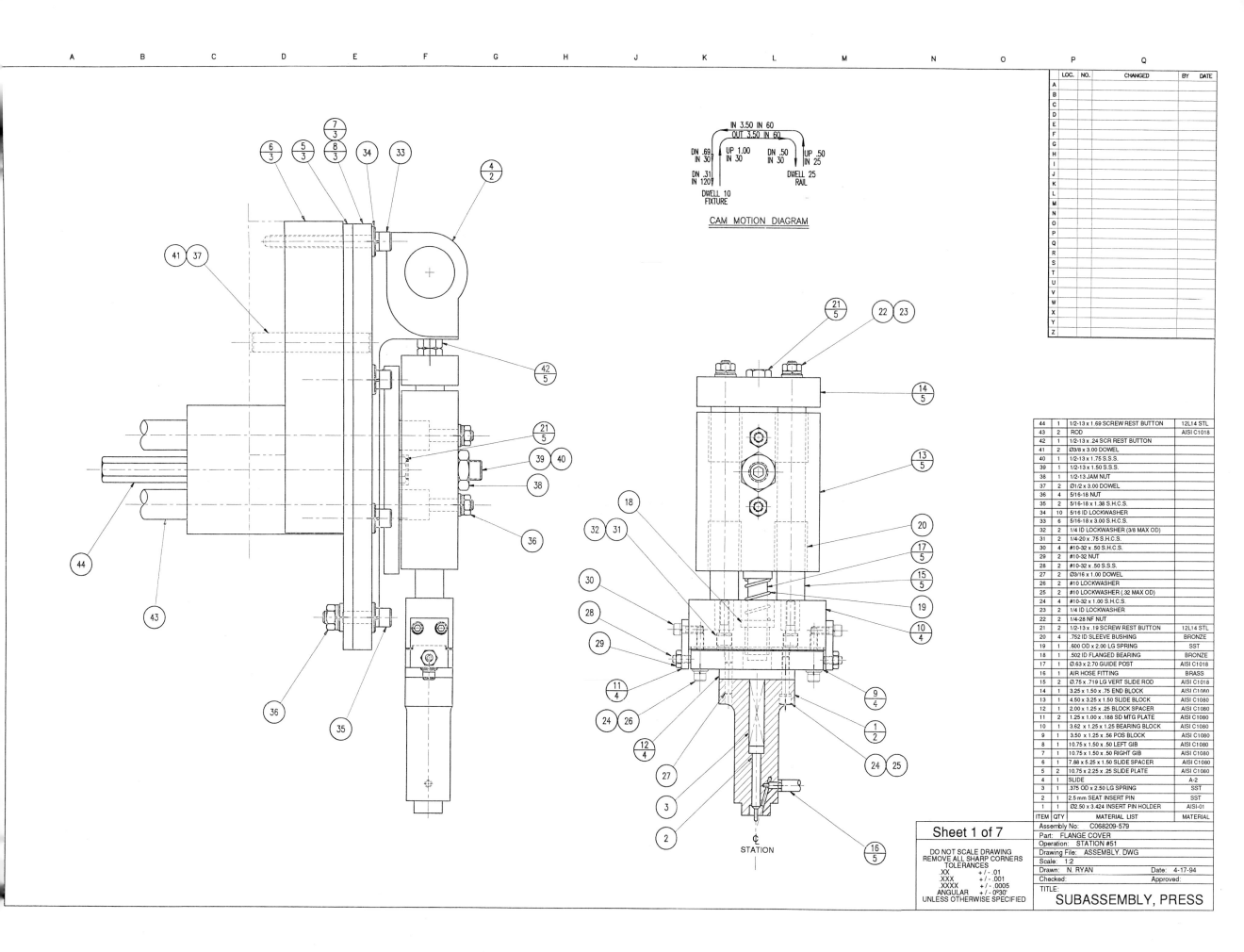

Appendices **385**

APPENDIX A
HOLE/DRILL SIZE CHART

DRILL SIZE	DECIMAL DIA	DRILL SIZE	DECIMAL DIA	DRILL SIZE	DECIMAL DIA	DRILL SIZE	DECIMAL DIA	DRILL SIZE	DECIMAL DIA
80	.0135	50	.0700	22	.1570	17/64	.2656	1/2	.5000
79	.0145	49	.0730	21	.1590	H	.2660	33/64	.5156
1/64	.0156	48	.0760	20	.1610	I	.2720	17/32	.5312
78	.0160	5/64	.0781	19	.1660	J	.2770	35/64	.5469
77	.0180	47	.0785	18	.1695	K	.2811	9/16	.5625
76	.0200	46	.0810	11/64	.1719	9/32	.2812	37/64	.5781
75	.0210	45	.0820	17	.1730	L	.2900	19/32	.5937
74	.0225	44	.0860	16	.1770	M	.2950	39/64	.6094
73	.0240	43	.0890	15	.1800	19/64	.2968	5/8	.6250
72	.0250	42	.0935	14	.1820	N	.3020	41/64	.6406
71	.0260	3/32	.0937	13	.1850	5/16	.3125	21/32	.6562
70	.0280	41	.0960	3/16	.1875	O	.3160	43/64	.6719
69	.0292	40	.0980	12	.1890	P	.3230	11/16	.6875
68	.0310	39	.0995	11	.1910	21/64	.3281	45/64	.7031
1/32	.0313	38	.1015	10	.1935	Q	.3320	23/32	.7187
67	.0320	37	.1040	9	.1960	R	.3390	47/64	.7344
66	.0330	36	.1065	8	.1990	11/32	.3437	3/4	.7500
65	.0350	7/64	.1093	7	.2010	S	.3480	49/64	.7656
64	.0360	35	.1100	13/64	.2031	T	.3580	25/32	.7812
63	.0370	34	.1110	6	.2040	23/64	.3594	51/64	.7969
62	.0380	33	.1130	5	.2055	U	.3680	13/16	.8125
61	.0390	32	.1160	4	.2090	3/8	.3750	53/64	.8281
60	.0400	31	.1200	3	.2130	V	.3770	27/32	.8437
59	.0410	1/8	.1250	7/32	.2187	W	.3860	55/64	.8594
58	.0420	30	.1285	2	.2210	25/64	.3906	7/8	.8750
57	.0430	29	.1360	1	.2280	X	.3970	57/64	.8906
56	.0465	28	.1405	A	.2340	Y	.4040	29/32	.9062
3/64	.0469	9/64	.1406	15/64	.2344	13/32	.4062	59/64	.9219
55	.0520	27	.1440	B	.2380	Z	.4130	15/16	.9375
54	.0550	26	.1470	C	.2420	27/64	.4219	61/64	.9531
53	.0595	25	.1495	D	.2460	7/16	.4375	31/32	.9687
1/16	.0625	24	.1520	1/4	.2500	29/64	.4531	63/64	.9844
52	.0635	23	.1540	F	.257	15/32	.4687	1	1.0000
51	.0670	5/32	.1562	G	.2610	31/64	.4843		

386 Appendices

APPENDIX A (continued)
INCHES/METRIC DECIMAL EQUIVALENTS

FRACTION	DECIMAL		FRACTION	DECIMAL	
	INCH	METRIC (MM)		INCH	METRIC (MM)
1/64	.015625	0.3969	33/64	.515625	13.0969
1/32	.03125	0.7938	17/32	.53125	13.4938
3/64	.046875	1.1906	35/64	.546875	13.8906
1/16	.0625	1.5875	9/16	.5625	14.2875
5/64	.078125	1.9844	37/64	.578125	14.6844
3/32	.09375	2.3813	19/32	.59375	15.0813
7/64	.109375	2.7781	39/64	.609375	15.4781
1/8	.1250	3.1750	5/8	.6250	15.8750
9/64	.140625	3.5719	41/64	.640625	16.2719
5/32	.15625	3.9688	21/32	.65615	16.6688
11/64	.171875	4.3656	43/64	.671875	17.0656
3/16	.1875	4.7625	11/16	.6875	17.4625
13/64	.203125	5.1594	45/64	.703125	17.8594
7/32	.21875	5.5563	23/32	.71875	18.2563
15/64	.234375	5.9531	47/64	.734375	18.6531
1/4	.2500	6.3500	3/4	.7500	19.0500
17/64	.265625	6.7469	49/64	.765625	19.4469
9/32	.28125	7.1438	25/32	.78125	19.8438
19/64	.296875	7.5406	51/64	.796875	20.2406
5/16	.3125	7.9375	13/16	.8125	20.6375
21/64	.328125	8.3384	53/64	.828125	21.0344
11/32	.34375	8.7313	27/32	.84375	21.4313
23/64	.359375	9.1281	55/64	.859375	21.8281
3/8	.3750	9.5250	7/8	.8750	22.2250
25/64	.390625	9.9219	57/64	.890625	22.6219
13/32	.40625	10.3188	29/32	.90625	23.0188
27/64	.421875	10.7156	59/64	.921875	23.4156
7/16	.4375	11.1125	15/16	.9375	23.8125
29/64	.453125	11.5094	61/64	.953125	24.2094
15/32	.46875	11.9063	31/32	.96875	24.6063
31/64	.484375	12.3031	63/64	.984375	25.0031
1/2	.5000	12.7000	1	1.0000	25.4000

APPENDIX B
TAP DRILL SIZES FOR MACHINE SCREWS

SIZE OF SCREW		NUMBER OF THREADS PER INCH	TAP DRILLS	
NUMBER OR DIAMETER	DECIMAL EQUIVALENT		DRILL SIZE	DECIMAL EQUIVALENT
0	.060	80	3/64	.0469
1	.073	64 72	53 53	.0595 .0595
2	.086	56 64	50 50	.0700 .0700
3	.099	48 56	47 45	.0785 .0820
4	.112	40 48	43 42	.0890 .0935
5	.125	40 44	38 37	.1015 .104
6	.138	32 40	36 33	.1065 .1130
8	.164	32 36	29 29	.1360 .1360
10	.190	24 32	25 21	.1495 .1590
12	.216	24 28	16 14	.1770 .1820
1/4	.250	20 28	7 3	.2010 .213
5/16	.3125	18 24	F I	.2570 .2720
3/8	.375	16 24	5/16 Q	.3125 .3320
7/16	.4375	14 20	U 25/64	.3680 .3906
1/2	.500	13 20	27/64 29/64	.4219 .4531

APPENDIX C
CLEARANCE HOLE SIZES FOR MACHINE SCREWS

CLEARANCE HOLE DRILLS					
SIZE OF SCREW		CLOSE FIT		TAP DRILLS	
NUMBER OR DIAMETER	DECIMAL EQUIVALENT	DRILL SIZE	DECIMAL EQUIVALENT	DRILL SIZE	DECIMAL EQUIVALENT
0	.060	52	.0635	50	.0700
1	.073	48	.0760	46	.0810
2	.086	43	.0890	41	.0960
3	.099	37	.1040	35	.1100
4	.112	32	.1160	30	.1285
5	.125	30	.1285	29	.1360
6	.138	27	.1440	25	.1495
8	.164	18	.1695	16	.1770
10	.190	9	.1960	7	.2010
12	.216	2	.2210	1	.228
1/4	.250	F	.2570	H	.2660
5/16	.3125	P	.3230	Q	.3320
3/8	.375	W	.3860	X	.3970
7/16	.4375	29/64	.4531	15/32	.4686
1/2	.500	33/64	.5156	17/32	.5312

Appendices   **389**

APPENDIX D
WORD ABBREVIATIONS ON DRAWINGS

GENERAL RULES:

1. Abbreviations of word combinations should not be separated for single use.
2. Single abbreviations may be combined when necessary.
3. Spaces between word combination abbreviations may be filled with a hyphen (-) for clarity.
4. The same abbreviations should be used for all tenses, the positive case, singular, and plural forms of a given word.

WORD	ABBREVIATION	WORD	ABBREVIATION	WORD	ABBREVIATION
Absolute	ABS	August	AUG	Brinnell hardness	
Accessory	ACCESS.	Authorize	AUTH	number	BHN
Accumulate	ACCUM	Automatic		British thermal unit	BTU
Actual	ACT.	frequency control	AFC	Bronze	BRZ
Adapter	ADPT	Automatic		Brown	BRN
Addendum	ADD.	gain control	AGC	Brown and	
Addition	ADD.	Automatic		Sharpe (gage)	B&S
Adjust	ADJ	volume control	AVC	Building	BLDG
After	AFT.	Auxiliary	AUX	Burnish	BHN
Alignment	ALIGN.	Auxiliary power unit	APU	Bushing	BUSH.
Allowance	AllOW.	Average	AVG	Buzzer	BUZ
Alloy	ALY	Avoirdupois	AVDP	Bypass	BYP
Alteration	ALT	Azimuth	AZ	Cadmium	CAD.
Alternate	ALT	Balance	BAL	Calibrate	CAL
Alternating		Ball bearing	BB	Camber	CAM.
current	AC	Bandpass	BP	Capacitor	CAP.
Ambient	AMB	Bandwidth	B	Capacity	CAP.
American		Base line	BL	Carbon	C
Wire Gage	AWG	Basic	BSC	Carbon steel	CS
Amount	AMT	Battery (electrical)	BAT.	Carton	CTN
Ampere	AMP	Bearing	BRG.	Castellate	CTL
Ampere		Beat-frequency	BF	Casting	CSTG
(combination form)	A	Beat frequency		Cast iron	CI
Ampere hour	AMP HR	oscillator	BFO	Cathode-ray tube	CRT
Amplifier	AMPL	Berylium	BE.	Center	CTR
Amplitude modulation	AM.	Between	BET.	Center line	CL or ¢
And	&	Between centers	BC	Center of gravity	CG
Annunciator	ANN	Bill of material	B/M	Center tap	CT
Anodize	ANOD	Binding	BIND.	Center to center	C TO C
Antenna	ANT.	Black	BLK	Centigrade	C
Application	APPL	Blank	BLK	Centimeter	CM
Approved	APPD	Blue	BLU	Ceramic	CER
Approximate	APPROX	Board	BD	Chamfer	CHAM
April	APR	Bolt circle	BC	Change	CHG
Arc Weld	ARC W	Both faces	BF	Change notice	CN
Armature	ARM.	Both sides	BS	Change order	CO
Assemble	ASSEM	Bottom face	BF	Chassis	CHAS
Assembly	ASSY	Bracket	BRKT	Chromium	Cr
Attention	ATTN	Brass	BRS	Circuit	CKT
Attenuation,		Brazing	BRZG	Circuit breaker	CKT BKR
Attenuator	ATTEN	Bridge	BRDG	Circular	CIR
Audio frequency	AF	Brinnell hardness	BH	Circular pitch	CP

APPENDIX D (continued)
WORD ABBREVIATIONS ON DRAWINGS

WORD	ABBREVIATION	WORD	ABBREVIATION	WORD	ABBREVIATION
Circumference	CIRC	Decibel	DB	External	EXT
Class	CL	Decimal	DEC	Fahrenheit	F
Clearance	CL	Deep-drawn	DD	Farad	F
Clockwise	CW	Degree	DEG or A7	Far side	FS
Coaxial	COAX.	Delay	DLY	Fastener	FASTNR
Coefficient	COEF	Delay line	DL	February	FEB
Cold-drawn steel	CDS	Depth	D	Federal	FED.
Cycles per minute	CPM	Detail	DET	Federal specification	FS
Cycles per second	CPS	Deviation	DEV	Federal stock number	FSN
Cold-rolled	CR	Diagram	DIAG	Figure	FIG.
Cold-rolled steel	CRS	Diameter	DIA	Filament	FIL
Collector	COLL	Diameter bolt circle	DBC	Filament center tap	FCT
Color code	CC	Diametral pitch	DP	Fillister head	FIL H
Commercial	COML	Dimension	DIM.	Finish	FIN.
Common	COM	Direct-current	DC	Finish all over	FAO
Company	CO	Direct-current volts	VDC	Fixed	FXD
Composition	COMP	Direct current		Flange	FLG
Concentric	CONC	working volts	VDCW	Flat head	FH
Condition	COND	Disconnect	DISC.	Foot	FT or "feet"
Conductor	COND	Division	DIV	Foot-pound	FT LB
Contact	CONT	Door	DR	For example	EG
Continue	CONT	Double-pole	DP	Four-pole	4P
Continuous wave	CW	Double-pole		Frequency	FREQ
Control	CONT	double throw	DPDT	Frequency modulation	FM
Copper	COP	Double-pole		Front	FR
Corporation	CORP	single throw	DPST	Gage	GA
Corrosion	CORR	Double throw	DT	Gallon	GAL
Corrosion-resistant	CRE	Down	DN	Galvanize	GALV
Corrosion-resistant		Drawing	DWG	Half-hard	1/2 H
steel	CRES	Drawing list	DL	Half-round	1/2 RD
Cotangent	COT.	Drill	DR	Hard	H
Counterbore	CBORE	Drill rod	DR	Harden	HDN
Counterclockwise	CCW	Drive	DR	Hardware	HDW
Counterdrill	CDRILL	Drive fit	DF	Head	HD
Countersink	CSK	Duplicate	DUP	Headless	HDLS
Countersink		Each	EA	Heater	HTR
other side	CSKO	Eccentric	ECC	Height	HGT
Cross section	XSECT	Effective	EFF	Henry (electrical)	H
Crystal	XTAL	Electric	ELEC	Hertz	HZ
Cubic	CU	Electrolytic	ELECT.	Hexagon	HEX
Cubic centimeter	CC	Element	ELEM	Hexagonal head	HEX HD
Cubic feet	CU FT	Eliminate	ELIM	High frequency	HF
Cubic feet per minute	CFM	Elongation	ELONG	High-frequency	
Cubic feet per second	CFS	Emergency	EMER	oscillator	HFO
Cubic inch	CU IN.	End to end	E to E	Horizon, horizontal	HORIZ
Cubic meter	CU M	Engineer	ENGR	Horizontal center line	HCL
Cubic millimeter	CU MM	Engineering	ENGRG	Horsepower	HP
Current	CUR.	Equipment	EQUIP.	Hot-rolled steel	HRS
Cycle	CY	Equivalent	EQUIV	Hour	HR
Cycles per minute	CPM	Estimate	EST	Identification	IDENT
Cycles per second	CPS	Et cetera	ETC	Impeller	IMP.
Datum	D	Example	EX	Impulse	IMP.
Decalcomania	DECAL	Exclusive	EXCL	Inch	IN. or
December	DEC	Extension	EXT		"inches"

APPENDIX D (continued)
WORD ABBREVIATIONS ON DRAWINGS

WORD	ABBREVIATION	WORD	ABBREVIATION	WORD	ABBREVIATION
Inches per second	IPS	Megacycle	MC	Origin, original	ORIG
Inch-pound	IN LB	Megacycle per second	MCS	Oscillator	OSC
Include	INCL	Megohm	MEGO	Ounce	OZ
Inclusive	INCL	Meter	M	Ounce-inch	OZ IN
Incorporated	INC	Microfarad	UF or µF	Output	OUPT
Indicate	IND	Microhenry	UH or µH	Outside diameter	OD
Indicator	IND	Microhm	UOHM or	Outside radius	OR.
Information	INFO		µOHM	Oval head	OV HD
Inside diameter	ID	Microinch	UIN or µIN	Page	P
Inside radius	IR	Micro micro (10^{-12})	UU or µµ	Pair	PR
Insulation, Insulator	INS	Micromicrohenry	UUH or	Panel	PNL
Integrated circuit	IC	µµH		Paragraph	PARA
Intermediate		Micromicrofarad	UUF or µµF	Parallel	PAR.
frequency	IF	Micromicron	UU or µµ	Part	PT
Internal	INT	Micron (.001 mm)	U or µ	Part number	PN
Irregular	IRREG	Milliampere	MA	Passivate	PASS.
Issue	ISS	Millihenry	MH	Per	/
Jack	J	Millimeter	MM	Per centum	PCT
January	JAN	Millivolt	MV	Perpendicular	PERP
July	JUL	Milliwatt	MW	Phase	PH
Junction	JCT	Minimum	MIN	Phenolic	PHEN
Junction box	JB	Minor	MIN	Phillips head	PHL H
June	JUN	Miscellaneous	MISC	Phosphor bronze	PH BRZ
Keyway	KWY	Model (for general use)	MOD	Piece	PC
Kilocycle	KC	Modification	MOD	Pitch	P
Kilocycles per second	KP/S	Modify	MOD	Pitch circle (thread)	PC
Left	L	Modulator	MOD	Pitch diameter	PD
Left hand	LH	Mount	MT	Plastic (thread)	PLSTC
Length	LG	Mounting	MTG	Plate (electron tube)	P
Length overall	LOA	National coarse		Plug	PL
Light	LT	(thread)	NC	Plus or minus	±
Linear	LIN	National extra		Point	PT
Liquid	LIQ	fine (thread)	NEF	Point of intersection	PI
Load limiting resister	LLR	National fine (thread)	NF	Point of tangency	PT
Local oscillator	LO	National special		Pole	P
Lock washer	LK WASH.	(thread)	NS	Position	POS
Long	LG	Nickel	NI	Positive	POS
Loudspeaker	LS	Nominal	NOM	Pound	LB
Low frequency	LF	Normal	NORM.	Pound-foot	LB FT
Low-frequency		Normally closed	NC	Pounds per	
oscillator	LFO	Normally open	NO	square inch	PSI
Low pass	LP	Not applicable	NA	Power	PWR
Machine screw	MS	Not to scale	NTS	Power amplifier	PA
Maintenance	MAINT	November	NOV	Power supply	PWR SUP
Major	MAJ	Number	NO.	Preamplifier	PREAMP
Manual	MAN	Nylon	N	Preferred	PFD
Manufacture	MFR	Obsolete	OBS	Preliminary	PRELIM
Manufactured	MFD	October	OCT	Prepare	PREP
Manufacturing	MFG	Ohm (for use		Primary	PRIM.
March	MAR	on diagrams)	Ω	Project	PROJ
Master oscillator	MO	On center	OC	Quantity	QTY
Material	MATL	Opposite	OPP	Quarter-hard	1/4 H
Maximum	MAX	Optional	OPT	Quartz	QTZ
Mechanical	MECH	Orange	ORN	Radio frequency	RF

392   Appendices

APPENDIX D (continued)
WORD ABBREVIATIONS ON DRAWINGS

WORD	ABBREVIATION	WORD	ABBREVIATION	WORD	ABBREVIATION
Radius	R or RAD	double throw	SPDT	Toggle	TGL
Rate	RT	Single pole,		Tolerance	TOL
Reactor	REAC	single throw	SPST	Torque	TOR
Received	RECD	Single throw	ST	Transformer	XMFR
Receiver	RCVR	Slotted	SLOT.	Transistor	TSTR
Receptacle	RECP	Small	SM	Transmitter	XMTR
Rectifier	RECT	Socket	SOC	Transmitter receiver	TR
Reference	REF	Socket head	SCH	Triple pole	3P
Reference line	REF L	Solenoid	SOL.	Triple throw	3T
Relay	REL	Spacer	SPR	True position	TP
Release	REL	Speaker	SPKR	Tubing	TUB.
Relief	REL	Special	SPL	Twisted	TW
Remove	REM	Specification	SPEC	Typical	TYP
Required	REQD	Spectrum analyzer	SA	Unfinished	UNFIN
Resistor	RES.	Spherical	SPHER	Unified coarse thread	UNC
Revision (thread)	REV	Split ring	SR	Unified extra fine thread	UNEF
Rheostat	RHEO	Spot face	SF	Unified fine thread	UNF
Right	R	Spot weld	SW	Unified special thread	UNS
Right hand	RH	Spring	SPG	United States Air Force	USAF
Rivet	RIV	Square	SQ	Upper	UP.
Rochwell hardness	RH	Square foot	SQ FT	Used with	U/W
Rotary	ROT.	Square inch	SQ IN.	Variable	VAR
Round	RD	Stainless steel	SST	Variable-frequency	
Roundhead	RH	Standard	STD	oscillator	VFO
Schedule	SCH	Steel	STL	Vertical	VERT.
Schematic	SCHEM	Stranded	STRD	Vertical center line	VCL
Screw	SCR	Surface	SURF.	Very-high frequency	VHF
Seamless	SMLS	Switch	SW	Very-low frequency	VLF
Seamless steel tubing	SSTU	Symbol	SYM	Video	VID
Section	SECT.	Tangent	TAN.	Viscosity	VIS
Selector	SEL	Tapping	TAP.	Volt	V
September	SEP	Tempered	TEMP	Washer	WASH.
Serial	SER	Terminal	TERM.	Watt	W
Servo	SVO	That is	i.e.	Weight	WT
Set screw	SS	Thermal	THRM	White	WHT
Sheet	SH	Thermistor	TMTR	Wire-wound	WW
Shield	SHLD	Thick	THK	With (abbreviate only	
Shoulder	SHLD	Thread	THD	in conjunction with	
Silver	SIL	Threads per inch	TPI	other abbreviations)	W/
Similar	SIM	Three-pole	3P	Without	W/O
Single pole	SP	Through	THRU	Yellow	YEL
Single pole,		Time delay	TD	Zinc	Zn

APPENDIX E
STANDARD SCREW THREAD CHARTS
EXTERNAL THREADS

NOMINAL SIZE AND THREADS PER INCH	SERIES	CLASS DESIG-NATION	ALLOW-ANCE	MAJOR DIA LIMITS			PITCH DIA LIMITS			MINOR DIA
				MaxA	Min	MinB	Max	Min	Tol	
1-64	UNC	2A	.0006	.0724	.0686	—	.0623	.0603	.0020	.0532
		3A	.0000	.0730	.0692	—	.0629	.0614	.0015	.0538
2-56	UNC	2A	.0006	.0854	.0813	—	.0738	.0717	.0021	.0635
		3A	.0000	.0860	.0819	—	.0744	.0728	.0016	.0641
3-48	UNC	2A	.0007	.0983	.0938	—	.0848	.0825	.0023	.0727
		3A	.0000	.0990	.0945	—	.0855	.0838	.0017	.0734
4-40	UNC	2A	.0008	.1112	.1061	—	.0950	.0925	.0025	.0805
		3A	.0000	.1120	.1069	—	.0958	.0939	.0019	.0813
5-40	UNC	2A	.0008	.1242	.1191	—	.1080	.1054	.0026	.0935
		3A	.0000	.1250	.1199	—	.1088	.1059	.0019	.0943
6-32	UNC	2A	.0008	.1372	.1312	—	.1169	.1141	.0028	.0989
		3A	.0000	.1380	.1320	—	.1177	.1156	.0021	.0997
6-40	UNF	2A	.0008	.1372	.1321	—	.1210	.1184	.0026	.1065
		3A	.0000	.1380	.1329	—	.1218	.1198	.0020	.1073
8-32	UNC	2A	.0009	.1631	.1571	—	.1428	.1399	.0029	.1248
		3A	.0000	.1640	.1580	—	.1437	.1415	.0022	.1257
8-36	UNF	2A	.0008	.1632	.1577	—	.1452	.1424	.0028	.1291
		3A	.0000	.1640	.1585	—	.1460	.1439	.0021	.1299
10-24	UNC	2A	.0010	.1890	.1818	—	.1619	.1586	.0033	.1379
		3A	.0000	.1900	.1828	—	.1629	.1604	.0025	.1389
10-32	UNF	2A	.0009	.1891	.1831	—	.1688	.1658	.0030	.1508
		3A	.0000	.1900	.1840	—	.1697	.1674	.0023	.1517
12-24	UNC	2A	.0010	.2150	.2078	—	.1879	.1845	.0034	.1639
		3A	.0000	.2160	.2088	—	.1899	.1863	.0026	.1649
12-28	UNF	2A	.0010	.2150	.2088	—	.1918	.1886	.0032	.1712
		3A	.0000	.2160	.2095	—	.1928	.1904	.0024	.1722
1/4-20	UNC	1A	.0011	.2489	.2367	—	.2164	.2108	.0056	.1876
		2A	.0011	.2489	.2408	.2367	.2164	.2127	.0037	.1876
		3A	.0000	.2500	.2419	—	.2175	.2147	.0028	.1887
1/4-28	UNF	1A	.0010	.2490	.2392	—	.2258	.2208	.0050	.2052
		2A	.0010	.2490	.2425	—	.2258	.2225	.0033	.2052
		3A	.0000	.2500	.2435	—	.2268	.2243	.0025	.2062

APPENDIX E (continued)
STANDARD SCREW THREAD CHARTS
EXTERNAL THREADS

NOMINAL SIZE AND THREADS PER INCH	SERIES	CLASS DESIG- NATION	ALLOW- ANCE	MAJOR DIA LIMITS			PITCH DIA LIMITS			MINOR DIA
				MaxA	Min	MinB	Max	Min	Tol	
5/16-18	UNC	1A	.0012	.3113	.2892	—	.2752	.2691	.0061	.2431
		2A	.0012	.3113	.3026	.2982	.2752	.2712	.0040	.2431
		3A	.0000	.3125	.3038	—	.2764	.2734	.0030	.2443
5/16-24	UNF	1A	.0011	.3114	.3006	—	.2843	.2788	.0055	.2603
		2A	.0011	.3114	.3042	—	.2843	.2806	.0037	.2603
		3A	.0000	.3125	.3053	—	.2854	.2827	.0027	.2614
3/8-16	UNC	1A	.0013	.3737	.3595	—	.3331	.3266	.0065	.2970
		2A	.0013	.3737	.3643	.3595	.3331	.3287	.0044	.2970
		3A	.0000	.3750	.3656	—	.3344	.3311	.0033	.2983
3/8-24	UNF	1A	.0011	.3739	.3631	—	.3468	.3411	.0057	.3228
		2A	.0011	.3739	.3667	—	.3468	.3430	.0038	.3228
		3A	.0000	.3750	.3678	—	.3479	.3450	.0029	.3239
7/16-14	UNC	1A	.0014	.4361	.4205	—	.3897	.3826	.0071	.3485
		2A	.0014	.4361	.4258	.4206	.3897	.3850	.0047	.3485
		3A	.0000	.4375	.4272	—	.3911	.3876	.0035	.3499
7/16-20	UNF	1A	.0013	.4362	.4240	—	.4037	.3975	.0062	.3749
		2A	.0013	.4362	.4281	—	.4037	.3995	.0042	.3749
		3A	.0000	.4375	.4294	—	.4050	.4019	.0031	.3762
1/2-13	UNC	1A	.0015	.4985	.4822	—	.4485	.4411	.0074	.4041
		2A	.0015	.4985	.4876	.4822	.4485	.4435	.0050	.4041
		3A	.0000	.5000	.4891	—	.4500	.4463	.0037	.4056
1/2-20	UNF	1A	.0013	.4987	.4865	—	.4662	.4598	.0064	.4374
		2A	.0013	.4987	.4906	—	.4662	.4619	.0043	.4374
		3A	.0000	.5000	.4918	—	.4675	.4643	.0032	.4387

Appendices 395

APPENDIX E
STANDARD SCREW THREAD CHARTS
INTERNAL THREADS

NOMINAL SIZE AND THREADS PER INCH	SERIES DESIG-NATION	CLASS	MINOR DIA LIMITS		PITCH DIA LIMITS			MINOR DIA MIN.
			Min	Max	Min.	Max.	Tol.	
1-64	UNC	2B	.0561	.0623	.0629	.0655	.0026	.0730
		3B	.0561	.0623	.0629	.0648	.0119	.0730
2-56	UNC	2B	.0667	.0737	.0744	.0772	.0028	.0860
		3B	.0667	.0737	.0744	.0765	.0021	.0860
3-48	UNC	2B	.0764	.0845	.0855	.0885	.0030	.0990
		3B	.0764	.0845	.0855	.0877	.0022	.0990
4-40	UNC	2B	.0849	.0939	.0958	.0991	.0033	.1120
		3B	.0849	.0939	.0958	.0982	.0024	.1120
5-40	UNC	2B	.0979	.1062	.1088	.1121	.0033	.1250
		3B	.0979	.1062	.1088	.1113	.0025	.1250
6-32	UNC	2B	.1040	.1140	.1177	.1214	.0037	.1380
		3B	.1040	.1140	.1177	.1204	.0027	.1380
6-40	UNF	2B	.1110	.1190	.1218	.1252	.0034	.1380
		3B	.1110	.1186	.1218	.1243	.0025	.1380
8-32	UNC	2B	.1300	.1390	.1437	.1475	.0038	.1640
		3B	.1300	.1389	.1437	.1465	.0028	.1640
8-36	UNF	2B	.1340	.1420	.1460	.1496	.0036	.1640
		3B	.1340	.1416	.1460	.1487	.0027	.1640
10-24	UNC	2B	.1450	.1560	.1629	.1672	.0043	.1900
		3B	.1450	.1555	.1629	.1661	.0032	.1900
10-32	UNF	2B	.1560	.1640	.1697	.1736	.0039	.1900
		3B	.1560	.1641	.1697	.1726	.0029	.1900
12-24	UNC	2B	.1710	.1810	.1889	.1933	.0044	.2160
		3B	.1710	.1807	.1889	.1922	.0033	.2160
12-28	UNF	2B	.1770	.1860	.1928	.1970	.0042	.2160
		3B	.1770	.1857	.1928	.1959	.0031	.2160
1/4-20	UNC	1B	.1960	.2070	.2175	.2248	.0073	.2500
		2B	.1960	.2070	.2175	.2223	.0048	.2500
		3B	.1960	.2067	.2175	.2211	.0036	.2500
1/4-28	UNF	1B	.2110	.2200	.2268	.2333	.0065	.2500
		2B	.2110	.2200	.2268	.2311	.0043	.2500
		3B	.2110	.2190	.2268	.2300	.0032	.2500
5/16-18	UNC	1B	.2520	.2650	.2764	.2843	.0079	.3125
		2B	.2520	.2650	.2764	.2817	.0053	.3125
		3B	.2520	.2630	.2764	.2903	.0039	.3125

APPENDIX E (continued)
STANDARD SCREW THREAD CHARTS
INTERNAL THREADS

NOMINAL SIZE AND THREADS PER INCH	SERIES DESIG-NATION	CLASS	MINOR DIA LIMITS		PITCH DIA LIMITS			MINOR DIA MIN.
			Min	Max	Min.	Max.	Tol.	
5/16-24	UNF	1B	.2670	.2770	.2854	.2925	.0071	.3125
		2B	.2670	.2770	.2854	.2902	.0048	.3125
		3B	.2670	.2754	.2854	.2890	.0036	.3125
3/8-16	UNC	1B	.3070	.3210	.3344	.3429	.0085	.3750
		2B	.3070	.3210	.3344	.3401	.0057	.3750
		3B	.3070	.3182	.3344	.3387	.0043	.3750
3/8-24	UNF	1B	.3300	.3400	.3479	.3553	.0074	.3750
		2B	.3300	.3400	.3479	.3528	.0049	.3750
		3B	.3300	.3372	.3479	.3516	.0037	.3750
7/16-14	UNC	1B	.3600	.3760	.3911	.4003	.0092	.4375
		2B	.3600	.3760	.3911	.3972	.0061	.4375
		3B	.3600	.3717	.3911	.3957	.0046	.4375
7/16-20	UNF	1B	.3830	.3950	.4050	.4131	.0081	.4375
		2B	.3830	.3950	.4050	.4104	.0054	.4375
		3B	.3830	.3916	.4050	.4091	.0041	.4375
1/2-13	UNC	1B	.4170	.4340	.4500	.4597	.0097	.5000
		2B	.4170	.4340	.4500	.4565	.0065	.5000
		3B	.4170	.4284	.4500	.4548	.0048	.5000
1/2-20	UNF	1B	.4460	.4570	.4675	.4759	.0084	.5000
		2B	.4460	.4570	.4675	.4731	.0056	.5000
		3B	.4460	.4537	.4675	.4717	.0042	.5000

APPENDIX F
PINS

SQUARE END STRAIGHT DOWEL PIN

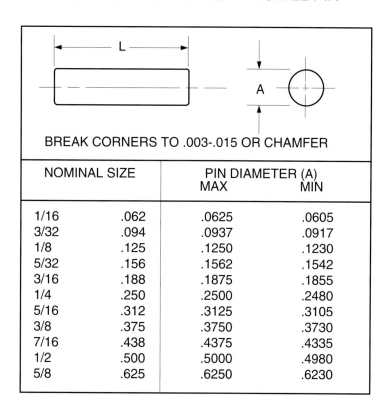

BREAK CORNERS TO .003-.015 OR CHAMFER

NOMINAL SIZE		PIN DIAMETER (A) MAX	MIN
1/16	.062	.0625	.0605
3/32	.094	.0937	.0917
1/8	.125	.1250	.1230
5/32	.156	.1562	.1542
3/16	.188	.1875	.1855
1/4	.250	.2500	.2480
5/16	.312	.3125	.3105
3/8	.375	.3750	.3730
7/16	.438	.4375	.4335
1/2	.500	.5000	.4980
5/8	.625	.6250	.6230

CLEVIS PIN

(ANSI B 18.8.1-1972-R1977)

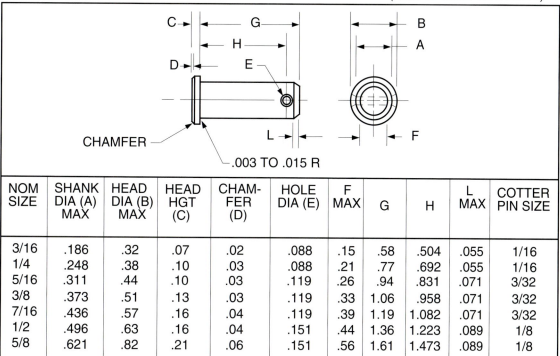

NOM SIZE	SHANK DIA (A) MAX	HEAD DIA (B) MAX	HEAD HGT (C)	CHAMFER (D)	HOLE DIA (E)	F MAX	G	H	L MAX	COTTER PIN SIZE
3/16	.186	.32	.07	.02	.088	.15	.58	.504	.055	1/16
1/4	.248	.38	.10	.03	.088	.21	.77	.692	.055	1/16
5/16	.311	.44	.10	.03	.119	.26	.94	.831	.071	3/32
3/8	.373	.51	.13	.03	.119	.33	1.06	.958	.071	3/32
7/16	.436	.57	.16	.04	.119	.39	1.19	1.082	.071	3/32
1/2	.496	.63	.16	.04	.151	.44	1.36	1.223	.089	1/8
5/8	.621	.82	.21	.06	.151	.56	1.61	1.473	.089	1/8

APPENDIX F (continued)
PINS

COTTER PIN

(ANSI B 18.8.1-1972-R1977)

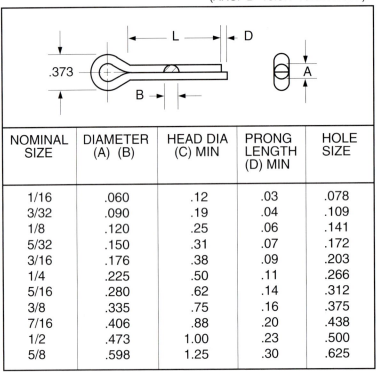

NOMINAL SIZE	DIAMETER (A) (B)	HEAD DIA (C) MIN	PRONG LENGTH (D) MIN	HOLE SIZE
1/16	.060	.12	.03	.078
3/32	.090	.19	.04	.109
1/8	.120	.25	.06	.141
5/32	.150	.31	.07	.172
3/16	.176	.38	.09	.203
1/4	.225	.50	.11	.266
5/16	.280	.62	.14	.312
3/8	.335	.75	.16	.375
7/16	.406	.88	.20	.438
1/2	.473	1.00	.23	.500
5/8	.598	1.25	.30	.625

TAPER PIN

(ASA B5.20-1958)

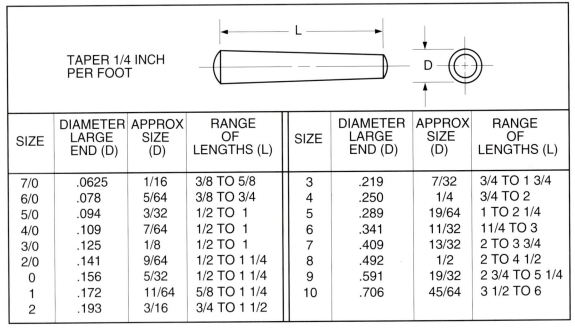

TAPER 1/4 INCH PER FOOT

SIZE	DIAMETER LARGE END (D)	APPROX SIZE (D)	RANGE OF LENGTHS (L)	SIZE	DIAMETER LARGE END (D)	APPROX SIZE (D)	RANGE OF LENGTHS (L)
7/0	.0625	1/16	3/8 TO 5/8	3	.219	7/32	3/4 TO 1 3/4
6/0	.078	5/64	3/8 TO 3/4	4	.250	1/4	3/4 TO 2
5/0	.094	3/32	1/2 TO 1	5	.289	19/64	1 TO 2 1/4
4/0	.109	7/64	1/2 TO 1	6	.341	11/32	1 1/4 TO 3
3/0	.125	1/8	1/2 TO 1	7	.409	13/32	2 TO 3 3/4
2/0	.141	9/64	1/2 TO 1 1/4	8	.492	1/2	2 TO 4 1/2
0	.156	5/32	1/2 TO 1 1/4	9	.591	19/32	2 3/4 TO 5 1/4
1	.172	11/64	5/8 TO 1 1/4	10	.706	45/64	3 1/2 TO 6
2	.193	3/16	3/4 TO 1 1/2				

APPENDIX F (continued)
PINS

GROOVED PIN

(ASA B5.20-1958)

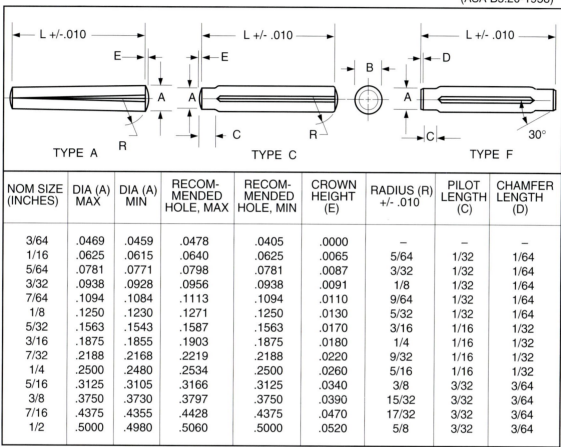

NOM SIZE (INCHES)	DIA (A) MAX	DIA (A) MIN	RECOM-MENDED HOLE, MAX	RECOM-MENDED HOLE, MIN	CROWN HEIGHT (E)	RADIUS (R) +/- .010	PILOT LENGTH (C)	CHAMFER LENGTH (D)
3/64	.0469	.0459	.0478	.0405	.0000	–	–	–
1/16	.0625	.0615	.0640	.0625	.0065	5/64	1/32	1/64
5/64	.0781	.0771	.0798	.0781	.0087	3/32	1/32	1/64
3/32	.0938	.0928	.0956	.0938	.0091	1/8	1/32	1/64
7/64	.1094	.1084	.1113	.1094	.0110	9/64	1/32	1/64
1/8	.1250	.1230	.1271	.1250	.0130	5/32	1/32	1/64
5/32	.1563	.1543	.1587	.1563	.0170	3/16	1/16	1/32
3/16	.1875	.1855	.1903	.1875	.0180	1/4	1/16	1/32
7/32	.2188	.2168	.2219	.2188	.0220	9/32	1/16	1/32
1/4	.2500	.2480	.2534	.2500	.0260	5/16	1/16	1/32
5/16	.3125	.3105	.3166	.3125	.0340	3/8	3/32	3/64
3/8	.3750	.3730	.3797	.3750	.0390	15/32	3/32	3/64
7/16	.4375	.4355	.4428	.4375	.0470	17/32	3/32	3/64
1/2	.5000	.4980	.5060	.5000	.0520	5/8	3/32	3/64

APPENDIX F (continued)
PINS

GROOVED PIN

(ABA B5.20-1958)

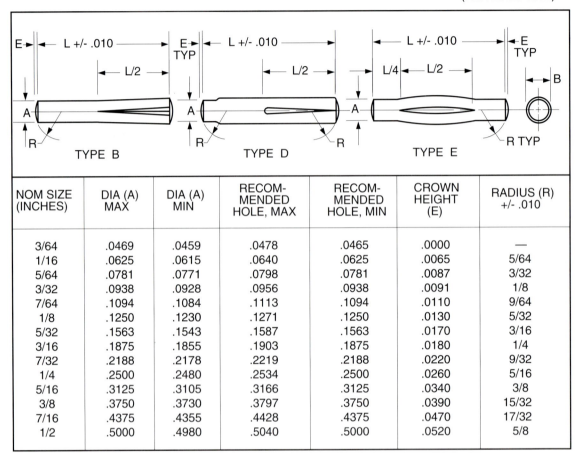

NOM SIZE (INCHES)	DIA (A) MAX	DIA (A) MIN	RECOM- MENDED HOLE, MAX	RECOM- MENDED HOLE, MIN	CROWN HEIGHT (E)	RADIUS (R) +/- .010
3/64	.0469	.0459	.0478	.0465	.0000	—
1/16	.0625	.0615	.0640	.0625	.0065	5/64
5/64	.0781	.0771	.0798	.0781	.0087	3/32
3/32	.0938	.0928	.0956	.0938	.0091	1/8
7/64	.1094	.1084	.1113	.1094	.0110	9/64
1/8	.1250	.1230	.1271	.1250	.0130	5/32
5/32	.1563	.1543	.1587	.1563	.0170	3/16
3/16	.1875	.1855	.1903	.1875	.0180	1/4
7/32	.2188	.2178	.2219	.2188	.0220	9/32
1/4	.2500	.2480	.2534	.2500	.0260	5/16
5/16	.3125	.3105	.3166	.3125	.0340	3/8
3/8	.3750	.3730	.3797	.3750	.0390	15/32
7/16	.4375	.4355	.4428	.4375	.0470	17/32
1/2	.5000	.4980	.5040	.5000	.0520	5/8

APPENDIX G
BENDS IN SHEET METAL

BENDS IN SHEET METAL — 90-DEGREE DEVELOPED LENGTH

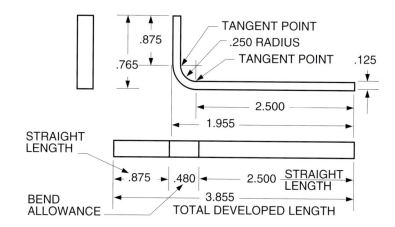

To determine the developed length of an object with a 90-degree bend:
1. Locate the tangent points at the inside of the .250-inch radius. The tangent point is where the straight length meets the radius.
2. Find the material thickness on the drawing: .125.
3. Refer to the chart below where the horizontal row at the top indicates the inside radius and the left-hand column indicates the material thickness. The chart shows that the .250-inch radius and the .125-inch sheet thickness meet at .480 inch (bold letters). This dimension is called the bend allowance and is the length that must be added to the straight lengths to determine the total developed length.

BEND ALLOWANCE FOR 90-DEGREE BENDS (INCH)

RADII THK	.031	.063	.094	.125	.156	.188	.219	**.250**	.281	.313	.344	.375	.438	.500
.013	.058	.108	.157	.205	.254	.304	.353	.402	.450	.501	.549	.598	.697	.794
.016	.060	.110	.159	.208	.256	.307	.355	.404	.453	.503	.552	.600	.699	.796
.020	.062	.113	.161	.210	.259	.309	.358	.406	.455	.505	.554	.603	.702	.799
.022	.064	.114	.163	.212	.260	.311	.359	.408	.457	.507	.556	.604	.703	.801
.025	.066	.116	.165	.214	.263	.313	.362	.410	.459	.509	.558	.607	.705	.803
.028	.068	.119	.167	.216	.265	.315	.364	.412	.461	.511	.560	.609	.708	.805
.032	.071	.121	.170	.218	.267	.317	.366	.415	.463	.514	.562	.611	.710	.807
.038	.075	.126	.174	.223	.272	.322	.371	.419	.468	.518	.567	.616	.715	.812
.040	.077	.127	.176	.224	.273	.323	.372	.421	.469	.520	.568	.617	.716	.813
.050		.134	.183	.232	.280	.331	.379	.428	.477	.527	.576	.624	.723	.821
.064		.144	.192	.241	.290	.340	.389	.437	.486	.536	.585	.634	.732	.830
.072			.198	.247	.296	.346	.394	.443	.492	.542	.591	.639	.738	.836
.078			.202	.251	.300	.350	.399	.447	.496	.546	.595	.644	.743	.840
.081			.204	.253	.302	.352	.401	.449	.498	.548	.598	.646	.745	.842
.091			.212	.260	.309	.359	.408	.456	.505	.555	.604	.653	.752	.849
.094			.214	.262	.311	.361	.410	.459	.507	.558	.606	.655	.754	.851
.102				.268	.317	.367	.416	.464	.513	.563	.612	.661	.760	.857
.109				.273	.321	.372	.420	.469	.518	.568	.617	.665	.764	.862
125				.284	.333	.383	.432	**.480**	.529	.579	.728	.677	.776	.873
.156					.355	.405	.453	.502	.551	.601	.650	.698	.797	.895
.188						.427	.476	.525	.573	.624	.672	.721	.820	.917
.203								.535	.584	.634	.683	.731	.830	.928
.218								.546	.594	.645	.693	.742	.841	.938
.234								.557	.606	.656	.705	.753	.852	.950
.250								.568	.617	.667	.716	.764	.863	.961

Appendices 401

APPENDIX G (continued)
BENDS IN SHEET METAL

BENDS IN SHEET METAL — GREATER THAN 90-DEGREE DEVELOPED LENGTH

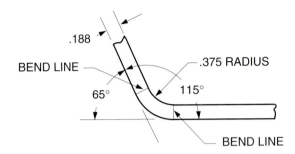

For the above illustration, material thickness equals .188 inch and the inside radius equals .375 inch. Where these two dimensions meet on the chart (in bold letters, reading horizontally and vertically) equals .00802. If the inside bend equals 115 degrees, the bend allowance (B/A) equals 180 degrees minus 115 degrees equals 65 degrees times .00802 equals .5213 inch. This is the bend allowance, which is the length of material required from bend line to bend line given the illustrated conditions. The .5213 inch needs to be added to the straight lengths to determine the total developed length.

BEND ALLOWANCE FOR EACH 1 DEGREE OF BEND

RADII THK	.031	.063	.094	.125	.156	.188	.219	.250	.281	.313	.344	**.375**	.438	.500
.013	.00064	.00120	.00174	.00228	.00282	.00338	.00392	.00446	.00500	.00556	.00610	.00664	.00774	.00883
.016	.00067	.00122	.00176	.00231	.00285	.00342	.00395	.00449	.00503	.00559	.00613	.00667	.00777	.00885
.020	.00069	.00125	.00179	.00233	.00287	.00343	.00397	.00452	.00506	.00561	.00616	.00670	.00780	.00888
.022	.00071	.00127	.00181	.00235	.00289	.00345	.00399	.00453	.00508	.00563	.00617	.00672	.00782	.00890
.025	.00074	.00129	.00184	.00238	.00292	.00348	.00402	.00456	.00510	.00566	.00610	.00674	.00784	.00892
.028	.00076	.00132	.00186	.00240	.00294	.00350	.00404	.00458	.00512	.00568	.00622	.00676	.00786	.00894
.032	.00079	.00134	.00189	.00243	.00297	.00353	.00407	.00461	.00515	.00571	.00625	.00679	.00789	.00897
.038	.00084	.00140	.00194	.00248	.00302	.00358	.00412	.00466	.00520	.00576	.00630	.00684	.00794	.00902
.040	.00085	.00141	.00195	.00249	.00303	.00359	.00413	.00468	.00522	.00577	.00632	.00686	.00796	.00904
.050		.00149	.00203	.00258	.00312	.00368	.00422	.00476	.00530	.00586	.00640	.00694	.00804	.00912
.064		.00160	.00214	.00268	.00322	.00378	.00432	.00486	.00540	.00596	.00650	.00704	.00814	.00922
.072			.00220	.00274	.00328	.00384	.00438	.00492	.00546	.00602	.00656	.00710	.00820	.00929
.078			.00225	.00279	.00333	.00389	.00443	.00497	.00551	.00607	.00661	.00715	.00825	.00933
.081			.00227	.00281	.00335	.00391	.00445	.00499	.00554	.00609	.00664	.00718	.00828	.00936
.091			.00235	.00289	.00343	.00399	.00453	.00507	.00561	.00617	.00671	.00725	.00835	.00944
.094			.00237	.00291	.00346	.00401	.00456	.00510	.00564	.00620	.00674	.00728	.00838	.00946
.102				.00298	.00352	.00408	.00462	.00516	.00570	.00626	.00680	.00734	.00844	.00952
.109				.00303	.00357	.00413	.00467	.00521	.00575	.00631	.00685	.00739	.00849	.00958
.125				.00316	.00370	.00426	.00480	.00534	.00588	.00644	.00698	.00752	.00862	.00970
.156					.00394	.00450	.00504	.00558	.00612	.00688	.00722	.00776	.00886	.00994
.188						.00475	.00529	.00583	.00637	.00693	.00747	**.00802**	.00911	.01019
.203								.00595	.00649	.00704	.00759	.00813	.00923	.01031
.218								.00606	.00660	.00716	.00770	.00824	.00934	.01042
.234								.00619	.00673	.00729	.00783	.00837	.00947	.01055
.250								.00631	.00685	.00741	.00795	.00849	.00959	.01068

APPENDIX H
SHEET METAL GAGES

NUMBER GAGE	*AMERICAN WIRE GAGE OR BROWN & SHARPE GAGE	** UNITED STATES STANDARD GAGE	MACHINE AND WOOD SCREW GAGE
6/0	.5800	-----	-----
5/0	.5165	-----	-----
4/0	.4600	.4063	-----
3/0	.4096	.3750	-----
2/0	.3648	.3438	-----
0	.3249	.3125	.060
1	.2893	.2813	.073
2	.2576	.2656	.086
3	.2294	.2500	.099
4	.2043	.2344	.112
5	.1819	.2188	.125
6	.1620	.2031	.138
7	.1443	.1875	.151
8	.1285	.1719	.164
9	.1144	.1563	.177
10	.1019	.1406	.190
11	.0907	.1250	.203
12	.0808	.1094	.216
13	.0720	.0938	-----
14	.0641	.0781	.242
15	.0571	.0703	-----
16	.0508	.0625	.268
17	.0453	.0563	-----
18	.0403	.0500	.294
19	.0359	.0438	-----
20	.0320	.0375	.320
21	.0285	.0344	-----
22	.0253	.0313	-----
23	.0226	.0281	-----
24	.0201	.0250	.372
25	.0179	.0219	-----
26	.0159	.0188	-----
27	.0142	.0172	-----
28	.0126	.0156	-----
29	.0113	.0141	-----
30	.0100	.0125	.450
31	.0089	.0109	-----
32	.0080	.0102	-----
33	.0071	.0094	-----
34	.0063	.0086	-----
35	.0056	.0078	-----
36	.0050	.0070	-----

*American Wire or Brown & Sharpe Gage: copper wire, brass, copper alloys, and nickel silver wire and sheet, also aliminum sheet, rod, and wire.

**United States Standard Gage: steel and monel metal sheets.
The use of decimals of an inch for specifying sheet and wire gages is recommended.

APPENDIX I
KEYS

SQUARE, FLAT, PLAIN TAPER, AND GIB HEAD KEYS

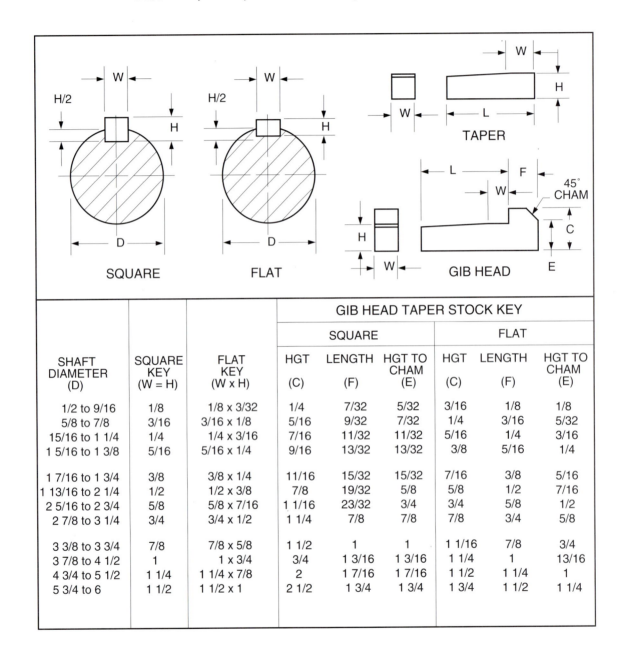

SHAFT DIAMETER (D)	SQUARE KEY (W = H)	FLAT KEY (W x H)	GIB HEAD TAPER STOCK KEY					
			SQUARE			FLAT		
			HGT (C)	LENGTH (F)	HGT TO CHAM (E)	HGT (C)	LENGTH (F)	HGT TO CHAM (E)
1/2 to 9/16	1/8	1/8 x 3/32	1/4	7/32	5/32	3/16	1/8	1/8
5/8 to 7/8	3/16	3/16 x 1/8	5/16	9/32	7/32	1/4	3/16	5/32
15/16 to 1 1/4	1/4	1/4 x 3/16	7/16	11/32	11/32	5/16	1/4	3/16
1 5/16 to 1 3/8	5/16	5/16 x 1/4	9/16	13/32	13/32	3/8	5/16	1/4
1 7/16 to 1 3/4	3/8	3/8 x 1/4	11/16	15/32	15/32	7/16	3/8	5/16
1 13/16 to 2 1/4	1/2	1/2 x 3/8	7/8	19/32	5/8	5/8	1/2	7/16
2 5/16 to 2 3/4	5/8	5/8 x 7/16	1 1/16	23/32	3/4	3/4	5/8	1/2
2 7/8 to 3 1/4	3/4	3/4 x 1/2	1 1/4	7/8	7/8	7/8	3/4	5/8
3 3/8 to 3 3/4	7/8	7/8 x 5/8	1 1/2	1	1	1 1/16	7/8	3/4
3 7/8 to 4 1/2	1	1 x 3/4	3/4	1 3/16	1 3/16	1 1/4	1	13/16
4 3/4 to 5 1/2	1 1/4	1 1/4 x 7/8	2	1 7/16	1 7/16	1 1/2	1 1/4	1
5 3/4 to 6	1 1/2	1 1/2 x 1	2 1/2	1 3/4	1 3/4	1 3/4	1 1/2	1 1/4

APPENDIX I
KEYS

WOODRUFF KEYS

ANSI B 17.2-1967-R1978

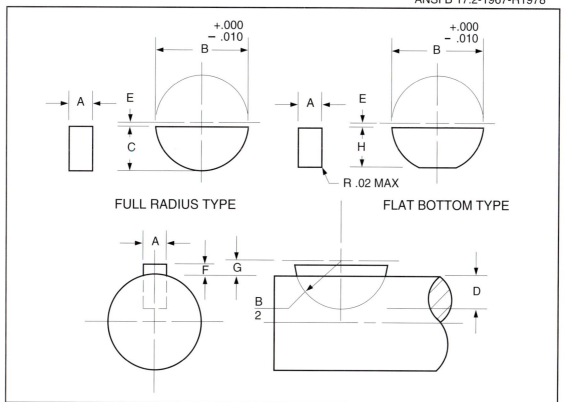

FULL RADIUS TYPE   FLAT BOTTOM TYPE

KEY NO.	NOMINAL SIZE	\	\	MAXIMUM DIMENSIONS	\	\	\
	A x B	E	F	G	H	D	C
204	1/16 x 1/2	3/64	1/32	5/64	.194	.1718	.203
304	3/32 x 1/2	3/64	3/64	3/32	.194	.1561	.203
305	3/32 x 5/8	1/16	3/64	7/64	.240	.2031	.250
404	1/8 x 1/2	3/64	1/16	7/64	.194	.1405	.203
405	1/8 x 5/8	1/16	1/16	1/8	.240	.1875	.250
406	1/8 x 3/4	1/16	1/16	1/8	.303	.2505	.313
505	5/32 x 5/8	1/16	5/64	9/64	.240	.1719	.250
506	5/32 x 3/4	1/16	5/64	9/64	.303	.2349	.313
507	5/32 x 7/8	1/16	5/64	9/64	.365	.2969	.375
606	3/16 x 3/4	1/16	3/32	5/32	.303	.2193	.313
607	3/16 x 7/8	1/16	3/32	5/32	.365	.2813	.375
608	3/16 x 1	1/16	3/32	5/32	.428	.3443	.438
609	3/16 x 1 1/8	5/64	3/32	11/64	.475	.3903	.484
807	1/4 x 7/8	1/16	1/8	3/16	.365	.2500	.375

Note: All dimensions are given in inches. The key numbers indicate nominal key dimensions.

APPENDIX I
KEYS

PRATT AND WHITNEY ROUND-END KEYS

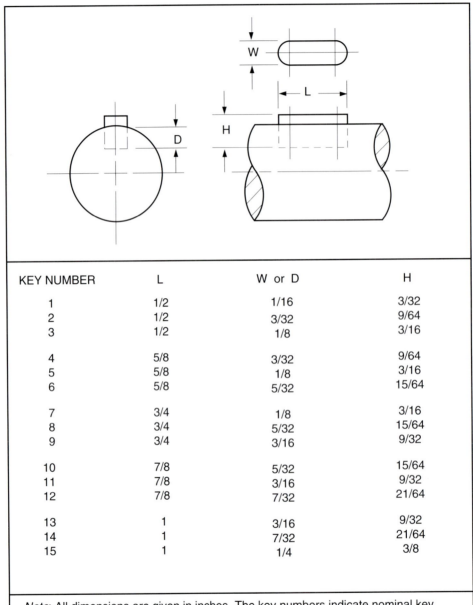

KEY NUMBER	L	W or D	H
1	1/2	1/16	3/32
2	1/2	3/32	9/64
3	1/2	1/8	3/16
4	5/8	3/32	9/64
5	5/8	1/8	3/16
6	5/8	5/32	15/64
7	3/4	1/8	3/16
8	3/4	5/32	15/64
9	3/4	3/16	9/32
10	7/8	5/32	15/64
11	7/8	3/16	9/32
12	7/8	7/32	21/64
13	1	3/16	9/32
14	1	7/32	21/64
15	1	1/4	3/8

Note: All dimensions are given in inches. The key numbers indicate nominal key dimensions.

INDEX

A

adhesive bonding, 213
American National Standards Institute
(ANSI), 4, 14, 37, 39, 49, 84, 89,
104, 131, 168, 196, 206, 250, 277,
300
auxiliary views, 103, 112-117
front, 114
partial, 117
primary, 113-114
secondary 113, 114-117
side, 114
top, 114

B

balloon numbers, 39-40
belt drives, 246
crossed belt, 246
open belt, 246
open belt with idler, 246
quarter-twist, 246
belts, 244-248
flat, 246-247
V, 246, 248
bevel gear, 234, 236-238
hypoid, 237
spiral, 237
straight tooth, 237
zero, 237
bevel gear nomenclature, 237-238
addendum angle, 237
apex, 236, 237
dedendum angle, 237
face angle, 237
face width, 238
mounting distance, 238
pitch angle, 238
pitch cone radius, 238
root angle, 238
whole depth, 238
bill of material, 39
bit, 360
blueline diazo process, 1
bolt, 200-201
brazing, 211-213
breaks, 112
elongated objects, 112
byte, 360

C

CAD, 1, 3, 4, 51, 353-370
CAD applications, 363-370

calipers, 15-19
dial, 15
digital, 15
electronic, 15
hermaphrodite, 15
spring, 15, 18
vernier, 15, 18-19
cam, 240-244
cylindrical, 242
disk, 242
drum, 242
plate, 242
radial, 242
cam follower, 240-242
flat, 242
offset, 242
point, 242
roller, 242
spherical, 242
swinging, 242
yoke, 242
cam nomenclature, 242-244
base circle, 242, 244
cam surface, 242, 244
displacement diagram, 242, 244
dwell, 242, 244
fall, 242, 244
pitch curve, 244
pitch point, 244
pressure angle, 244
prime circle, 244
rise, 242, 244
trace point, 244
central processing unit (CPU), 4, 355, 360,
361
chain—types, 250-251
bead, 250
detachable, 250
offset, 250
pintle, 250
roller, 250
silent, 250-251
coatings, 283-286
metallic, 283-285
nonmetallic, 285-286
polymer, 286
computer-aided drafting, 1, 3, 4, 51,
353-370
computer applications, 364-370
booting, 363-364
command menu, 364
creating a design file, 365
creating text, 366
layers, 366
line menu, 365

logging-on, 364
plotting, 366-370
computer system components, 354-361
application programs, 355
central processing unit (CPU), 360, 361
hardware, 354
input devices, 355-358
memory, 360
operating system software, 355
output devices, 358-359
processing devices, 360-361
computer systems, 361
analog, 361
desktop, 362
digital, 361
laptop, 362
mainframe, 361
mini, 362
micro, 362
portable, 362
CPU. *See* central processing unit

D

developments, 327-335
developable, 329
parallel line, 330-333
radial line, 330,333
triangulation, 330,335
non-developable, 329
dimensional measuring, 11
dimensioning, 131-145, 336
datum, 133-135, 338
progressive, 338
dimensions, 12, 131-132, 162-163, 167
dimensioning systems—types, 132-133
aligned, 132-133
unidirectional, 132
dimensioning terms, 133-143
actual size, 133
arc, 21, 141
basic size, 133
blind hole, 141
chamfer, 143
counterbore, 142
counterdrill, 142
countersink, 142
datum, 133-135
diameter, 137-138
dimension line, 135-136
extension line, 136
feature, 135
foreshortened radius, 138
keyway, 144-145
leader, 136-137

408   **Index**

dimensioning terms, *continued*
  limit dimensions, 133
  nominal size, 133
  radius, 138
  reference dimension, 135
  slotted hole, 142
  spotface, 142-143
  true position, 133
disk, 364
  formatting, 364
drafting, 3-4
  computer-aided, 3, 4
  manual, 3-4
drawings, 1-3, 35-36, 37, 82-84, 85, 89, 132,
    299-305, 327, 335-340
  assembly, 3, 300, 301-305
  design layout, 3, 301
  detail, 3, 301-302
  detail assembly, 305
  development, 327
  expanded assembly, 3, 305
  exploded view, 3, 305
  general assembly, 3, 305
  inseparable, 305
  mono-detail, 300-301
  multi-detail, 301
  multiview, 82-84, 85, 89
  one-view, 84
  principal view, 132
  production, 2, 300
  separable, 305
  sheet metal, 327, 335-340
  three-view, 85
  two-view, 85
  working, 2, 300
drawing field, 37
drawing medium, 35-36, 61-62
  erasability, 35-36
  permanency, 35
  strength, 35
  translucency, 35
drawing medium—types, 35-36, 61-62
  cloth, 35-36
  polyester film, 36
  vellum, 36
drawing notes, 53
  general, 53
  local, 53
dual-dimensioning system, 14

E

engineering change notice (ECN), 39
  revision area, 39
  revision block, 39
engineering drawing, 1-3
equipment enclosures, 335-337
  cabinet, 336-337
  chassis, 336
  console, 337
  rack, 335-336

F

fasteners, 198-199
  permanent, 198-199
  removable, 198-199
  semipermanent, 198-199
fasteners—threaded, 199-201
  bolts, 200-201
  nuts, 200-201
  screws, 200-201
fasteners—unthreaded, 202-205
  pins, 202-203
  rings, 202-204
  rivets, 204-205
font, 51

G

gear broaching, 235
gear hobbing, 235
gears, 233-240
  bevel, 234, 236-238
  pinion, 234-235
  spur, 234-236
  worm, 234, 239-240
gear shaping, 235
gear train, 234

H

hard copy, 1, 358, 366
hardware, 4, 354
hieroglyphics, 3

I

idler, 246
input, 354
input devices, 4, 355-358
  alphanumeric keyboard, 355
  cursor, 355-356
  fax/modem, 356
  graphics tablet, 356
  menu tablet, 356
  modem, 356
  mouse, 357
  prompt, 357
  puck, 357
  screen menu, 357
  touch pen, 357
  voice-data entry, 358
interactive, 4
International System of Units (SI), 5, 12, 13
involute curve, 234

J

joining, 205-213
  adhesive bonding, 2133
  brazing, 211-213
  soldering, 213
  welding, 206-211

L

lettering, 51-53
  commercial Gothic-style, 51-53
  templates, 53
  transfer-type, 53
lettering style, 51-53
line, 49
line conventions, 49-51, 104-105
  break line, 50
  center line, 50
  cutting plane line, 51, 104
  dimension line, 50
  extension line, 50
  hidden line, 51
  leader line, 50
  outline, 51
  phantom line, 50-51
  section lining, 51, 104-105
  visible line, 51
lines, 82, 85, 87, 89, 104
  coincidence of, 89
  crosshatching, 104
  foreshortened, 138
  hatching, 104
  miter, 85
  precedence of, 89
  projection, 82

M

measurement systems, 12-14
  decimal-inch, 12, 12-13, 21
  metric, 12, 13
measuring, 11, 14-24
measuring tools—nonprecision type, 15-18
  inside caliper, 15
  outside caliper, 15
  protractor, 15, 21
  scale, 15
  steel rule, 15
measuring tools—precision type, 15, 18-19
   21-24
  micrometer, 15, 21-24
  vernier caliper, 15, 18-19
memory, 360
memory device, 355
metallic coatings, 283-285
  electroless plating, 283
  electroplating, 283
  immersion plating, 283-285
metric equivalents, 13
metrology, 12
micrometer, 15, 21-24
  blade, 22
  digital electronic, 22
  disc, 22
  hub, 22
  outside, 15, 22
  screw thread, 22
millimeter, 13
monitor, 1, 354

Index   **409**

N

networking, 361
nonmetallic coatings, 285-286
   anodize, 285-286
   chemical conversion, 285
   chromate coating, 286
   phosphate coating, 286
nut, 200-202
   acorn, 202
   castle, 202
   hexagon, 202
   slotted, 202
   square, 202

O

output, 354
output devices, 4, 358-359
   CRT, 358
   plotter, 358
   printer, 358

P

part-numbering system, 40
parts list, 39
pins, 202-203
   clevis, 202
   cotter, 202-203
   dowel, 203
   grooved, 203
   roll, 203
   spring, 203
   taper, 203
point, 85-87
plotter, 1, 359
polymer coatings, 286
   epoxies, 286
   urethanes, 286
   vinyls, 286
power transmission, 233
power transmission elements, 233-252
   belts, 244-248
   cams, 240-244
   chains, 249-252
   gears, 233-240
print, 1, 51
processing device, 360-361
   CPU, 360
   floppy disk drive, 360-361
   hard disk drive, 360-361
   memory, 360
projection, 81-84, 85
   first-angle, 81, 82
   orthographic, 81
   third-angle, 81-82
protective coatings, 283-286
protractor, 21
   bevel, 21
   combination set bevel, 21
   plate, 21

   universal bevel, 21
   vernier, 21
pulley, 244, 246-247
   driven, 247
   driver, 247

R

radian, 21
RAM (random access memory), 371
retaining ring, 202, 204
   external, 204
   internal, 204
rib, 110
rivet, 204-205
ROM (read-only memory), 371

S

screws, 193-198, 200-201
screw thread application, 196-198
   decimal-inch, 196-198
   designation, 196-198
   metric, 198
   representation, 196
screw thread—forms, 194-195
   acme, 194
   American National, 194
   buttress, 195
   knuckle, 195
   sharp V, 195
   square, 194
   Unified National, 194-195
   worm, 195
screw thread—terminology, 195-196
   crest, 195
   depth, 195
   external, 195
   internal, 195
   left-hand, 195
   major diameter, 195
   minor diameter, 195
   nominal thread size, 195
   pitch, 196
   pitch diameter, 196
   right-hand, 196
   root, 196
   thread class, 196
   threads per inch (TPI), 196
   series, 196
sectional views, 103-112
   broken-out, 109
   full, 106
   half, 106
   offset, 109-110
   removed, 110
   revolved, 109
   unlined, 110-112
set screw points, 201
   cone, 201
   cup, 201

   flat, 201
   full dog, 201
   half dog, 201
   oval, 201
set screws, 201
sheave, 244, 248
sheet metal, 327-340
sheet metal—terms, 327-340
   base line, 333
   bead, 339-340
   bend allowance, 337
   bend angle, 337
   bend line, 337
   bend relief, 339
   blank, 327
   brake, 337-338
   drawn parts, 340
   leg, 337
   noncircular openings, 339
SI. *See* International System of Units
silver solder, 213
sketching, 66-69
   arcs, 69
   circles, 69
   rectangles, 66-68
   squares, 66-68
   triangles, 66-68
software, 4, 354-355
soldering, 213
spoke, 110
sprocket, 249, 251-252
sprocket nomenclature, 252
   bore, 252
   bottom diameter, 252
   hub length, 252
   outside diameter, 252
   pitch, 252
   pitch diameter, 252
spur gear nomenclature, 234-235
   addendum, 235
   center distance, 235
   clearance, 235
   circular pitch, 235
   dedendum, 235
   outside diameter, 235
   pitch circle, 235
   pressure angle, 235
   root diameter, 235
   tooth space, 235
   tooth thickness, 235
surface, 89, 278
surface finish, 278
surface irregularities, 278
surface texture, 278-282
   center line, 279
   measured profile, 279
   microinch, 279
   micrometer, 279
   nominal profile, 279
   peak, 279
   profile, 279

Index

surface texture, *continued*
    roughness sampling length, 279
    roughness width sampling, 279
    sampling length, 279
    spacing, 279
    traversing length, 279
    valley, 279
    waviness height, 279
    waviness spacing, 279
surface texture nomenclature, 278-279
    flaws, 279
    lay, 279
    measured surface, 278
    nominal surface, 278
    roughness, 278
    waviness, 278-279
Systeme internationale d'unites. See International System of Units (SI)

T

technical sketches, 61, 62-67
    axonometric, 63
    dimetric, 63
    isometric, 63
    multiview, 62-63
    oblique, 63-64
    perspective, 64-66
    trimetric, 63
technical sketching, 61-70
text, 51
    computer-generated, 51, 55
    freehand, 51
    mechanical, 51
text presentation, 51
threaded fasteners, 199-201
    cap screw, 200

carriage bolt, 200-201
eyebolt, 201
lag screw, 201
machine screw, 200
set screw, 201
stove bolt, 200
title block, 37-38
tolerance, 162
tolerancing, 161-176
    accumulation of, 164-165
    allowance, 162
    angularity, 171-172
    bilateral tolerance, 162, 171
    circularity, 170
    circular runout, 176
    clearance allowance, 162
    coaxiality, 176
    concentricity, 176
    cylindricity, 170-171
    datum identification symbol, 166-167
    feature control frame, 165-166
    feature control symbol, 165
    fit, 162
    geometric characteristic symbol, 165
    geometric tolerance, 162
    interference allowance, 162
    least material condition (LMC), 168
    limits of size, 162, 163
    mating parts, 162
    maximum material condition (MMC), 167
    modifier, 167-168
    parallelism, 175
    perpendicularity, 174
    positional tolerance, 176
    profile of a line, 171
    profile of a surface, 171
    projected tolerance zone, 168

regardless of feature size (RFS), 168
roundness, 170
straightness, 168
tolerance zone, 167
total runout, 176
unilateral tolerance, 162, 171

V

views, 81-84, 103, 106-117
    auxiliary, 103, 112-117
    placement of, 84
    sectional, 103, 106-112
    selection of, 84
visual copy, 358

W

welding, 206-211
    basic symbols, 206
    common types, 206
    projection, 211
    resistance, 208-211
    seam, 211
    spot, 208-211
    supplementary symbols, 209
    symbol locations, 208
web, 110
workstation, 362
worm, 239
worm gear, 234, 239-340
worm wheel, 239
worm wheel nomenclature, 239-240
    lead, 240
    pitch, 240
    throat diameter, 240
    throat radius, 240